POST-YIELD FRACTURE MECHANICS

POST-YIELD
FRACTURE MECHANICS

Edited by

D. G. H. LATZKO

*Delft University of Technology,
Delft, The Netherlands*

APPLIED SCIENCE PUBLISHERS LTD
LONDON

APPLIED SCIENCE PUBLISHERS LTD
RIPPLE ROAD, BARKING, ESSEX, ENGLAND

British Library Cataloguing in Publication Data

Post-yield fracture mechanics.
1. Fracture mechanics   2. Welded steel
structures
I. Latzko, D G H
671.5'20422   TA409

ISBN 0-85334-775-1

WITH 6 TABLES AND 165 ILLUSTRATIONS

© APPLIED SCIENCE PUBLISHERS LTD 1979

Printed in Great Britain by Galliard (Printers) Ltd, Great Yarmouth

Contents

Contributors

J. A. BEGLEY
   Department of Metallurgical Engineering, The Ohio State University, 116W 19th Avenue, Columbus, Ohio 42310, USA.

T. K. HELLEN
   Research Department, Berkeley Nuclear Laboratories, Central Electricity Generating Board, Berkeley, Gloucestershire GL13 9PB, UK.

J. D. LANDES
   Westinghouse Electric Corporation, Research and Development Center, 1310 Beulah Road, Pittsburgh, Pennsylvania 15235, USA.

D. G. H. LATZKO
   Laboratory for Nuclear Engineering, Department of Mechanical Engineering, Delft University of Technology, Department of Mechanical Engineering, Laboratory for Thermal Power Engineering, Rotterdamsweg 139a, Delft, The Netherlands.

C. E. TURNER
   Department of Mechanical Engineering, Imperial College of Science and Technology, Exhibition Road, London SW7 2BX, UK.

Main Symbols and Their Units

A	area	(m^2)
a	crack length for edge crack	(m)
	half length of centre crack	
	minor axis of semi-elliptical embedded or surface crack	
B	specimen thickness	(m)
b	remaining ligament thickness in cracked specimen	(m)
	$W - a$	
C	compliance	(N/m)
c	major axis of semi-elliptical embedded or surface crack	(m)
D	gauge length (may be less than length L, unless L is redefined)	(m)
d	clip gauge opening	(m)
E	Young's modulus	(MN/m^2)
E'	E for plane stress	(MN/m^2)
	$E/(1 - v^2)$ for plane strain	
F	force	(N)
f	restraining stress in the Dugdale model plastic zone (may be σ_Y but not necessarily so)	(MN/m^2)
G	crack extension force	(MN/m)
	strain energy release rate	
J	line integral	(MN/m)
K	stress intensity factor	(MN/m$^{3/2}$)
k	elastic stress concentration factor	(—)
L	length (of specimen)	(m)
M	bending moment	(Nm)
m	constrained yield factor, $m\sigma_Y$ (*not* strictly 'constraint factor')	(—)
N	constant in relation between J (or G) and work done	(—)
	number of load cycles	(—)
	work hardening exponent, σ^N	(—)
N	specimen depth	(m)

P	potential energy	(MN/m)
p	pressure	(MN/m)
Q	flaw shape parameter (as used in ASME XI, app. A)	(—)
q	load line displacement	(m)
R	elastic–plastic equivalent of K, $R^2 = E'J$	(MN/m$^{3/2}$)
r	radius	(m)
r_Y	plastic zone adjustment at crack tip	(m)
S	span in bending	(m)
T_i	element of traction vector	(MN/m^2)
t	thickness of structural member (not specimen)	(m)
U	strain energy	(MN/m)
u, v, w	displacements in cartesian x, y, z directions	(m)
u_i	element of displacement vector	(m)
x, y, z	cartesian coordinates	(m)
	x—parallel to crack	
	y—perpendicular to crack (in-plane)	
	z—thickness direction (out-of-plane)	
Y	shape factor, $K = Y\sigma\sqrt{a}$ as in ASTM 410	(—)
Y^*	shape factor, $R = Y^*\sigma\sqrt{a}$	
Δ	referring to an increment of the variable	(—)
σ	crack (-tip) opening displacement	(m)
ε	strain	(—)
ε_{ij}	strain tensor	(—)
η	constant in relation between G and work done $(= 1/N)$	(—)
σ	normal stress	(MN/m^2)
σ_Y	(effective) yield stress	(MN/m^2)
σ_{UTS} or σ_o	ultimate tensile stress	(MN/m^2)
σ_{ij}	stress tensor	(MN/m^2)
v	Poisson constant	(—)
τ	shear stress	(MN/m)
γ	non-dimensional clip gauge opening	(—)
θ	rotation angle of bend bar	(—)
ω	work done	(MN/m)
ϕ	non-dimensional COD	(—)

Prefixes

r	remote from crack
l	local to crack

Subscripts
i   initiation
m   maximum
I, II, III referring to crack opening modes

Note
A number of additional symbols are defined as they occur in the text.

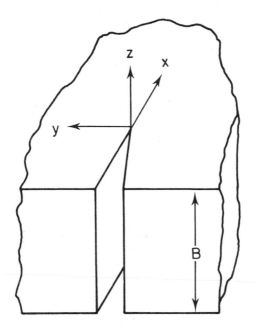

Crack tip coordinate system.

Preface

For the prevention of failure in structures, designed, manufactured or operated under their responsibility, engineers have traditionally looked for quantitative criteria permitting prediction of the entire structure's behaviour from simple laboratory tests on small specimens. In the case of brittle fracture, the failure mode characterised by rapid propagation of flaws occurring at overall stresses well below the yield stress and involving very little plastic deformation, such tests have long been exclusively based on the use of notched specimens subject to impact loading at various temperatures, preferably covering the entire range of possible service temperatures. The single-parameter criterion thus obtained is the so-called transition temperature, marking a change between relatively low and high energy absorption by the specimens in fracture. Apart from being of little use for materials lacking a clear-cut transition temperature or absorbing little energy even above the transition temperature, this criterion ignores the influences of flaw size and shape and of the stresses in the structure.

A number of serious brittle fracture failures in the welded hulls of 'Liberty' ships and T2 oil tankers mass-produced during the Second World War, and the introduction of high-strength, low-toughness materials in several critical aerospace applications, precipitated large-scale research aimed at providing quantitative design rules against brittle fracture based on an improved understanding of the fracture process. Starting from the Griffith approach, formulated as early as 1920, which stated that an existing crack would propagate if the total energy of the system would thereby be lowered, this research led to the proof by Irwin, in 1957, that this energy approach is equivalent to a stress-intensity approach according to which fracture occurs if the stresses near the crack tip reach a critical value, characteristic of the material. This Griffith–Irwin concept was subsequently expanded into the field of *fracture mechanics*. This offspring of a propitious mating of the theory of elasticity and experimental materials engineering has succeeded in providing quantitative expressions for the fracture stress in terms of crack shape and size, component geometry and a material property defined as fracture toughness. Chapter 1 presents a user-oriented summary of these relationships together with the underlying principles and the experimental determination of fracture toughness for the fracture safety assessment of structures.

The validity of this approach is predicated upon the assumption of linear elastic behaviour of the material surrounding the crack tip. For the more common structural and pressure vessel steels serving at and above room temperature this condition becomes unrealistic for the actual structure and difficult to achieve in test specimens, due to the spread of plastic behaviour around the crack tip. The aim of the present book is to present a concise review of the work undertaken to date to extend the single-parameter fracture mechanics approach into the realm of elasto-plastic material behaviour. Chapter 2 reviews the main concepts, with particular emphasis on the crack opening displacement (COD) and J integral as one-parameter criteria that have reached the stage of practical application. Care is taken to emphasise the relationship between these and other concepts, notably the stress intensity factor used in linear elastic fracture mechanics.

Experimental methods for establishing the critical COD and J values at which stable crack growth will start for a given material are discussed and evaluated in Chapter 3, with emphasis on the latter concept. Care is taken to point out the limitations and remaining uncertainties of the various methods.

Chapter 4 covers the computation of J integral values using the finite element method. In keeping with the user-oriented character of the book the emphasis is on results, while for readers less familiar with finite element techniques an outline of the method is presented in a separate Appendix.

While the authors feel that the wealth of knowledge on post-yield fracture mechanics generated over the past years justifies their effort to present the gist in book form, they are at the same time well aware that the field is still very much in flux and that important information is still unknown or awaiting confirmation.

An example in point is the limitation of the book to the determination and hence the prevention of initiation of crack extension, to the total exclusion of unstable crack propagation and crack arrest. While plasticity effects are of known relevance to such important areas of concern as running cracks in pressurised pipelines, where they will affect the flow strength and critical stress intensity in the zone around the crack tip, it was felt that the present state of knowledge did not warrant the inclusion of crack dynamics in a textbook such as this. The book's closing chapter, therefore, briefly identifies what the authors consider to be the main areas requiring further investigation and discusses their effect on the present-day application of post-yield fracture mechanics.

Any book authored by five individuals at five separate locations will incorporate contributions, both spiritual and material, from a number of people far too large to be acknowledged by name. In expressing the collective gratitude of all the authors, the editor wishes to point out that the foundations of this book were laid in December 1975 at the Winterthur meeting of the IAEA's International Working Group on 'Reliability of Reactor Pressure Components' under the inspiring chairmanship of Dr R. W. Nichols.

D. G. H. LATZKO

1

Linear Elastic Fracture Mechanics: A Summary Review

D. G. H. LATZKO

Delft University of Technology, Delft, The Netherlands

1.1   INTRODUCTION

If a flawed structural member such as a bar is exposed to a tensile force of increasing magnitude, one can visualise two extreme modes of failure: separation without any previous elongation and change in cross-section, or a gradual reduction in cross-sectional area to the vanishing point. The former might be termed a perfect *brittle fracture*, while the second is called *rupture*. This book is concerned with the prevention of fracture in engineering materials containing cracks. Such materials, e.g. steels and other structural metals, will usually exhibit some plastic deformation prior to fracture. This chapter is concerned with cases where plastic deformation preceding fracture is sufficiently small to use the assumption of linear elastic material behaviour as a basis for fracture control. As the resulting approach, *linear elastic fracture mechanics*, further referred to as LEFM, does not form the subject of this book, this present chapter presents its principles, methods and application in summary only. Prior to this, a note on the fracture process will try to emphasise some phenomena essential to the understanding of the types of fracture discussed in the remainder of the book.

1.2   THE FRACTURE PROCESS

Understanding of the inhomogeneous deformation process called fracture should start on the atomic scale (10^{-10} m). Here fracture occurs when bonds between atoms are broken, either by *cleavage* perpendicular to or by *shear* along the fracture plane. The question which of the two forms will predominate at the microscopic scale (10^{-6} to 10^{-5} m), i.e. in a single crystal under tensile load, will, broadly speaking, depend on the mobility of the lattice irregularities called *dislocations*. As the mobility decreases, e.g.

by local pile-ups of dislocations or by low temperature, plastic deformation is increasingly impeded and the stress is raised to the point where cleavage occurs. Propagation of such a cleavage across the grain boundaries in polycrystalline materials will require a further increase in stress to connect cleavages in adjacent grains by local areas of plastic deformation: see Fig. 1.1a. Thus cleavage fractures, while appearing smooth to the eye on a macroscopic scale, are generally discontinuous on a microscopic scale.

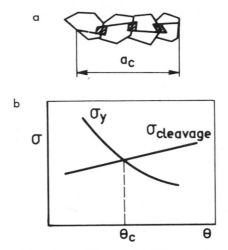

FIG. 1.1.   (a) Fracture path in polycrystalline material; (b) temperature dependence of yield and cleavage stresses in single crystals.

Dislocation mobility within a grain will, in addition to temperature (cf. Fig. 1.1b), be controlled by lattice arrangement and by the presence of interstitial and substitutional atoms. An example of the former is the difference in yield behaviour between face-centred cubic (fcc) materials such as austenitic stainless steels and body-centred cubic (bcc) materials such as the ferritic steels, while the second influence is evident in solid solution hardening and dispersion hardening.

On a macroscopic scale the predominance of cleavage results in fracture with flat separation surfaces, usually called *brittle* fracture and occurring at overall stresses well below yield.[1] With increasing shear effects, i.e. plastic deformation along planes of atoms having a low resistance to shear and called *slip planes*, fracture becomes increasingly *ductile* and the appearance

---

[1] The converse does not generally apply, i.e. cleavage on the microscale is not a prerequisite for brittle fracture.

of the fracture surfaces changes from flat to slanting (cf. Fig. 1.10). Localised plastic deformation leads to the formation and subsequent coalescence of voids. Extensive void formation with shearing of the remaining material leads to normal rupture typified by the cup-and-cone appearance in round bars of high-ductility materials fractured in tensile tests. The progress from brittle fracture to ductile rupture may thus be viewed as being associated with a continuing decrease in *plastic constraint*, a fact of the utmost importance throughout this book.

For further reading on the fracture process, reference is made to the pertinent literature, e.g. Yokobori [1], Tetelman & McEvily [2] and Liebowitz [3].

1.3   BASIC PRINCIPLES OF LEFM

1.3.1 Fracture Modes
Figure 1.2 shows schematically three distinct possible modes of crack extension under external load. In engineering practice the importance of the opening mode I far exceeds that of the other modes. Hence, further

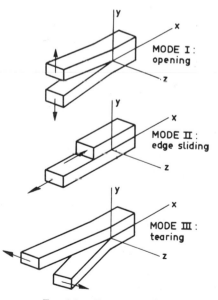

FIG. 1.2.   Fracture modes.

discussion throughout the book will be limited to mode I unless specifically stated otherwise.

1.3.2 Energy Considerations

Crack extension will require energy, which may be supplied from the work done by the applied external load and from the strain energy stored in the structure. The latter supply of energy will increase with crack extension, as the resulting increase in compliance (compliance = deformation/unit load) will reduce the structure's strain energy storage capacity.

A critical condition arises if the decrease in energy storage capacity per unit increase of crack area becomes at least equal to the energy absorbed per unit increase of crack area: crack propagation can then take place without the need for additional work to be done by the external load. Denoting the instantaneous value of the rate of energy supply per unit of crack area extension by G, the *strain energy release rate* [4], the critical condition may be defined by

$$G = G_c \qquad (1.1)$$

The strain energy release rate, G, can be expressed (see for example Bueckner, *Trans. ASME*, 1958, p. 1225) in terms of the structure's compliance C:

$$G = \frac{1}{2} F^2 \frac{dC}{da} \qquad (1.2a)$$

with

$$C = \frac{u}{F} \qquad (1.2b)$$

Measurements of G and G_c may hence be made from compliance measurements on cracked specimens, using load–displacement recordings as schematically shown in Fig. 1.3. This approach will be expanded upon in Chapter 3 for application to post-yield fracture mechanics. In LEFM it has been found possible to relate G uniquely to the stress distribution around the crack tip and thus to base the prediction of crack extension on stress considerations.

1.3.3 The Elastic Crack Tip Stress Field

Using Westergaard's complex stress functions, Irwin [4] demonstrated that the elastic stress field near the tip of the crack in an infinite sheet could be

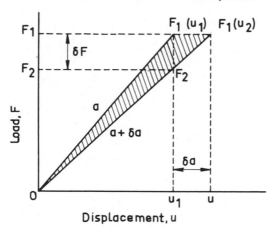

FIG. 1.3.   Elastic loading curves.

described for mode I (similar expressions were subsequently derived for modes II and III) by:

$$\sigma_x = \frac{K_I}{\sqrt{2\pi r}} \cos\frac{\theta}{2}\left[1 - \sin\frac{\theta}{2}\sin\frac{3\theta}{2}\right] \tag{1.3a}$$

$$\sigma_y = \frac{K_I}{\sqrt{2\pi r}} \cos\frac{\theta}{2}\left[1 + \sin\frac{\theta}{2}\sin\frac{3\theta}{2}\right] \tag{1.3b}$$

$$\tau_{xy} = \frac{K_I}{\sqrt{2\pi r}} \sin\frac{\theta}{2}\cos\frac{\theta}{2}\cos\frac{3\theta}{2} \tag{1.3c}$$

where the radius, r, and angle, θ, are defined as in Fig. 1.4.

The corresponding displacements for plane strain ($\varepsilon_z = 0$) become:

$$u = \frac{K_I}{\mu}\sqrt{\frac{r}{2\pi}}\cos\frac{\theta}{2}\left[1 - 2v + \sin^2\frac{\theta}{2}\right] \tag{1.4a}$$

$$v = \frac{K_I}{\mu}\sqrt{\frac{r}{2}}\sin\frac{\theta}{2}\left[2 - 2v - \cos^2\frac{\theta}{2}\right] \tag{1.4b}$$

$$w = 0 \tag{1.4c}$$

This analysis indicates that the stress and deformation fields in the vicinity of the crack tip can be uniquely characterised by a single parameter,[2] K, called the *stress intensity factor*.

---

[2] For consideration of the non-singular terms omitted from eqns. (1.3) see, for example, Larsson & Carlsson [5].

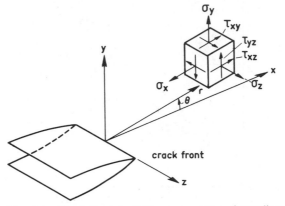

FIG. 1.4.   Crack tip stress field: components and coordinates.

In order to relate K to the strain energy release rate, G, Irwin considered a small crack extension in the x direction and calculated the energy released by the displacement of the crack surfaces in the y direction. He found

$$K^2 = E'G \qquad (1.5)$$

where $E' = E$ for plane stress and $E' = E/(1 - |v|^2)$ for plane strain. The critical value of K_I corresponding to G_c is called *fracture toughness*, K_{Ic}. (In accordance with general practice, K_{Ic} will hereafter be considered as specifically referring to plane strain considerations. For all other cases, fracture toughness is usually denoted by K.)

From dimensional considerations of eqns. (1.3) it follows that K_I is given by:

$$K_I = \text{constant} \cdot \sigma\sqrt{a} = Y \cdot \sigma\sqrt{a} \qquad (1.6)$$

where σ = applied stress in y direction, remote from the crack. Prediction of the extension of a given crack in a given stress field will therefore consist of

computation of the stress intensity factor, i.e. of the constant in eqn. (1.6) and

determining a unique value for the fracture toughness of the pertinent material.

1.3.4 Stress Intensity Factors
In the determination of stress intensity factors by elastic stress analysis

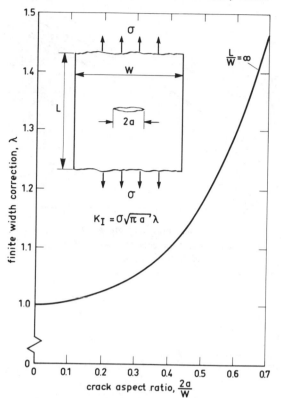

Fɪɢ. 1.5. Elastic stress intensity factor K_I for central crack in infinitely long rectangular sheet subjected to uniform uniaxial tension [12].

three levels of complexity may be distinguished according to the type and location of the flaw in question:

(a) through thickness cracks, for which the above *two-dimensional* approach is essentially valid, though requiring correction factors in eqn. (1.6) for

the finite size of the structure and

bulging if the cracked structure is pressurised (as, for example, in pressure vessels and piping); and

(b) surface or embedded flaws requiring a *three-dimensional* analysis, further to be subdivided into

(*i*) cases where the load on the crack may be approximated by simple tension, and

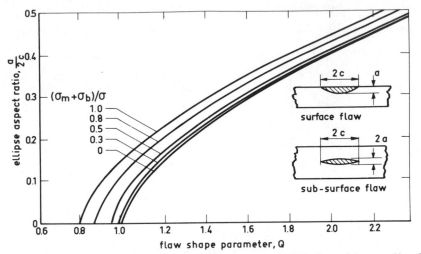

FIG. 1.6. Shape factors for surface and embedded flaws (*ASME Boiler and Pressure Vessel Code*, Section XI, App. A, 1974).

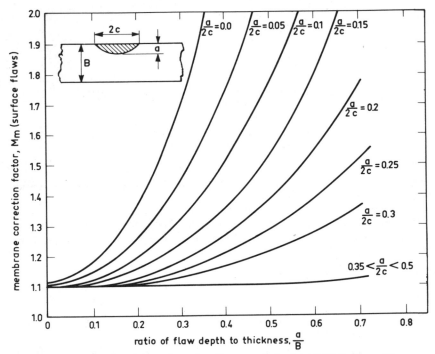

FIG. 1.7a. Magnification factors for surface flaws, tensile load (*ASME Boiler and Pressure Vessel Code*, Section XI, App. A., 1974).

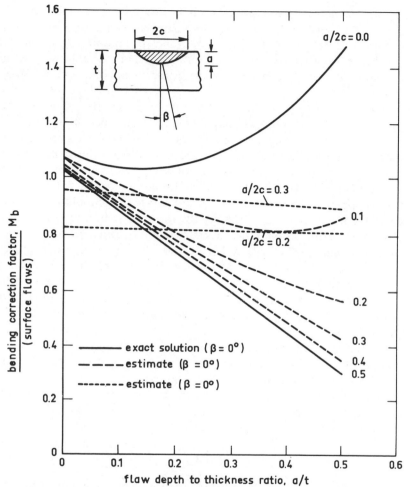

FIG. 1.7b.　Magnification factors for surface flaws, bending load (*ASME Boiler and Pressure Vessel Code*, Section XI, App. A., 1974).

(*ii*) cases where the crack is located in a complex stress field (e.g. at the corner of two intersecting structural members such as in pressure vessel nozzles and T-branches in piping or tubular structures).

In most engineering structures outside the aerospace field, notably in power generating and process equipment, the three-dimensional situations mentioned under (b) will prevail.

Both the actual computations and the compilation of the results are beyond the scope of this introductory chapter. The reader is referred, for example, to Refs. [6–8] for the former and to handbooks such as Refs. [9–11] for the latter purpose. The following remarks should only serve to indicate lines of approach and provide an impression of the present state of the art.

For cases involving through thickness cracks the basic elasticity equation, $\nabla^2 (\nabla^2 \phi) = 0$, has been solved for a number of geometries by using either appropriate algebraic or transcendental closed forms for the stress function ϕ or by describing it by polynomial expressions with constants obtained by matching the stress distribution given by the function to the relevant boundary conditions: the so-called boundary collocation method. Cases in point are sheets, discs, tubes and bars for which solutions giving K as a function of geometry—usually the ratio of crack length a and some characteristic dimension of the structure such as width—are to be found in the above handbooks for both tensile and bending loads. An example in point is given in Fig. 1·5 for a centre-cracked sheet in tension. These data are based on work by Isida [12] using the boundary collocation method. Additional data as well as verifications for the aforementioned solutions have been obtained using the finite element method discussed in Chapter 4. In thin-walled pressurised structures containing through cracks the stress intensity factor will be influenced by bulging of the material around the crack. Although such situations are inconsistent with plane strain and hence with the fracture toughness concept as discussed in Section 1·5, it is customary to account for bulging by multiplying the stress intensity by a factor obtained from purely elastic considerations. For cylinders the following correction is suggested [13]:

$$K_{\mathrm{I}_{\text{bulg.}}} = K_{\mathrm{I}} \left[1 + C \frac{t}{R} \left(\frac{a}{t} \right)^{\mu} \right] \tag{1.7}$$

where R is the cylinder radius while C and μ are experimentally determined constants.

Unfortunately information on part-through surface cracks, forming the bulk of engineering flaws, is much more limited because of the added complexity due to three-dimensionality. One approach open to the designer is that given in Appendix A to Section XI of the *ASME Boiler and Pressure Vessel Code*, 1974 Edition, which defines the stress intensity factor for part-through and embedded flaws, assumed to be of elliptical shape, by the general equation:

$$K_{\mathrm{I}} = \sigma_{\mathrm{m}} M_{\mathrm{m}} \sqrt{\pi} \sqrt{\frac{a}{Q}} + \sigma_{\mathrm{b}} M_{\mathrm{b}} \sqrt{\pi} \sqrt{\frac{a}{Q}} \tag{1.8}$$

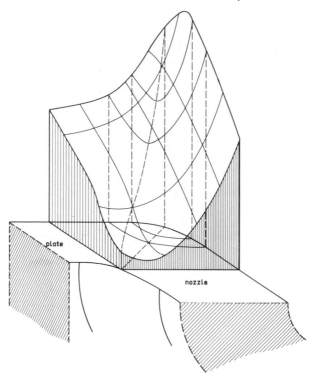

FIG. 1.8.   Typical stress distribution perpendicular to the crack plane in the nozzle transition region for uncracked nozzle-on-flat plate model (uniaxial loading).

where σ_m and σ_b are the gross[3] tensile (membrane) and bending stresses, respectively, in the flawed cross-section. Q is a shape factor to be obtained from Fig. 1.6, based on work presented by Tiffany & Masters [14]; $\sqrt{\pi/Q}$ is thus comparable to the constant in eqn. (1.6). M_m and M_b are elastic magnification factors lumping the effects of both front and back surfaces for tensile and bending loads, respectively; they are plotted for surface flaws in Figs. 1.7a and 1.7b, respectively, the latter being taken from Ref. [13]. The information contained in these graphs is the result of extensive computations in an area which is under continuing development, as witnessed, for example, Refs. [15–17]. Of the various computational methods reviewed in Ref. [7] the Schwarz alternating technique and the finite element method using three-dimensional elements have been used most frequently in the more recent literature.

---

[3] i.e. not corrected for cross-section of the flaw.

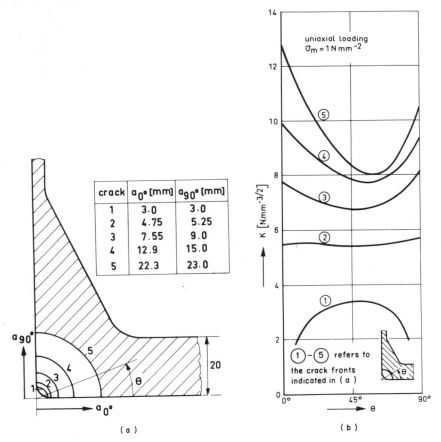

FIG. 1.9.   (a) Dimensions of crack fronts for nozzle N5 model computations; (b) three-dimensional finite element K-factor distribution along the crack fronts of nozzle N5.

An even more difficult problem is posed by the flaws grouped under (*ii*) above, notably those located at intersections such as nozzle corners. This is due to the complex stress fields already prevailing in the uncracked structure, as demonstrated in Fig. 1.8. Recent computational and experimental results, the former obtained by both analytical and finite element techniques presented by Broekhoven [18], show significant variation of K values along the crack front—Fig. 1.9. This fact, since confirmed by both experiment [19] and computation [20] with shapes varying according to the intersection's geometry, is potentially important for fatigue crack growth predictions.

1.4 FRACTURE TOUGHNESS TESTING

The LEFM approach described so far is predicated upon the existence and measurability of a *unique* fracture toughness value for the material of concern. However, Fig. 1.10, obtained for Al 7075-T6 [21], but qualitatively typical for many structural metals, shows the value for G— or K— at fracture to be strongly thickness-dependent. As shown in the figure

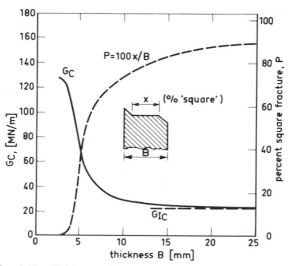

FIG. 1.10.   Thickness effect on fracture energy and appearance.

this dependence corresponds to a change in fracture appearance from fully slanted for thin sections to flat for the thick sections where the $G_c(K_c)$ curve flattens out to its minimum value. The cause of this thickness dependence is the plastic zone near the crack tip, neglected in the preceding discussion and schematically shown in Fig. 1.11. In the region remote from the free surfaces, where a state of plane strain is maintained, this zone is kept small by the constraint of the surrounding elastic material, favouring flat fracture. As the thickness is increased the plane strain area becomes predominant and the fracture toughness reaches its minimum value, the *plane strain fracture toughness*, K_{Ic}, which is a material constant. The transition from plane strain to plane stress should actually be shown as a fairly sharp demarcation in Fig. 1.11 because the increase in plastic zone size caused by relaxation of σ_z will in turn cause a further relaxation of σ_z.

D. G. H. Latzko

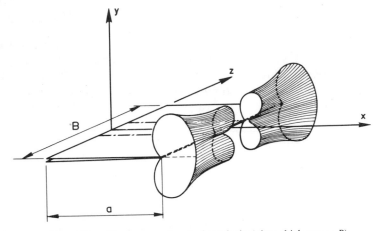

FIG. 1.11.   Plastic zone ahead of crack tip (plate thickness $= B$).

It follows that valid fracture toughness measurements require a minimum degree of plane strain. According to Section 1.5 the size of the plastic zone is proportional to $(K_{Ic}/\sigma_y)^2$. Hence, it appears logical to couch the requirement for an adequate degree of plane strain in terms of this ratio, as done in the empirical ASTM limit:

$$B \geq 2 \cdot 5 \left(\frac{K_{Ic}}{\sigma_y} \right)^2 \qquad (1.9)$$

The ASTM standard [22] containing this limit[4] also prescribes the specimen configurations, namely three-point bend and compact tension, their dimensions and the clip gauge for measuring displacements during the test, all according to Fig. 1.12. The test itself consists of a load–displacement recording, used to define the load at fracture, and of accurate measurement of the crack length at fracture. From these two data the fracture toughness value is computed by means of formulae also given in [22]. Figure 1.13 shows typical results for a Mn–Mo ferritic pressure vessel steel widely used in plates for nuclear reactor vessels. Both K_{Ic} and the yield stress are shown as functions of temperature. The figure illustrates the fundamental problem of fracture testing of medium and low strength steels: to meet the plane strain validity criterion of eqn. (1.9) at room temperature and above, plate thicknesses in excess of $0 \cdot 2$ m are required.

---

[4] Also valid for the crack length a.

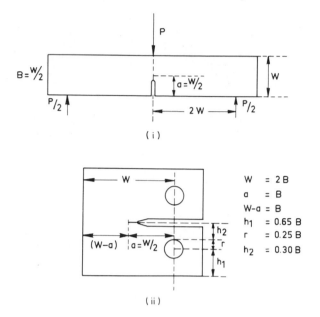

FIG. 1.12. (i and ii) Standard test pieces for fracture toughness testing [22]: (i) three-point bend specimen; (ii) C(ompact) T(ension) specimen. (iii) Clip gauge for measuring crack opening displacement [22].

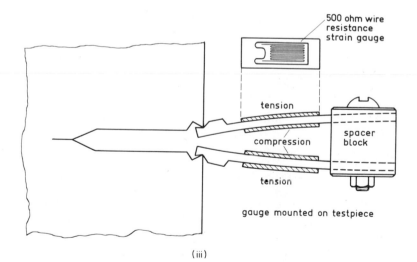

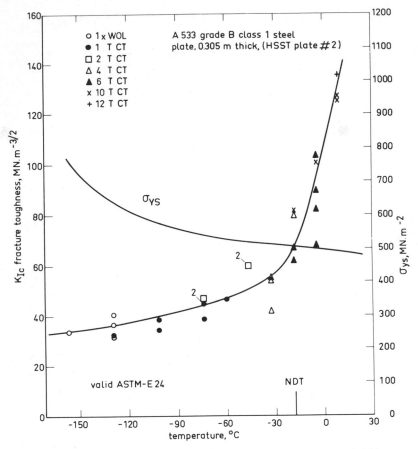

F<small>IG.</small> 1.13.   Temperature dependence of plane strain fracture toughness K_{Ic} and yield stress σ_Y
for Mn–Mo pressure vessel steel [23].

1.5   PLASTIC ZONE SIZE

Figure 1.14a indicates schematically the effect of small-scale yielding on the
elastic stress distribution near the crack tip described by eqns. (1.3). By
'small-scale' we indicate a yielded zone small in comparison with both the
crack length ($2a$) and the thickness (B) and other dimensions of the cracked
structure. Analytical prediction of the size and shape of this plastic zone is
extremely difficult for mode I (and II) of crack extension. As a first simple
approximation, use may be made of the concept of a '*notional crack*' [24]

whose tip is supposed to be located at the point where the actual stress σ_y equals the yield stress σ_Y (cf. Fig. 1.14a). It follows from eqn. (1.3b) that the distance r_Y from this point to the actual crack tip must be

$$r_Y = \frac{1}{2\pi}\left(\frac{K_I}{\sigma_Y}\right)^2 \tag{1.10a}$$

Arguments for using this adjustment for the actual crack length are based on analogy with the mode III elastic–plastic solution, predicting a circular

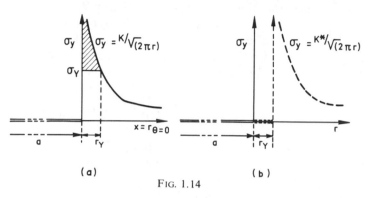

(a)   (b)

FIG. 1.14

plastic zone boundary with radius of $1/2\pi\,(K_{III}/k)^2$, where k is the resistance to plastic shear [25].

This adjustment holds for plane stress. From Fig. 1.6 and the associated arguments of Section 1.4 it follows that under plane strain conditions the adjustment must be significantly smaller. Reduction by a factor of 3 is most commonly assumed, i.e.

$$r_{Y\,\text{plane strain}} = \frac{1}{6\pi}\left(\frac{K_I}{\sigma_Y}\right)^2 \tag{1.10b}$$

The result of these adjustments is a notional crack with length $2(a + r_Y)$, to which the concepts of LEFM outlined above are again applicable, with a revised elastic stress distribution as shown schematically in Fig. 1.14b.

A more general expression for plastic zone size would be:

$$r_Y = \frac{1}{2\pi}\left(\frac{K_I}{m\sigma_Y}\right)^2 \tag{1.10c}$$

where m indicates the amount of constraint, with $m = 1$ for plane stress and $m = \sqrt{3}$ for plane strain.

It should be emphasised that the adjustments of eqns. (1.10) are approximations lacking a firm theoretical basis and excluding such effects as work hardening and large strains. They should not be mistaken, therefore, for alternatives to elasto-plastic crack tip modelling described in Chapters 2 and 4.

1.6   FATIGUE CRACK GROWTH

Cracks that are non-propagating under static loads will grow under cyclic loads through the formation of new voids due to the pile-up of dislocations and may ultimately reach the critical length a_c corresponding to $K_I = K_{Ic}$. A failure thus occurring after a certain number of load cycles (N) is termed a fatigue failure. Its prevention requires the capability to predict the crack growth per cycle. The latter is given for many materials of engineering interest, such as steels, by

$$\frac{da}{dN} = C_0(\Delta K_I)^m \qquad (1.11)$$

where ΔK_I is the stress intensity factor range $K_{I\,max} - K_{I\,min}$, while C_0 and m are experimentally determined material constants. This correlation is often referred to as Paris' Law because of its origin [27].[5] The plot of crack growth per cycle versus stress intensity factor range, shown in Fig. 1.15 for the same pressure vessel steel (ASTM A533 grade B, class 1) as used for Fig. 1.13, confirms the validity of eqn. (1.11) while indicating some spread in the constant C_0.

Writing eqn. (1.6) as

$$\Delta K_I = \Delta\sigma C\sqrt{a} = C_1\sqrt{a} \qquad (1.12)$$

and combining it with eqn. (1.11) yields:

$$\frac{da}{dN} = C_0 \cdot C_1^m \cdot a^{m/2} \qquad (1.13)$$

Strictly speaking the 'constant' C_1 will vary with crack length, as borne out by Figs. 1.6, 1.7a and 1.7b. However, by assuming conservative values for the various constituents of C_1 and subsequently checking after completion of the analysis, C_1 may be kept constant and eqn. (1.13) be integrated to yield:

$$a_f = (a_i^{(2-m)/2} + C_2 N_{tot})^{2/(2-m)} \qquad (1.14)$$

[5]Various more complex relationships, proposed in the literature, take account of additional effects such as the ratio between the maximum and minimum values of K during a load cycle.

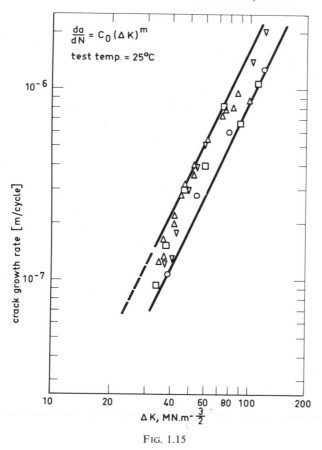

$$\frac{da}{dN} = C_0 (\Delta K)^m$$

test temp. = 25°C

crack growth rate [m/cycle]

ΔK, MN.m$^{-\frac{3}{2}}$

FIG. 1.15

where a_i and a_f are the initial and final crack length, respectively, N_{tot} is the number of relevant load cycles of a certain amplitude and C_2 is a constant $= [(2 - m)/m]C_0C_1^m$.

If the final crack length thus obtained approaches the critical value, a repeat analysis using the incremental approach, i.e. step-wise variation of C_1 with crack length, is recommended.

Usually a structure is exposed to various load cycles of different amplitudes. In view of the underlying assumption of linear elasticity the crack length increases obtained for each distinct cycle by the method outlined above may in such cases be added algebraically to yield the total crack growth.

1.7 CONCLUDING REMARKS

Two limitations to the application of LEFM, both due to plastic zone extension, emerge from the above:

the size of test pieces required for valid K_{Ic} measurements for tough materials at temperatures near those prevailing in the actual structure under load, and

the occurrence of extensive yielding in cracked structures of highly ductile materials.

These limitations have spawned the need for post-yield fracture mechanics, the subject to be discussed in the remainder of this book.

REFERENCES

1. Yokobori, T. (1968). *An interdisciplinary approach to fracture and strength of solids*, Wolters-Noordhoff, Groningen.
2. Tetelman, A. S. & McEvily, A. J. (1967). *Fracture of structural materials*, John Wiley, New York.
3. Liebowitz, H. (ed.) (1968). *Fracture*, vol. 1, Academic Press, New York.
4. Irwin, G. R. (1957). Analysis of stresses and strains near the end of a crack traversing a plate, *Trans. ASME, J. Appl. Mech.*, **24**, pp. 361–4.
5. Larsson, S. G. & Carlsson, A. J. (1973). Influence of non-singular stress terms and specimen geometry on small-scale yielding at crack tips in elastic–plastic materials, *J. Mech. Phys. Solids*, **21**, pp. 263–77.
6. Paris, P. C. & Sih, G. C. (1965). Stress analysis of cracks, pp. 30–83, in *Fracture toughness testing and its applications*, ASTM–STP 381, Philadelphia.
7. Sih, G. C. (1973). *Methods of analysis and solutions of crack problems*, Noordhoff Int. Publ., Leyden.
8. Kassir, M. & Sih, G. C. (1975). *Three-dimensional crack problems*, Noordhoff Int. Publ., Leyden.
9. Sih, G. C. (1973). *Handbook of stress intensity factors for researchers and engineers*, Inst. Fract. & Sol. Mech., Lehigh Univ.
10. Tada, H., *et al.* (1974). *The stress analysis of cracks handbook*, Del Research Corp., Hellertown (Pa).
11. Rooke, D. P. & Cartwright, D. J. (1976). *Compendium of stress intensity factors*, HMSO, London.
12. Isida, M. (1971). Effect of width and length on stress intensity factors of internally cracked plates under various boundary conditions, *Int. J. Fract. Mech.*, **7**, pp. 301–7.
13. Wilhem, D. P. (1970). 'Fracture mechanics guidelines for aircraft structural applications', AFFDL-TR-69-111, Feb.
14. Tiffany, C. F. & Masters, J. N. (1965). Applied fracture mechanics, pp. 249–89,

in *Fracture toughness testing and its applications*, ASTM–STP 381, Philadelphia.

15. Swedlov, J. L. (ed.) (1972). *The surface crack: physical problems and computational solutions*, ASME, New York.
16. Kobayashi, A. S., *et al.* (1975). Stress intensity factors for elliptical cracks, pp. 525–44, in *Prospects of fracture mechanics* (G. C. Sih *et al.*, eds.), Noordhoff Int. Publ., Leyden.
17. Nishimura, A. *et al.*, (1977). Stress intensity factor for a semi-elliptical crack in an internally pressurized cylinder, *3rd Int. Conf. Press. Vessel Technol.* (Tokyo), pp. 517–33.
18. Broekhoven, M. J. G. (1977). Fatigue and fracture behaviour of cracks at nozzle corners: comparison of theoretical predictions with experimental data, *3rd Int. Conf. Press. Vessel Technol.* (Tokyo), pp. 839–52.
19. Smith, C. W., *et al.*, (1977). Geometric influence upon stress intensity distributions along reactor vessel nozzle cracks, SMIRT-4 (San Francisco), Paper G4/3.
20. Kobayashi, A. S., *et al.*, (1977). Stress intensity factors of corner cracks in two nozzle–cylinder intersections, SMIRT-4 (San Francisco), Paper G4/4.
21. Brown, W. F. & Srawley, J. E. (1965). Fracture toughness testing, pp. 133–98, in *Fracture toughness testing and its applications*, ASTM–STP 381, Philadelphia.
22. *Standard method of test for plane strain fracture toughness of metallic materials*, ASTM E 399–74, 1974.
23. Wessel, E. T. (1969). Linear elastic fracture mechanics for thick-walled, welded pressure vessels: materials property considerations, in *Practical fracture mechanics for steel* (R. W. Nichols *et al.*, eds.), Chapman & Hall, London.
24. ASME Boiler of Pressure Vessel Code, Section XI: Rules for in service inspection of nuclear power plant components (1974 Edition).
25. Knott, J. F. (1973). *Fundamentals of fracture mechanics*, Butterworth, London.
26. McClintock, F. A. & Irwin, G. R. (1965). Plasticity aspects of fracture mechanics, pp. 84–113, in *Fracture toughness testing and its applications*, ASTM–STP 381, Philadelphia.
27. Paris, P. C. & Erdogan, F. (1963). A critical analysis of crack propagation laws, *Trans. ASME, J. Basic Engng*, **85**, pp. 528–34.

2

Methods for Post-yield Fracture Safety Assessment

C. E. TURNER

Imperial College of Science and Technology,
London, UK

2.1 INTRODUCTION

Failure of a cracked component loaded beyond the range to which LEFM is applicable may occur in one of several different ways. A centre-cracked panel loaded in tension is taken as a simple example. A stress very near to the crack tip (not here specifically defined but taken as representative of the local stress conditions) is denoted by $_1\sigma$, the uniaxial yield stress by σ_Y, net section stress by σ_n, and the uniform stress remote from the crack by σ. It is useful to distinguish four regimes, which can loosely be described as follows:

(a) $_1\sigma > \sigma_Y > \sigma_n > \sigma$: yielding limited to a zone in the *immediate* vicinity of the crack, notionally of vanishingly small extent. This is the LEFM problem. If failure occurs it is usually by unstable rapid propagation of the crack.

(b) $_1\sigma > \sigma_Y \gtrsim \sigma_n > \sigma$: yielding is extensive, but does not spread to a lateral boundary of the structure and is thus contained. This is a regime that can be called elastic–plastic and to which yielding fracture mechanics can be applied, but for which LEFM, with correction for the extent of the plastic zone, may still give an acceptable answer. If failure occurs it is usually by unstable rapid propagation of the crack.

(c) $_1\sigma > \sigma_n \gtrsim \sigma_Y > \sigma$: yielding is very extensive and spreads to the lateral boundary ahead of the crack, and is thus uncontained. This is a regime that can be called gross yield, to which yielding fracture mechanics must be applied. For configurations with little lateral constraint and low hardening, tough materials may fail by plastic collapse of the net section, whilst for less tough materials a crack may spread by stable or unstable growth.

23

(d) $_1\sigma > \sigma_n > \sigma > \sigma_Y$: since the applied stress σ is greater than the yield
      stress, extensive plasticity develops along the components as well as
      across the section, implying work-hardening of the net section. This
      is a regime that can be called general yield. Failure may well be by
      plastic collapse, and limit analysis the preferred tool, but tearing or,
      in some materials, crack propagation may still be the failure mode.

These four conditions are shown schematically in Fig. 2.1. It must be
clear that the stress levels used to describe the four states merge from one
condition to the other and are affected by configuration, induced biaxial
and triaxial stresses and work-hardening. The terms elastic–plastic, and
gross and general yield, while descriptive are by no means universally
adopted. Clearly, the regimes starting in (b) and extending through (c) and,
for some materials, into (d) are those that constitute the realm of yielding
fracture mechanics of interest here.

As noted, the so-called elastic–plastic regime (b) forms a logical extension
of LEFM and is perhaps the condition in which many components, with
regions of stress concentration or unintentional defects, are likely to
operate under normal loading. In so far as representative large-scale test
pieces can be used, avoidance of failure under conditions of similar severity
overall might be judged adequate. But if, for obvious reasons of economy
and convenience, small-scale test pieces are used, they may well have to
experience overall deformations at the upper end of regime (b) or well into
regime (c) of gross yield in order to simulate the severity of crack tip
deformation in the real (large) component. Accident or fault conditions
may also take real components into this regime or beyond it into general
yield (d), particularly if the defect is small and the net section work-hardens
appreciably. Test pieces may correspondingly be taken into this regime,
and reaching it implies a level of ductility likely to be found only in the lower
strength–weight-ratio materials or alloys developed with the particular
demands of toughness in mind. *It must be emphasised that components
likely to reach gross or general yield, even if analysed satisfactorily by
fracture mechanics, must be checked against plastic collapse by an
appropriate design procedure.*

Section 2.2 describes in some detail the two yielding fracture mechanics
concepts that have received most attention: crack opening displacement
(COD) and the J contour integral (J). These have become philosophies in
their own right, to form the mainstream of fracture mechanics. Their
application to design problems, although rather limited so far, is included
in the discussion, and the final part of this section deals with the relationship

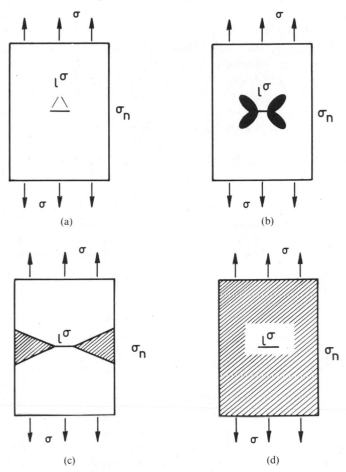

FIG. 2.1. Schematic representation of increasing degrees of yielding: (a) LEFM, (b) elastic–plastic, (c) gross yield, (d) general yield.

between these two concepts. It can be said at once that they are by no means in direct conflict, although the details of analysis and proposed methods of application differ considerably.

Section 2.3 describes several procedures that either have evolved along empirical lines or have been developed with specific applications in mind. Some appear to be more relevant to the elastic–plastic regime and some to gross or general yield. The present status of these methods differs from that of the COD and *J* concepts inasmuch as none of them has, to the author's knowledge, reached the stage of being considered for incorporation in national standards. Nevertheless, some may be propounded and applied in specific fields by very knowledgeable and responsible users, and backed by experiment or experience very relevant to a particular field.

The final section, 2.4, deals with a number of aspects and problems of post-yield fracture behaviour, notably slow stable crack growth, the onset of unstable crack growth or tearing and of size effects that are as yet not fully understood or, despite the use of computers, are too complex to analyse usefully.

The number of concepts and procedures described in this chapter is indicative of the still developing nature of post-yield fracture mechanics pointed out in the Preface. Their concurrent development may also at a first glance seem to evidence a lack of communication between the various schools of thought. However, here it should be remembered from Chapter 1 that the development of yielding fracture mechanics has resulted from two distinct requirements. The first is to permit the use of relatively small test pieces in laboratory testing from which the fracture behaviour of large components can be predicted. The second requirement is to design in circumstances of limited ductility yet of an extent apparently too great for the use of LEFM. This requires the measurement of a fracture toughness parameter applicable after yielding. In addition to differences in emphasis resulting from these distinct requirements the very reasons for considering yielding in fracture control vary with the field of application.

In the high-pressure components of the power and chemical process industries the main reason stems from the use of materials that, although ductile in conventional tests, are liable to fracture with only limited ductility when used in thick sections, in situations of high constraint (triaxial stresses) or perhaps after some form of metallurgical damage from welding, fabrication or even environmental embrittlement. The contribution to toughness of the so-called shear lip effect, whereby plastic flow in the free surface allows at least locally ductile behaviour, may be negligible or entirely absent if a defect is deeply embedded or penetrates into the wall of a thick component. At the other end of the range, in thin-sheet aerospace structures, much of the toughness is attributable to the development of shear lips by plastic deformation, particularly for cracks penetrating completely through the thickness.

While the authors of the present book are far more familiar with the former, plane-strain type of situation, methods specifically valid for thin sheets and through-cracks have been included in this chapter for the sake of completeness.

2.2 SOME CONCEPTS OF ELASTIC–PLASTIC FRACTURE MECHANICS

2.2.1 Crack Opening Displacement (COD)

2.2.1.1 *Outline of the Method*

The method is based on the assumption—put forward independently by Wells [1], Cottrell [2] and Barenblatt [3]—that where significant plasticity occurs the fracture process will be controlled primarily by the intense deformations adjacent to the crack tip, and that the separation of the crack faces, or crack opening displacement, will be a measure of the intense deformation. Crack extension will then begin at some critical value of this crack opening displacement, according to the particular micro-mode of fracture that occurs. Hence, the method requires analytical prediction of displacements near the crack tip.

For the elastic case the displacement in the y direction relevant to mode I fracture follows from formula (1.4):

$$v = \frac{2K}{E'} \sqrt{\frac{2r}{\pi}} \tag{2.1}$$

From the elastic solution for the crack of length $2(a + r_\text{Y})$, the displacement at a (i.e. r_Y behind the notional crack tip but right at the real crack tip) is found to be

$$v_a = \frac{2K}{E'} \sqrt{\frac{2r_\text{pl}}{\pi}} \tag{2.2}$$

Using the plane stress value for r_Y from eqn. (1.10a) gives

$$\delta = 2v_a = \frac{4K^2}{\pi E' \sigma_\text{Y}} \tag{2.3}$$

Wells [4] went on to suggest that an energy balance for a small increment of crack growth required

$$G = \sigma_\text{Y} \delta \tag{2.4}$$

since σ_Y is the stress acting across the plastic zone in the plane stress Irwin model. On the other hand, using the relationship (1.5) from linear mechanics

$$K^2 = E' G$$

together with eqn. (2.3) gives

$$G = \frac{\pi}{4}\sigma_Y \delta \qquad (2.5)$$

This discrepancy of $4/\pi$ in the argument is simply neglected and the crack tip opening, now called crack opening displacement, COD, or sometimes more explicitly, the crack tip opening displacement CTOD, became the first parameter to be considered seriously for fracture at a crack after extensive yielding.

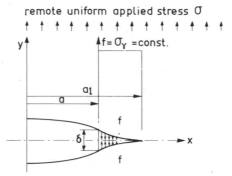

FIG. 2.2.   Crack-opening displacement in the BCS–Dugdale or 'strip-yield' model of crack tip plasticity.

A more refined model for defining the plastic zone than the Irwin correction used above, yet which avoids the complexities of a true elastic–plastic solution, was proposed by Dugdale [5]. As shown in Fig. 2.2, he considered an extended crack of length $2a_1$ acted upon by a remote uniform stress σ but, over the distances $(a_1 - a)$ corresponding to a plastic zone ahead of the real crack, a uniform stress, f, is developed by yielding of the plastic material. The laws of plasticity are not invoked, however, and the extent of the plastic zone in the axial direction is not considered.

Dugdale pointed out that the stress at the real crack tip could not now be singular (since it is limited to f) and suggested that this condition would be satisfied if the singularities from the local stress system, f, cancelled with the singularity from the applied system, σ. For this to be so, he showed the length of plastic region, $(a_1 - a)$, was

$$(a_1 - a) = a[\sec (\pi\sigma/2f) - 1] \qquad (2.6)$$

By this combination of elastic solutions, Burdekin & Stone [6] evaluated the displacement at the tip of the real crack for the line plasticity model and found

$$\delta = 2v_a = \frac{8fa}{\pi E} \ln\left(\sec\frac{\pi\sigma}{2f}\right) \qquad (2.7)$$

A similar result was obtained by Bilby *et al.* [7][1] from a dislocation model in anti-plane strain but the engineering applications of the Wells–Dugdale model are usually restricted to the mode I situation. If the ln (sec) term is expanded

$$\delta = \frac{8fa}{\pi E}\left[\frac{1}{2}\left(\frac{\pi\,\sigma}{2\,f}\right)^2 + \frac{1}{12}\left(\frac{\pi\,\sigma}{2\,f}\right)^4 + \frac{1}{45}\left(\frac{\pi\,\sigma}{2\,f}\right)^6 + \cdots\right] \qquad (2.8)$$

Taking the first term only, and using the plane stress relation $K^2 = EG = \sigma^2\pi a$:

$$\delta = \frac{\pi\sigma^2 a}{Ef} = \frac{G}{f} \qquad (2.9)$$

which, if f as taken as σ_Y, agrees with eqn. (2.4). If the second term in eqn. (2.8) is included, again with $f = \sigma_Y$, the result is closely similar to LEFM with a plane stress plastic zone correction differing only in the term $\pi^2/24$ instead of $\frac{1}{2}$. Thus, although Wells had general yielding in mind when developing the COD concept, the use of it would seem to be a logical extension of LEFM, at least for plane stress.

Burdekin & Stone also evaluated the overall strain, ε, over a gauge length, D, in the axial direction and encompassing it across the centre. They found, for the case $f = \sigma_Y$ representing plane stress,

$$\varepsilon/\varepsilon_Y = \frac{2}{\pi}\left\{2n\coth^{-1}\left[\frac{1}{n}\left(\frac{k^2 + n^2}{1 - k^2}\right)^{1/2}\right]\right.$$
$$\left. + (1 - v)\cot^{-1}\left(\frac{k^2 + n^2}{1 - k^2}\right)^{1/2} + v\cos^{-1}k\right\} \qquad (2.10)$$

where $n = a/y$, $k = \cos\pi\sigma/2f$ and $\varepsilon_Y = \sigma_Y/E$ (the yield strain). The origin of the y coordinate is at the centre of the crack. They then evaluated strain as a function of the non-dimensional COD, ϕ,

$$\phi = \delta/2\pi\varepsilon_Y a \qquad (2.11)$$

[1] Often referred to as the Bilby–Cottrell–Swinden or BCS model.

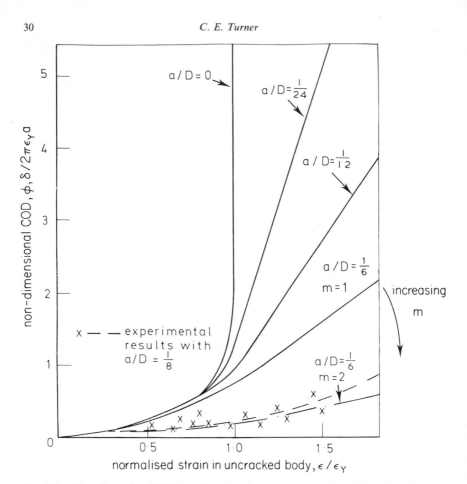

FIG. 2.3.   Non-dimensional COD as a function of strain over gauge length D, with early test results from Ref. [6] and effect of constrained yield factor m added from Ref. [47].

for a series of gauge lengths, expressed as multiples of a: Fig. 2.3. Any particular value of ϕ implies, of course, a corresponding value of σ/σ_Y for a given value of crack length, a.

Initial experimental results, on sheet aluminium alloy at fairly small overall strain, confirmed the curve of Fig. 2.3 for the relevant gauge length, but more extensive tests [6] using steel, the first of which was carried out on steel plates roughly 1 m square by 75 mm thick with a variety of gauge lengths and notch depths, showed no general trend [6]. As is evident from

Fig. 2.4, all the experimental results scattered around one part of the diagram.

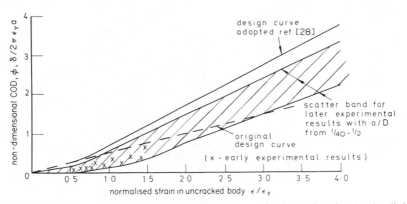

FIG. 2.4.   Non-dimensional experimental COD data as a function of strain, showing little effect of gauge length. (Based on Ref. [28].)

2.2.1.2 *The Fracture Criterion:* $\delta = \delta_c$

As in the case of LEFM the present approach assumes that fracture occurs if the parameter selected, δ, reaches a critical value, δ_c, which should be a material constant. On this latter vital point the evidence is as yet incomplete, partly because of uncertainties in measurement and partly because of ambiguities in the definition of δ.

At the time that much of the early data was collected, there was no standard experimental technique. At first, paddle COD-meters were used in which, for sawn slits, a transducer was operated by the movement of a spade or 'paddle'-shaped probe inserted into the tip of the slit and spring-loaded in torsion to bear against the sides of the slit as they opened. With the growing use of fatigue cracked pieces this technique became impracticable. For material property data the three-point bend test, in the full thickness of interest, was preferred, as described in Chapter 3, and test techniques developed in which the tip COD was inferred from clip gauge measurements at the mouth of a slit notch. The principle of the analysis is the assumption of a centre of rotation at $r(W - a)$ below the tip, about which the ligament hinges by plastic bending. The value of r depends on the degree of deformation but is in the range 0·3 to 0·4 for the preferred deep

32 C. E. Turner

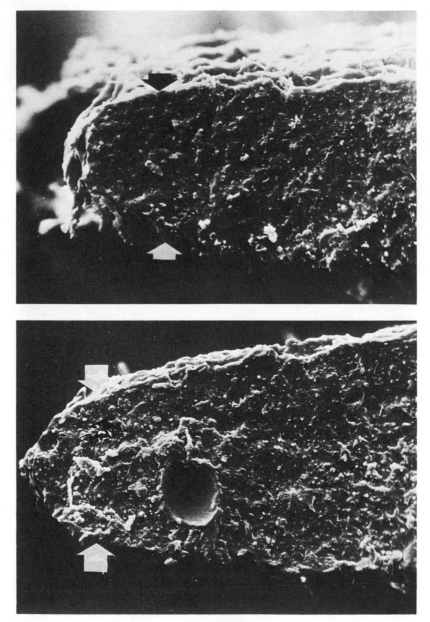

FIG. 2.5. Two examples of crack tip profiles from infiltration studies. Arrows show the
position at which COD was measured [12].

notch three-point bend pieces. Details are given in Chapter 3 and Ref. [8].[2] Naturally the early measurements to establish the validity of the concept did not use the present technique, so that care must be exercised in the interpretation of such data. Some studies have been made using micro-indentations spaced across the crack face from which tip displacement can be inferred by extrapolation of flank displacements to the tip [9]. Approximate values have also been inferred from notch–root contractions, angle of bend double-notch techniques and high speed photography. Some of these, carried out mainly for dynamic purposes, are described in Ref. [10]. More recently, crack infiltration methods have been used by Robinson & Tetelman [11]. They infiltrated the crack tip with a hardening silicone rubber and then broke open the specimen to obtain a cast of the crack from which COD could be measured. In a first generation of work predominantly on steels, aluminium and titanium alloys of high strength–toughness ratio and rather low work-hardening (but including the pressure vessel steel A533B) it was inferred from grid measurements that plane strain existed over the mid-thickness of 10-mm-sized pieces at least up to plastic deformation of 4° angle of bend. The values of COD so measured were at a fairly well defined shoulder about δ or $\delta/2$ back from the blunting tip (Fig. 2.5) and, at least in these tests, came from small pieces, mainly $a/W = 0.2$, and thus were not in accord with the proposed standard test method [8]. Clip gauge readings were taken and values of the rotational constant r inferred but a direct comparison of the infiltration COD and that for the recommended procedure was not made. The object of this work was to demonstrate a relation between clip gauge and true tip (infiltration) COD

---

[2] The COD test procedure [8] will probably be amended shortly. The main feature of the alteration is understood to be the relation of the elastic component (which is proportional to load) to K, leaving the plastic component derived in terms of COD as at present. This forces the result to degenerate to LEFM for small ductility in the spirit proposed [85] for J testing. Thus,

$$\delta = \frac{K^2}{2\sigma_Y^2 E} + \frac{V_{pl}}{1 + 2\cdot5\dfrac{a + z}{w - a}}$$

where V_{pl} = plastic component of clip gauge mouth opening displacement, and z = distance of clip gauge knife edge above the surface. The factor 2 in the term for K implies $m = 2$ in the relationship $G = m\sigma_Y\delta$. The factor 2·5 corresponds to a rotational factor 0·4 of the ligament width, for deep-notch three-point bend. K is calculated by standard LEFM formulae for the load at the point of interest on the load-clip gauge recording.

from which K_{Ic} could be inferred from a small test piece. A unique curve for the rotational factor r was found as a function of COD value for all the materials tested, and COD related to LEFM by

$$K_{Ic}^2/E' = G_{Ic} = m\sigma_Y\delta_c \tag{2.12}$$

where $m = 1 \pm 0.05$. The agreement with K_{Ic} tended to validate the COD concept, but only a narrow range of geometries was tested. In later work Robinson [12] tested an alloy steel and a low-strength structural steel. LEFM data were not available, but if inferred from values of the J contour integral (to be discussed in Section 2.2.2.) gave a value of $m \simeq 1$ for the alloy steel and $2\cdot 2$ for the structural steel. Perhaps more importantly in the present context, the values of δ (at initiation) were the same for three-point bend and centre-notched tension pieces and thus supported a degree of independence of geometry, although, as seen later, the J values were equally well supported.

The point just mentioned, viz. that critical values of COD (or J) refer to the onset of crack growth, introduces another uncertainty, namely, if this onset or initiation is followed by slow stable crack growth, how should this be allowed for in a critical measure of the fracture parameter? The phenomenon, occurring in tough materials such as most steels prior to fracture and sometimes referred to as slow tearing, has long been accepted in plane stress (for example, Ref. [13]) but has been less widely recognised in plane strain until recently, although reported on fairly frequently in COD literature (for example, Ref. [14]). Smith & Knott [15] reported slow growth determined by breaking open interrupted tests and extrapolating back to zero growth. In thick pieces, growth occurred from mid-thickness as a 'thumbnail' before maximum load but, for their tests, maximum load was attained just before slow growth was evident on the surface. Harrison & Fearnehough [16] reported that COD at initiation of ductile tearing was relatively independent of test piece dimension, but the amount of growth varied from one geometry to another. Tests carried out in the UK on bend bars [17], pressurised spheres [18] and pressurised cylinders [19] seem to confirm that COD at initiation of crack growth (slow tearing) from a sharp pre-existing crack, δ_i, is a material property, for a given temperature, strain rate and environment.

Further evidence as to the validity of COD as a fracture criterion is shown in Fig. 2.6 where Kanazawa et al. [20] infer tensile failure stresses from COD measured in three-point bending. The agreement is good for a range of steels. Similarly, Urbensky & Müncner [21] report agreement between fracture of pressure vessels and full-thickness bend tests.

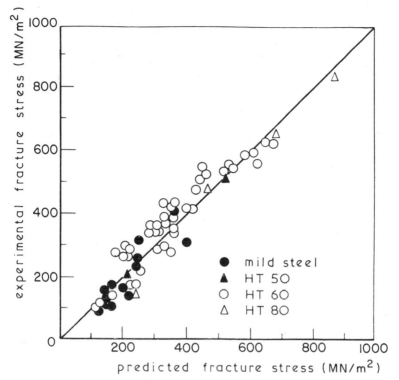

FIG. 2.6.   Estimation of fracture stress in tension using critical COD obtained from bending [20]. HT = yield stress (kg/mm^2).

Unfortunately, the precise treatment given to slow crack growth in these two studies is not clear.

The best established material property, in terms of COD, thus appears to be the initiation of slow tearing, δ_i. It may be objected that the validity of δ_i has not been demonstrated closely for a very wide range of geometries and thicknesses. However, provided deeply notched (e.g. $a/W = 0.5$ for three point bend) test pieces are used to obtain a severe yet realistic degree of constraint, such a value seems at least a conservative measure of the toughness of the lower strength structural steels in which slow stable crack growth occurs before final fracture. It has even been argued that use of δ_i in design is *unduly* conservative, but use of a maximum load COD value, δ_m, is certainly less well substantiated. If the value of δ_m occurs at a maximum load of the plastic collapse type with a slowly falling load–displacement curve, a direct relevance to some other configuration may be questionable, although the material has clearly shown ductile behaviour in the thickness of interest. If the value of δ_m occurs at a maximum on a still rising curve, by rapid fracture after some stable growth, relevance to a component of the same thickness may be more readily accepted, though not, to the writer's knowledge, either clearly demonstrated or theoretically founded. Experience shows a much wider experimental scatter on δ_m values than on δ_i so prudence dictates that a reasonable lower bound on δ_m must be established. This problem is discussed again in Section 2.4.1 in connection with more recent studies of stable crack growth and the final unstable fracture.

2.2.1.3 *Computational Verification*
An essential prerequisite for the COD—or any other quantitative— approach to fracture safety, is the possibility for accurate prediction of the selected parameter, namely δ.

This aspect of the validity of COD has been studied by computer methods, where results are from elastic–plastic finite element computations, not computations of the Dugdale model. The broad reasonableness of the concept is clearly demonstrated (e.g. Fig. 2.7), particularly in the three-point bend test [22], but detailed study by Sumpter [23] emphasises the difficulty of defining COD in general. For the bend test,

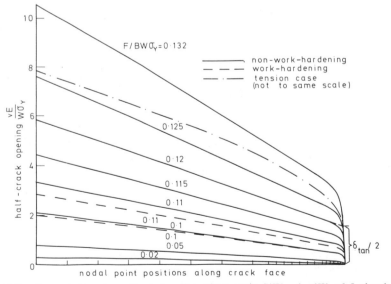

FIG. 2.7.   Computed crack profiles in bending, plane strain $S/W = 4$, $a/W = 0.5$, showing the tangent definition of the half COD and the more rounded profile in tension [22].

the near straight flank profile encourages the use of the tangent extrapolated to intercept the normal at the crack tip, δ_{tan}. In tension, the crack profile is more rounded and the point at which a tangent should be drawn is unclear. This subject has been discussed at some length [24, 25] based on data from two extremes, non-work-hardening bending with its clear definition of δ_{tan} and high work-hardening tension with a very

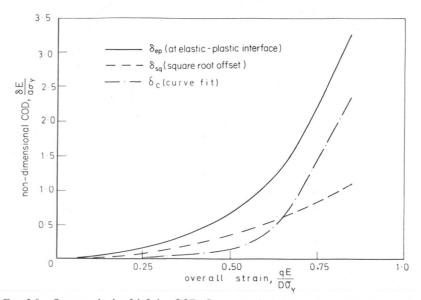

FIG. 2.8.   Some methods of defining COD. Centre-cracked plate, $2a/W = 0.5$, plane strain, non-work-hardening [23].

rounded profile. Other definitions of COD have been used in computer studies, such as the opening at the intercept on the profile of a chord at some particular angle from the tip, a 'square-root offset' that is exact for the linear elastic case and so on. Several of these are shown in Fig. 2.8. It will at once be seen that, in the small-scale yielding regime, differences by a factor of two between the various definitions can easily arise. Use of large displacement theory [26] gives a smooth tip for non-hardening material for which a meaningful δ can be defined (Fig. 2.9) but solutions with sharp corners at the tip can also be found so that the behaviour is not unique. One proposal that emerged was the use of the elastic–plastic interface, just behind the crack tip, to define the point at which COD is measured. As seen in Fig. 2.10, such a point may not be close to the tip and, although quite specific in

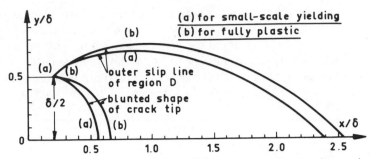

FIG. 2.9.   Deformation of crack tip region with large geometry changes, plane strain tension. Conventional solutions (small geometry change) predict a fan of slip lines centred on the tip. In the large-geometry-change solution, a non-centred fan focuses intense strain into region D, which extends ahead of the tip a distance of some 2δ [26].

theoretical or computational studies, it is not readily detectable in experimental studies. For extensive yielding all tend to reduce to straight lines not through the origin, found by taking a linear plastic regime and joining to the origin by a parabolic curve for the near LEFM regime (after eqn. (2.4) and Fig. 2.8). Thus

$$\delta = \delta_{pl} + \delta_{el} \tag{2.13a}$$

$$\delta_{pl} = a(\sigma/\sigma_Y) + b \tag{2.13b}$$

$$\delta_{el} = c\sigma^2 \tag{2.13c}$$

a, b, c are constants, in principle geometry-dependent. This is entirely

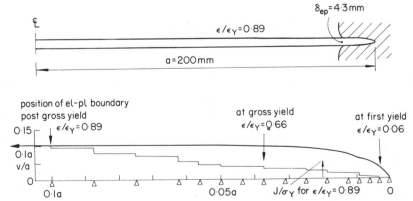

FIG. 2.10.   Crack profiles in the gross yield regime, to true scales. Centre-cracked plate, $D/W = 4$; $2a/W = 0.5$, plane strain, work-hardening, dimensionalised for $E/\sigma_Y = 250$; $a = 200\,mm$. (Data from Ref. [23].) The location of a possible measuring point for COD at J/σ_Y from the tip is indicated.

consistent with the COD design curve described below and indeed, to within a factor of about two-fold between the various definitions and cases there is no problem.

2.2.1.4 *Design Application*

As originally propounded [27, 28] the COD design method aimed to give the practitioner a very simple but assuredly conservative method for defining, for a given material, an acceptable combination of stress level and defect size. The material toughness, δ_c, was defined in a three-point bend test of full thickness material. The test method has since been standardised in BS DD19 ([8] and Chapter 3). The defect was originally envisaged as a *full through-thickness crack*, and the main feature of the design method was the provision of one curve relating the strain level in a component to the non-dimensional COD, $\phi = \delta/2\pi\varepsilon_Y a$ (eqn. (2.11)). The abscissa used was nominal strain, ε, in the absence of a crack, rather than stress, and the curve for ϕ was the safe bound of experimental results originally from plate and bend tests with full thickness defects, to which part through thickness crack data were later added. This master design curve has more recently been defined [29] by simple expressions and is clearly very similar in shape to that discussed in eqn. (2.13).

$$\phi = (\varepsilon/\varepsilon_Y)^2 \qquad \text{for} \qquad \varepsilon/\varepsilon_Y < 0.5 \qquad (2.14a)$$

$$\phi = (\varepsilon/\varepsilon_Y) - 0.25 \qquad \text{for} \qquad \varepsilon/\varepsilon_Y \geq 0.5 \qquad (2.14b)$$

The constants in eqn. (2.14) are chosen, as stated, as a bound to experimental evidence, but to a form guided by theory.

Although the obvious core of the method is the design curve itself defined by eqn. (2.14), this curve is proposed as part of a whole design package. The most important point is the establishment of a critical value of COD, δ_c. Reproducibility of δ_c measurements will depend both on the definition of that quantity, i.e. on *how* to test, and on representative test piece material, i.e. *what* to test. The former point has been discussed in Section 2.2.1.2, where the value at initiation of crack growth δ_i emerged as the most invariant measure of a material property in terms of COD, although considered by many as too conservative in the light of experience. A small arbitrary measure of crack growth can therefore be accepted, δ_c, or even a maximum load COD be used, δ_m, which may or may not include some crack growth.

This acceptance of some value of δ beyond initiation focuses attention on the real point at issue, i.e. on what part of the structure, parent plate, weld metal or heat affected zone shall the COD test be conducted? In short, the acceptance of the less conservative measure, δ_c or δ_m rather than δ_i, appears in a very different light if evaluated for the least tough zone in the structure. In many structures such zones are formed by weldments and their heat affected zones (HAZ), which also generally offer a higher likelihood of defects and are therefore prime targets for COD (and other fracture control methods). Indeed, the need to incorporate an assessment of the fracture resistance of weldments has been in the forefront of COD development in the UK, where the Welding Institute has been a leader in the assessment of fracture resistance by COD.

Unfortunately the toughness of weld metal or HAZ varies with welding procedure in a complicated way that does not seem to have been rationalised for the non-specialised metallurgist. Although the specialist can go a long way towards recommending good procedures for any given problem, a choice still has to be made between determining the minimum toughness of the worst region of the weldment, or the toughness of the region where cracks are most likely for the welding procedure in question. The former is often in the weld metal for medium-strength steels, though normally in the HAZ for the low-strength carbon–manganese structural steels that were predominant ten or more years ago and, indeed, are still widely used today. If the 'worst toughness' is being measured then it is desirable to use the standard DD19 test piece. If the 'realistic defect' concept is used, or indeed tests being conducted to evaluate an actual defect, then it may be desirable to match the defect size, shape and constraint as closely as possible so that testing of shallow notch pieces may be necessary. DD19 covers a range of *notch depths* from about $0.25 < a/W < 0.5$, although $a/W = 0.5$ is the preferred standard. Some shallow notch tests were discussed by Archer [30] and some hitherto unpublished data on mouth–tip opening calibrations in plane strain finite element computations are shown in Fig. 2.11 for shallow-notch pieces. This problem of toughness testing of weldments has recently been studied extensively by Dawes [31]. While the subject of toughness variation with welding process and welding parameters is obviously beyond the scope of this book (see for example, Refs. [32–39]) Figs. 2.12–2.14 are included to illustrate the sort of effects to be expected. It perhaps emphasises the orientation of the COD approach towards elastic–plastic design of a wide variety of components that more such weldment data seem to exist for COD than for other elastic–plastic methods.

To revert to the application of the method: once a suitable value of δ_c has been determined, the non-dimensional COD, ϕ, has to be assessed for the

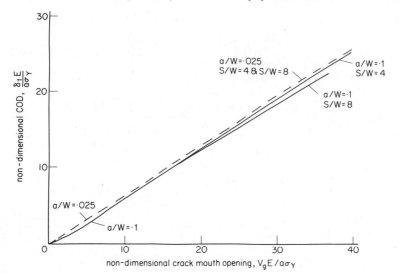

FIG. 2.11. Calibration curves for tip COD, δ_{tan} in terms of mouth opening V_g. Shallow-notch three-point bending; $S/W = 4$ and 8, $a/W = 0.1$ and 0.025, plane strain, work-hardening. The value of COD is defined by a tangent drawn to the crack profile. The degree of hardening used is described by $\sigma/\sigma_Y = 1.2$ at $\varepsilon/\varepsilon_Y = 10$; $\sigma/\sigma_Y = 1.4$ at $\varepsilon/\varepsilon_Y = 40$. The graphs can be represented approximately by the equation $\delta_{tan} = 0.66\, V_g(W - a)/W$, although the curves are strictly not linear.

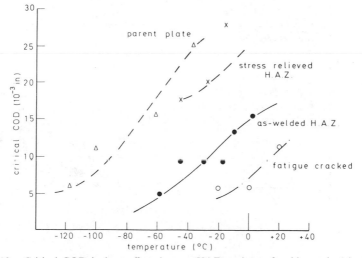

FIG. 2.12. Critical COD in heat-affected zone (HAZ) regions of weldment in 1-in-thick (25 mm) Mn–Cr–Mo–V steel (Burdekin, F. M. *et al.* (1968). *Brit. Weld. J.*, **15**, p. 59). Sawn notches 1.5 mm wide except as marked.

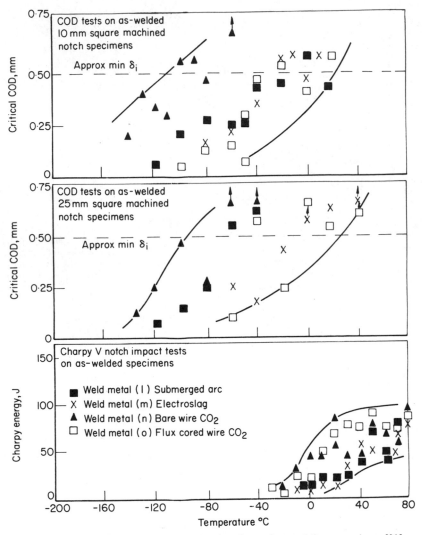

FIG. 2.13.   Critical COD and Charpy V energy for various welding procedures [31].

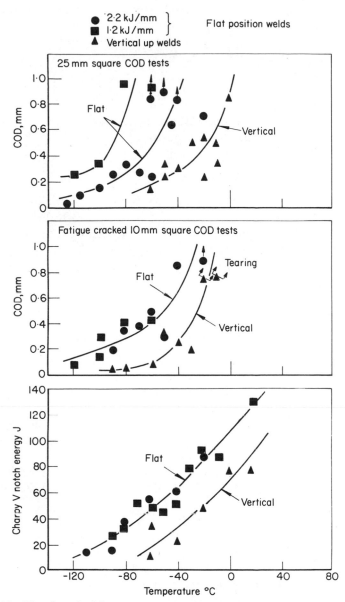

FIG. 2.14. The effect of welding procedures on critical COD and Charpy V transition curves [31].

structure from eqns. (2.14). The nominal strain, ε, is based on a simplified elastic assessment of the stresses acting in the absence of a crack, provided $\varepsilon/\varepsilon_Y < 2\cdot0$ when based on primary and secondary stresses alone. This procedure includes the effect of local peak stresses as causing a strain (including the basic strain) of

$$\varepsilon = k\sigma/E \qquad (2.15a)$$

where k is the elastic stress concentration factor and σ the nominal elastic stress. If there are residual stresses present of known magnitude, σ_r, then these are added

$$\varepsilon = (k\sigma + \sigma_r)/E \qquad (2.15b)$$

If the residual stresses are not known, a value $\sigma_r = \sigma_Y$ is allowed. If the primary and secondary stresses result in $\varepsilon/\varepsilon_Y > 2\cdot0$ then a closer determination of the nominal strain is recommended from experiments or computational methods. Thus, for a given applied nominal strain including the allowances for stress concentration and residual stress, ϕ is found from eqn. (2.14a) or (2.14b). The permissible crack size $\bar{a}$ then follows from:

$$\bar{a} = \delta_c/2\pi\varepsilon_Y\phi \qquad (2.16a)$$

$$= C\delta_c/\varepsilon_Y \qquad \text{(i.e. } C = 1/2\pi\,\phi) \qquad (2.16b)$$

This term is here called 'permissible' rather than 'critical' since eqns. (2.14) for ϕ are themselves a conservative bound to the data. The expression is phrased in terms of $\bar{a}$, an equivalent crack, because of the recent developments towards correction of the through thickness crack geometry assumed so far. Proposals to this end have been developed in the UK along the same lines as discussed in Chapter 1 for elastic conditions, though using somewhat different terminology and quantitative data [29–31]. The crack depth $\bar{a}$, corrected for finite thickness and free surface effects, may be read from Fig. 2.15. This figure is based on LEFM, with the COD method itself allowing for plasticity.

   For deep surface cracks $a/B > 0\cdot7$ or highly stressed situations, the ligament to the back face may yield. This point was initially [28] accounted for by taking $\bar{a} = c$, i.e. treating as if it were a through thickness crack of length $2c$, for $a/B > (\sigma_Y - \sigma)/\sigma_Y$; $c/B > 5\cdot0$. This has now been modified by Dawes [31] following tests by Egan [40] to

(i)   use of the above LEFM-based chart for

$$a/B < [(\sigma_Y + \sigma_u) - 2\sigma]/(\sigma_Y + \sigma_u) \qquad \text{and} \qquad a/B \le 0\cdot7$$

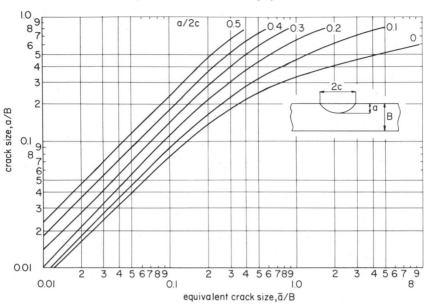

FIG. 2.15. Description of equivalent crack size, $\bar{a}$, for semi-elliptical surface cracks in COD design method [31, 120].

(ii) $\bar{a} = c$ for

$$a/B < [(\sigma_Y + \sigma_u) - 2\sigma]/(\sigma_Y + \sigma_u) \quad \text{and} \quad a/B > 0.7$$

(iii) $\bar{a} = 1.5\,c$ for

$$a/B > [(\sigma_Y + \sigma_u) - 2\sigma]/(\sigma_Y + \sigma_u)$$

Similar arguments can be made for the treatment of buried cracks; where the situation is dominantly elastic use LEFM shape factors and where there is yield of a small remaining ligament, treat as an enveloping crack. The preferred treatment of *oblique cracks* is again as indicated in Chapter 1, namely to resolve the stress normal to the crack and treat by mode I. This philosophy has been extended to adjacent cracks, clouds of defects and all the various circumstances that arise in interpreting NDT records of weldments. When applied to pressure vessels, an LEFM-based correction factor for bulging, such as eqn. (1.7), has to be added.

Attempts are being made to bring all these features together into a more finalised code. In so doing not only has the equivalent crack concept been extended to embrace the many practical situations that are found but the one curve (eqns. (2.14a and b)) treatment has also been amended. For steels,

C in eqn. (2.16b) is indeed still taken from eqns. (2.14a and b) but for other materials C is taken only from eqn. (2.16a) for all values of $\varepsilon/\varepsilon_Y$. This approach has been described in brief [41 and 42] and presented in a formal manner in a document of the International Institute of Welding [43] and in a BSI Draft for Public Comment [44]. Proposals for estimating lower bound K_{Ic} from Charpy data are also included so that it is perhaps a misconception to treat the document as the 'COD design curve method' although its origins can clearly be traced to the COD school of thought. This draft was withdrawn in the normal course of events and has not as yet been issued in any other form. It is perhaps worth quoting the preamble to again emphasise viewpoints rather than details of procedure. 'It is not intended that the final publication should be used in the way of a "specification" as it does not "specify" acceptance levels for defects in welds. The rules should be used by contract and agreement by all parties and then only if defects exceeding the specified quality control level are discovered, in order to determine whether any action is necessary'. And further, ' . . . a distinction has to be made between acceptance based on quality control and acceptance based on fitness for purpose. Quality control levels are, of necessity, both arbitrary and conservative . . . . If more severe defects are revealed . . . decisions on . . . rejection and/or repairs may be based on fitness for purpose . . . in the light of . . . experience . . . or on the basis of Engineering Critical Assessment (ECA). It is with the latter that this (draft) is concerned'.

2.2.1.5 Critique of the Method

Any method for quantitative fracture control must meet at least the following essential requirements:

> soundness of the underlying theoretical model

> unequivocal definition of the proposed parameter both for measurement and computational purposes

> demonstrable relevance of the chosen parameter to specified conditions of usage, preferably wide.

One uncertainty over the relevance of COD and indeed other elastic–plastic concepts, namely the effect of stable crack growth, has been discussed already, and will be taken up again in Section 2.4. Another source of uncertainty is that the design curve, defined by eqns. (2.14) is a bound to data obtained on cracked pieces, whereas the nominal strains calculated during design are for an uncracked body. The idea has been expressed [45] that for most practical applications where yield is exceeded due to residual stresses or at regions of stress concentration, the conditions are displacement- or strain-controlled and the design curve is directly applicable. For the relatively small crack to uncracked area ratio of interest in most applications (as distinct from the large a/W ratios of conventional test pieces) it does not seem that the difference in strain would be large, particularly for contained yielding. The many other sources of uncertainty in calculating a nominal design strain are probably more significant. Even so, the design curve is, by intent, a conservative bound to the original test data and is thought by some to err, if at all, on the side of over-conservatism.

Turning now to weaknesses in the model, the most obvious of these is the representation of plastic yielding by an elastic material with a line of plasticity only at its edge. As already remarked this is not unreasonable for plane stress, but is obviously unrepresentative for plane strain. For reasons that are not apparent, the usage of the Dugdale model described above has retained the plane stress assumption of $f = \sigma_Y$, but been applied to circumstances of at least partly developed plane strain. It has in fact been suggested [46, 47] that this inconsistency might be circumvented through the use of a constrained yield stress

$$f = m\sigma_Y \qquad (2.17)$$

where m might be $\sqrt{3}$ (as in the LEFM plastic zone correction) or even as high as 3 (strictly $(\sqrt{3}/2)\,(\pi/2)$) for a von Mises yielding material analysed by the Prandtl slip line field for contained yielding [48]. The problem with such a step is that if the constraint were maintained, the collapse load of the structure would be exceeded. The implication is that, at least for this model, the constraint must reduce as yielding spreads so that the choice of an appropriate value for m is difficult. Taking values such as 1·5, 2 or 3, arbitrarily, brings the curves of Fig. 2.3 down towards the right as shown schematically by the curved arrow, so that rather better agreement is found with some of the data from Fig. 2.4. This does not, however, seem to be the main cause for the discrepancy between theory and experiment. The model, as here described, ignores the effects of a finite plate width and of a work-hardening material. For a finite plate width it was suggested [6] that a net stress defined as $\sigma W/(W - 2a)$ be used, where σ is the gross section stress. Since $\delta \to \infty$ as $\sigma \to f$ (i.e. to σ_Y in most applications) this finite width correction factor cannot be used beyond net section yield, whereas the whole purpose of the model is to describe post-yield behaviour. However, the real effect of extensive yielding with work-hardening in a finite width plate is not represented adequately by eqn. (2.10), and this neglect is probably the main reason for the discrepancy between the experimental data and the trends of the theoretical curves shown in Fig. 2.3. The work-hardening effect, although omitted from the model, is included in the proposed design curve, as this is based on experimental results, on mild steel. The reliance of the latter on one material only has led to the cautioning by Turner & Burdekin [45] that care be exercised in employing the proposed approach to materials showing a high yield or proof to ultimate stress ratio. They suggested setting the limiting maximum yield to ultimate ratio at 75–80% until further research information is available.

A final, more general remark concerns the fact that more complex geometries were not treated analytically in the development of the COD design curve method, although an approximate method is now available as described later (Section 2.3.2).

Computational studies for COD can be made using the BCS–Dugdale model [49] or by finite element methods (as, for example, Refs. [22 and 23]). As pointed out in connection with Fig. 2.8, there are ambiguities of definition of COD for computations of models other than the Dugdale model. The model has itself been extended to finite plate geometries [50] but none of these methods has been used, as far as the writer is aware, in connection with the design curve, the whole object of which is to provide a reasonable estimate of non-dimensional COD, ϕ, in terms of a more or less normal design estimate for the uncracked body without recourse to rather complicated analyses of the cracked configuration.

2.2.2 Contour Integrals
2.2.2.1 *The J Contour Integral*
2.2.2.1.1 *Outline of the method*

It has already been argued that just as in LEFM a parameter is sought to characterise the singular stress and strain fields around a crack tip, so in yielding mechanics a characterising parameter is required for the elastic–plastic stresses and strains around a sharp crack. As outlined in Chapter 1, the LEFM approach was based on the concept of a balance of energy rates, leading to G, which was later shown to be uniquely related to the Irwin–Westergaard stress field intensity factor approach K. A rather similar pattern of events has developed in non-linear mechanics where it has been shown that a certain integral, now commonly called the J contour integral, describes the flow of energy into the tip region, while the dominant term in the description of the stress and strain singularities at the crack tip can be written in terms of J. Thus, J may be invoked in terms of either

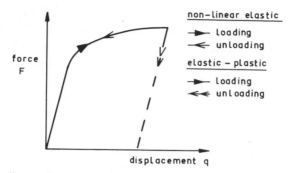

non-linear elastic

───► Loading
───◄ unloading

elastic – plastic

───► Loading
◄─◄ unloading

force
F

displacement q

FIG. 2.16. Loading and unloading paths for non-linear elastic and elastic–plastic behaviours.

energy or stress and strain state. These derivations of J are, however, strictly valid only for an elastic material, i.e. a material which even though non-linear in terms of load and deformation, on unloading, would return along the same load–deformation path, whereas with plasticity the unloading path would be roughly parallel to the original linear elastic path, Fig. 2.16. Conceptually this drawback appears insuperable at the present time. In practice, the use of J for plasticity-type materials can be supported in part by appeal to direct experimentation (as in Chapter 3) and in part by rather indirect analysis that will be described below.

Ignoring for the moment the problem of plasticity and considering only the non-linear aspects, the following relationships may be stated. By definition

$$J = \int_{\Gamma} \left\{ Z \, \mathrm{d}y - T_i \frac{\partial u_i}{\partial x} \, \mathrm{d}s \right\} \tag{2.18}$$

where Z is strain energy density, $Z = \int \sigma_{ij} \, \mathrm{d}\varepsilon_{ij}$, or density of stress working if the recoverable sense of the term strain energy is to be avoided, T_i are components of surface tractions as specified over an arbitrary part of the surface of a body (Fig. 2.17) and u_i are the components of displacements of

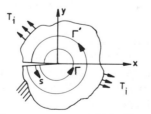

FIG. 2.17. Arbitrary contour over which J integral is evaluated.

the surface. A crack exists along the x axis, and s is the distance along any contour Γ traversed anti-clockwise from the lower face of the crack around the tip to the upper face. The crack faces are stress-free between the tip and where the contour cuts the faces. Mode I tensile opening is usually considered, although modes II and III can be discussed in a similar manner. It has been shown that this integral is independent of the path Γ, hence the common name '*the path-independent J contour integral*'.

It has also been shown that, for non-linear elastic material

$$J = -\frac{\partial P}{\partial a} \tag{2.19}$$

where P is potential energy and a is crack length. Thus for linear elastic material

$$J_{\text{linear}} = G \tag{2.20}$$

Equation (2.18) was first derived by Eshelby [51] as the energy–momentum tensor which is a measure of the energy flux into a singularity (or the several singularities) within the contour Γ. A similar expression was derived by Cherepynov [52] and separately by Rice [53]. This last work came at a time when the problem of elastic–plastic cracks was attracting much attention and together with an extensive review by Rice [54] led directly to the present applications of the concept.

Path independence requires that the density of stress working, Z, can be expressed as a function of the deformed state alone and not of the history of the loading path to that state. This is clearly not so for a body following the laws of incremental plasticity although, as seen later, numerical evidence may suggest the path dependence is small for cases loaded by applied mechanical forces. The energy balance does not include residual, inertial or thermal stress terms, nor, as usually formulated, loadings along the crack face itself between the tip and the points where the contour of integration cuts the crack flanks. The material should also be homogeneous in the direction of crack advance.

In three dimensions the contour implies a cylindrical region into which there is no flow of energy across the end faces so that application is usually restricted to two-dimensional plane stress or plane strain. A modified form of the expression for J, with the inclusion of an area integral, has been proposed [55], and other similar extensions are outlined in Section 2.2.2.2 for axially symmetric bodies.

When path independence is assured it seems that the conditions at the crack tip are unique, at least in angular average, with a $1/r$ singularity in energy density. It does not seem to be demonstrable that a near path independence in selected computations necessarily implies the same uniqueness in crack tip conditions and it is such restrictions on the generality of the results that has given rise to many of the extensive computational studies described in Section 2.2.2.1.3.

Another interpretation of J in characterising the singularity that has just been referred to is possible. The stresses around a crack in a non-linear elastic material obeying the stress–strain law

$$\frac{\varepsilon_{pl}}{\varepsilon_Y} = \alpha \left(\frac{\sigma}{\sigma_Y}\right)^N \tag{2.21}$$

where ε_Y and σ_Y are non-dimensionalising constants such that $\sigma_Y/\varepsilon_Y = E$, and α and N are constants, have been studied by Hutchinson [56] and Rice & Rosengren [57] (referred to jointly as *the HRR solutions*). Both papers considered an asymptotic solution for the stresses and strains near the crack tip and derived the power of the singularity in stress as $r^{-1/(N+1)}$ and of strain as $r^{-N/(N+1)}$ so that the singularity of energy density, $\sigma\varepsilon$, is r^{-1}. Higher order terms in powers of r would complete the solution of the stresses remote from the crack for any specific problem but, as in eqns. (1.3) describing the linear elastic singularity, only the leading term that dominates as $r \to 0$ is discussed. Hutchinson expressed his results in plastic stress and strain factors k_σ and k_ε. McClintock [58] pointed out that the HRR solutions could be expressed as in eqn. (2.22) although the precise form of terms differs since he did not use the normalising constants σ_Y and ε_Y in the stress–strain law of eqn. (2.21).

$$\frac{\sigma_{ij}}{\sigma_Y}(r,\theta) = \left(\frac{JE}{I\alpha\sigma_y^2 a}\right)^{1/(N+1)} \frac{1}{(r/a)^{1/(N+1)}} \tilde{\sigma}_{ij}(\theta) \tag{2.22a}$$

$$\frac{\varepsilon_{ij}}{\varepsilon_Y}(r,\theta) = \alpha \left(\frac{JE}{I\alpha\sigma_Y^2 a}\right)^{N/(N+1)} \frac{1}{(r/a)^{N/(N+1)}} \tilde{\varepsilon}_{ij}(\theta) \tag{2.22b}$$

$$\frac{u_i}{a}(r,\theta) = \alpha \left(\frac{JE}{I\alpha\sigma_Y^2 a}\right)^{N/(N+1)} (r/a)^{1/(N+1)} \tilde{u}_{ij}(\theta) \tag{2.22c}$$

I is a dimensionless factor introduced so that the $\tilde{\sigma}(\theta)$, $\tilde{\varepsilon}(\theta)$ and $\tilde{u}(\theta)$ terms have a maximum value of unity. In these expressions the only parameter that varies from problem to problem is J, which is related to Hutchinson's plastic stress and strain factors by $J = Ik_\sigma k_\varepsilon$. Linear elasticity is recovered and the singularity of r is $1/r^{1/2}$ for both stress and strain as expected [53], if

$$N = 1; \qquad (JE)^{1(N+1)} = (GE)^{1/2} = K \tag{2.23}$$

In the HRR derivations an infinite plate under uniaxial stress was considered so neither *biaxiality* nor the *effects of remote boundary conditions* were specifically included. It is often argued that provided a plate boundary is not too close to the crack region neither of these restrictions is essential to the derivation so that the stress and strain conditions at the tip of any crack in a power law hardening material are singular, of an amplitude characterised by the single parameter J. The writer is not aware of any specific statement in the published literature that supports or denies the suggestion that the characterisation is indeed unique for all remote stress states in a power law hardening elastic material. The point of biaxiality in plasticity is taken up in Section 2.2.2.1.3. The power of the singularity depends on $-1/(N+1)$ and $-N/(N+1)$ as already stated and the angular distribution is given by the functions $\tilde{\sigma}(\theta)$, $\tilde{\varepsilon}(\theta)$ and $\tilde{u}\theta$ of eqns. (2.22) as detailed in Refs. [56, 57 and 58]. For most purposes of fracture mechanics the value of these functions is irrelevant and only the value of J is required to compare the severity of stress or strain for different crack tip problems.

The foundation for the use of J is thus secure on both energetic and characterising grounds for a non-linear elastic material for externally applied loadings. The extension to plasticity is justified for materials following the laws of 'total' or 'deformation' plasticity (cf. Ref. [59] and Appendix 1), i.e. behaving exactly as does the non-linear elastic material in the case of monotonic loading. The difference is only apparent when the material 'unloads', either by reduction of the applied load or by crack advance (in which material adjacent to the crack is unloaded).

When metal deforms plastically, its behaviour is described by equations of incremental plasticity (Appendix 1) so that, strictly, even the loading event is not correctly described. However, it is generally agreed that for monotonic loading the differences between incremental and total theory are small so that a non-linear elastic formulation is accepted, with the proviso of no unloading, and the characterising role of J is carried over into plasticity. This does not mean that the energetic role of J defined by eqn. (2.19), i.e. as energy potentially available to grow a crack, would still be valid for plasticity, where part of the work done on the structure by the external load will be dissipated in plastic flow and, ultimately, mainly heat.

Since a growing crack implies unloading of the new crack face material, the postulate of monotonic loading is there violated. The practical consequence is that, with plasticity, the change in potential energy for a crack growing under load may differ from that found if the crack is unloaded, cut to longer length, and re-loaded. It has indeed been argued [60] that, in plasticity, materials in which, unlike the power law hardening description, the amount of work-hardening is finite (or of course zero for non-hardening), then for the limit $\delta a \to 0$ there will be *no* available rate of energy release over and above that required for continued plastic flow. Thus an analysis of energy rates on the Griffith concept, with the surface energy of the crack providing the balancing term between external work done (if any) and strain energy, is no longer feasible. This point is taken up in Appendix 2. In relation to J the subtlety is that eqn. (2.19) may be used to *evaluate J* by monotonic loading of cracks cut to successively longer lengths (as described in Chapter 3) although the physical meaning of 'potential energy available to propagate a crack' is not ascribed to the terms so evaluated and only the characterising meaning is attached to J.

Despite the apparent lack of a sound basis for the use of J in plasticity, both experimental and computational evidence support its application to fracture safety assessment. This will be elaborated upon in Sections 2.2.2.1.2 and 2.2.2.1.3, respectively.

2.2.2.1.2 *The fracture criterion:* $J = J_c$

A number of experimental studies for different materials indicate the existence of a geometry-independent critical J value at fracture. The first such results were published by Begley & Landes [61 and 62]. A general invariance of J_c with geometry was demonstrated, although clouded by uncertainty over slow stable crack growth (cf. also Chapter 3). Further confirming evidence is presented in Figs. 2.18 [63] and 2.19 [64], although it is again doubtful whether slow stable crack growth has been adequately allowed for in these early test results. The method of determining J for Fig. 2.18, was partly from computed values of J versus load and partly from J versus displacement assuming plane strain. The results refer to two configurations of bend test and compact K specimen (CKS) pieces and the difference in treatment (from load or deflection data) reflects the inadequacies of the computational results at that time. The broad agreement between critical J and COD as a function of temperature is clear and is further supported by the trend of J_c results in Fig. 2.19a. Figure 2.19b shows σ_c plotted against J_c for a 20-mm thick C–Mn steel of 300 MN/m^2 yield. The tests were conducted on several geometries: CKS, three-point bend and centre-cracked plates. The computational results by the finite element method are in plane stress. The precise definition of δ is not given. The conclusion is drawn that $J = \sigma_Y \delta$. Closer examination of Fig. 2.19b shows that for small J (low temperature results) $J_c \simeq 1 \cdot 2 \, \sigma_Y \delta_c$ and for large J (higher temperature results) $J_c \simeq 2 \sigma_Y \delta_c$. In short, the results may be used

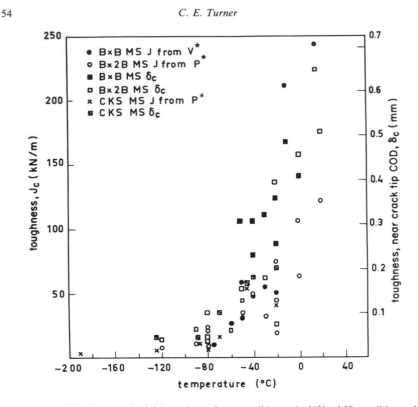

FIG. 2.18. Critical J and COD values for a mild steel [63]. MS = mild steel; B × B = square-section three-point-bend tests; B × 2B = rectangular-section three-point-bend tests; CKS = compact tension (K) tests; J from V^* = J evaluated from displacement; J from P^* = J evaluated from load; δ_c = COD tests.

to support a critical value of J at fracture for several geometries in a broad sense, but no conclusion can be drawn that establishes J as either more or less constant at fracture than is δ. Slow stable crack growth is not mentioned at all.

Data for J_i as a function of thickness was also derived by Griffis [65][3] and showed an independence of size above the thickness $B \simeq 50 J/\sigma_Y$. These results are described more fully in Chapter 3. These data specifically allow for slow crack growth but are restricted to one geometry ($a/W = 0.5$;

[3] J_i denotes J at initiation of crack growth, corresponding to the similar usage of δ_i for COD measured at initiation. The term is thus more specific than J_c and is discussed in more detail in Chapter 3.

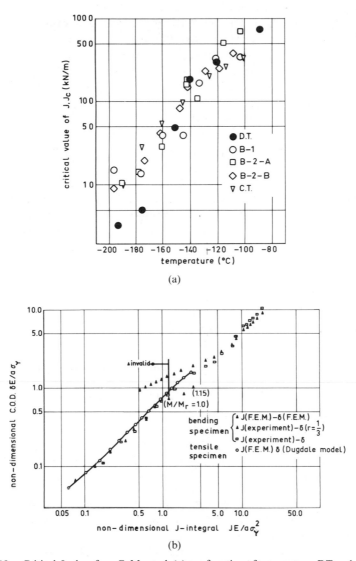

(a)

(b)

FIG. 2.19. Critical *J* values for a C–Mn steel: (a) as a function of temperature, DT = double-notch tension tests, B = various bending tests, CT = centre-cracked tension tests; (b) as a function of COD [64]. (FEM refers to computed finite element results.)

$S/W = 4$) although of various thicknesses and absolute sizes so that the effect of variation of in-plane constraint is not seen.

Robinson's results [12] supported a unique value of J for two geometries, three-point bend and centre-cracked pieces, in a low-strength structural steel (EN 32) and a medium-strength alloy steel (EN 24). It will be recalled from Section 2.2.1.2 that these data also supported a unique value of δ_i, since, to within experimental limits of some 10%, all tests showed $J \simeq 2\cdot2\,\sigma_Y\delta$ for both geometries so that either J or δ (as defined here by near-tip infiltration measurements) served as a fracture parameter for onset of crack growth. No specific comparison was made with the value of δ that would have been obtained from the standardised test procedure [8] conducted on full-thickness pieces.

Thus all the early data are broadly indicative of a reasonableness of J_c comparable to that for δ_c, but give little help in deciding whether one is preferable to the other. The importance of slow crack growth is clearly seen, and as emphasised elsewhere in this chapter the overall philosophy of dealing with effects of size, slow growth and degree of constraint appear more important than the precise criterion adopted for the measure of fracture.

2.2.2.1.3 Computational verification

Several partly interrelated questions requiring verification emerge from the outline of the method given in Section 2.2.2.1:

(i)   path independence of J if evaluated from eqn. (2.18) using the incremental laws of plasticity,

(ii)   effects of extensive yielding and of work-hardening,

(iii) energetic interpretation of J in plasticity and related use of load–displacement measurements for determining J values, and

(iv)   validity of crack tip characterising role of J, defined by eqns. (2.22) under the assumption of power law hardening, for different types of plastic behaviour.

In order to provide answers that are meaningful in an engineering sense the computations to be undertaken for verification should incorporate the incremental laws of plasticity.

(i) *Path independence.* A positive answer to the above question is prerequisite to the relevance of J to fracture analysis. In most of the finite element computations undertaken for verification, small deformation theory is adhered to (i.e. the stresses and strains are computed for the original shape of body and the shape is not redefined as deformation occurs). In what is described here, the incremental laws of plasticity are used with the von Mises' criterion of yield (cf. Appendix 1) and work-hardening is included or omitted at will. Unless specifically so stated, special crack tip elements incorporating a known stress singularity were not used in obtaining the results about to be discussed. A horizontal crack is thus represented by a line where elements 'above' the crack are not connected to those 'below' the crack. The tip is a nodal point which cannot itself deform, although all the elements around it can do so. Thus, as already remarked in connection with computations of COD, some uncertainty must remain on just how realistic is the representation of the crack tip region. If, as in most early studies, constant strain elements were used, then it seemed necessary, or at least desirable to use small elements, typically of the order of $a/100$ to $a/300$ close to the crack tip to ensure at least a plausible degree of realism. Elements have no absolute size factor, all the results being scaled to a dimension such as crack length, a, or plate width, W. Similarly, all stresses scale according to σ_Y and E.

The first such computations for J which the writer is aware of were made by Hayes [66], although earlier elastic–plastic computations of stress and strain fields had been made by Swedlow *et al.* [67] and Marcal & King [68]. Hayes found that, within numerical accuracy of some 7%, the results of integration along some eight or ten different paths were constant for a given overall geometry and condition of (uniaxial) loading provided paths within the three or four elements closest to the tip were ignored, for reasons associated with the obvious inaccuracy of using constant strain elements. Similar results were later obtained by Boyle [69], Sumpter [23] and Riccardella & Swedlow [70]. At least some of these results were based on completely different programs to that used by Hayes, Sumpter and originally Boyle, which was a direct descendant of that described by Marcal & King [68]. In these works, the strain energy term Z, eqn. (2.18), is computed from the integral $\int \sigma \, d\varepsilon$ and is thus a truly incremental term. Thermal or inertial stresses are not considered. Paths were chosen that were wholly within the plastic region (excluding the few elements closest to the tip), partly in the plastic region and partly in the elastic outer region and wholly in the outer elastic region. Provided a path did not cross a large change in element size, all results showed path independence. Sumpter [23] extended his calculations to more extensive plasticity than did Hayes. Results, taken from Sumpter [23] are shown in Fig. 2.20 for cases of both plane strain and plane stress. The extent of yielding is defined by the parameter $qE/D\sigma_Y$ where q is displacement and D is gauge length, i.e. by the parameter $\varepsilon/\varepsilon_Y$, and it is seen that the results for $\sigma/\sigma_Y \simeq 1$ are no better or

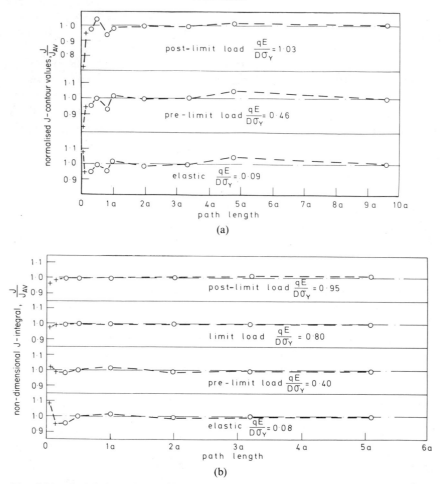

FIG. 2.20. Path independence of J contour values in incremental plasticity finite-element computations. Centre-cracked tension, $2a/W = 0.125$, non-work-hardening; (a) plane strain, (b) plane stress. (Data from Ref. [23].) ' + ' indicates very short path lengths omitted from J_{AV}, the averaged values of J.

worse than for the elastic case where it is rigorously known, that $J = G$ and must, therefore, be path independent. (Lyall-Saunders [71] had of course formulated a contour integral for the elastic case before J came into general usage, but the writer is not aware of any computational studies making use of it.)

Having thus obtained some measure of assurance that the contour integral J value was indeed definable in incremental plasticity computations, we may turn to the remaining questions formulated above.

(*ii*) *Evaluation of J for extensive yielding, with and without work-hardening.* Analogous to the relations (1.5) and (1.6) in linear mechanics it is convenient to write in yielding mechanics

$$R^2 = E'J \qquad (2.24)$$

and

$$R = Y^*\sigma\sqrt{a} \qquad (2.25)$$

where Y^* is a non-dimensional factor corresponding to the constant in eqn. (1.6).

For very small amounts of plasticity $Y^* \to Y$, $R \to K$ and $J \to G$. For significant amounts of plasticity Y^* is a function not only of geometry, but also of the applied stress σ/σ_Y. A typical curve for Y^* is shown in Fig. 2.21. Also shown, dotted, is the LEFM result including the plain strain plastic zone correction where

$$K = Y\sigma\sqrt{a + r_{pl}} = Y_{el}^*\sigma\sqrt{a} \qquad (2.26)$$

is taken to define an effective factor Y_{el}^*.

It will be seen that Y^* increases but little above the elastic value Y for small and moderate values of σ/σ_Y. In this region the LEFM-based Y_{el}^* is a good approximation to Y^*. As $\sigma/\sigma_Y \to 1$, Y^* increases rapidly and Y_{el}^* ceases to be a useful approximation to it. The corresponding load–deflection diagram is shown in Fig. 2.22 and illustrative maps of the extent of the plastic zone are presented in Fig. 2.23. Corresponding points can be identified on each diagram. Thus Y^* starts to increase rapidly at the 'elbow' of the load–deflection curve, B, where the plastic zone starts to spread to a free surface. This state represents the borderline between that described in Section 2.2.1 as 'elastic–plastic' and the subsequent 'gross yield'. It must be clearly recognised that a rigorous application of LEFM would be restricted to a regime as marked in Fig. 2.22, beyond which, if plotted to an enlarged extension axis, the load diagram is clearly non-linear. The positions for $r_{pl} = a/50$ and $a/10$ according to LEFM estimates in plane strain, are marked in Figs. 2.22 and 2.24. The degree of non-linearity and extent of the plastic zone up to A or B are, however, still small in comparison with the fully developed gross-yield region where plastic flow of the whole ligament may occur and a collapse-type situation develop with a limit load state reached for a non-hardening material. In short, the phrase 'extensive yielding' can be interpreted very differently when viewed in relation to the opposite extremes of LEFM or plastic collapse.

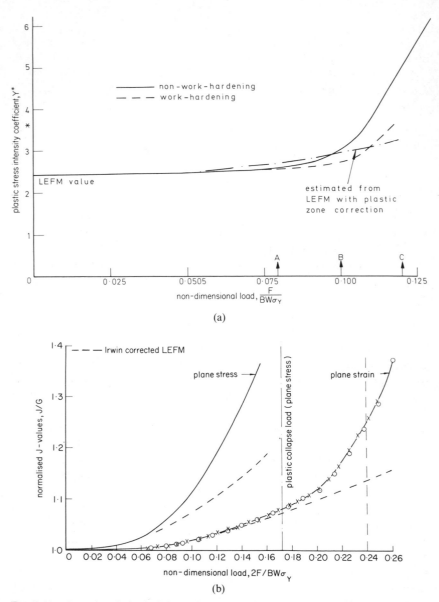

FIG. 2.21.   Intensity of plastic deformation at a crack tip as a function of applied load: (a) plastic stress intensity factor, Y^*, defined by $R = Y^*\sigma\sqrt{a}$, where $R^2 = E'J$, three-point bending, $a/W = 0.5$, $S/W = 4$, plane strain, work-hardening and non-work-hardening. (Refer to Figs. 2.22 and 2.23 for points A, B and C.) (b) J/G; compact tension specimen, $a/W = 0.5$, overall length $= 1.2W$, work-hardening $\sigma_u/\sigma_Y = 1.45$. (Refer to Fig. 2.23b for extent of yielding.) (Neale, B. K. (1975). CEGB Report RD/OB/N 3253.)

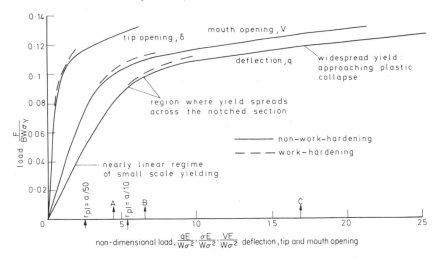

non-dimensional load, $\dfrac{qE}{W\sigma^2}, \dfrac{\sigma E}{W\sigma^2}, \dfrac{VF}{W\sigma^2}$ deflection, tip and mouth opening

FIG. 2.22. Load–deflection and load–crack opening curves. Three-point bending; $a/W = 0.5$, $S/W = 4$, plane strain, work-hardening and non-work-hardening. (Refer to Fig. 2.23 for extent of yielding associated with points A, B and C for non-work hardening case.) For the work-hardening case the degree of hardening is such that $\sigma/\sigma_Y = 1.4$ at $\varepsilon/\varepsilon_Y = 50$ and $\sigma/\sigma_Y = 1.6$ at $\varepsilon/\varepsilon_Y = 100$, representative of mild steel.

A position at the 'elbow' of the load deflection diagrams, where yield stress is first attained on the net section although a collapse mechanism has not yet formed, was found by Sumpter [23] to agree well with limit loads predicted from slip line constraint factors. The data were mainly non-hardening. Load continued to rise thereafter, particularly in plane–strain bending, even for no hardening, and it may well be that finite elements are overstiff to model a true collapse behaviour. This is discussed further in Chapter 4, and some evidence is shown later on the realism of predicted loads in both notched and unnotched problems.

The corresponding plot of J versus deflection q is shown in Fig. 2.24, non-dimensionalised in terms of $JE/W\sigma_Y^2$ and $qE/W\sigma_Y$. The start of this curve, if calculated from LEFM, is parabolic. The linear slope may be calculated from rigid–plastic collapse. Such an estimate is shown dotted in Fig. 2.25. The elastic calculation was continued up to the load causing collapse and the linear slope regime then grafted on, a technique that differs slightly from that originally suggested in Ref. [72]. Subject to certain additional features to be described in Section 2.2.2.1.4, Figs. 2.21 and 2.24 are representative in form of all the elastic–plastic computations known to the writer.

The effect of a particular degree of work-hardening, representative of a

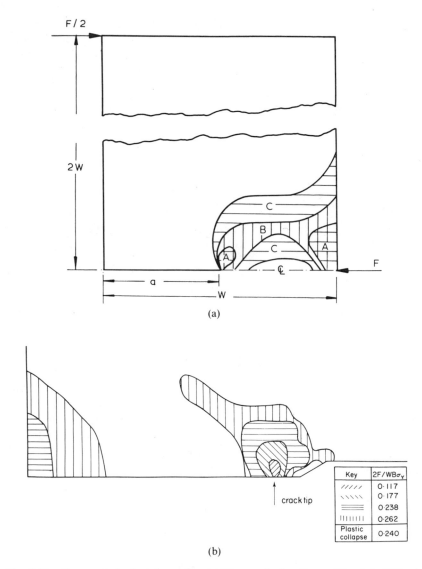

(a)

(b)

FIG. 2.23.   Extent of plastic deformation. (a) Three-point bending; $a/W = 0.5$, $S/W = 4$, plane strain, non-work-hardening. (b) Compact tension; $a/W = 0.5$ as in Fig. 2.21b. Refer to Figs. 2.21, 2.22 and 2.24 for points A, B and C.

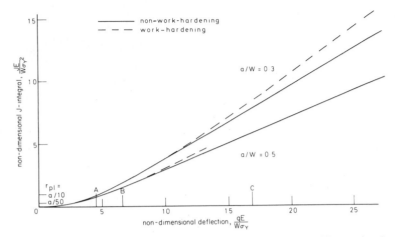

FIG. 2.24. Non-dimensionalised curves of *J* against deflection. Three-point bending; $a/W = 0.5$ and $a/W = 0.3$; $S/W = 4$, plane strain, work-hardening and non-work-hardening. Refer to Figs. 2.22 and 2.23 for points A, B and C.

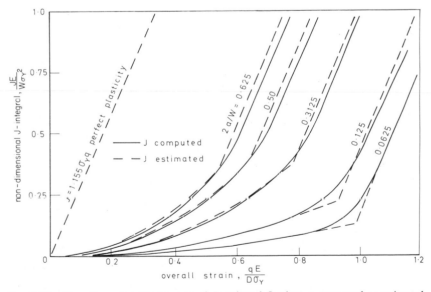

FIG. 2.25. Non-dimensionalised curves of *J* against deflection, *q*, computed or estimated from LEFM, plus rigid–plastic behaviour. Centre-cracked tension; various $2a/W$, $D/W = 2.5$; plane strain, non-work-hardening ([23] and Sumpter, J. D. G. *et al.* (1973). ICF-3, (I), 433.)

C. E. Turner

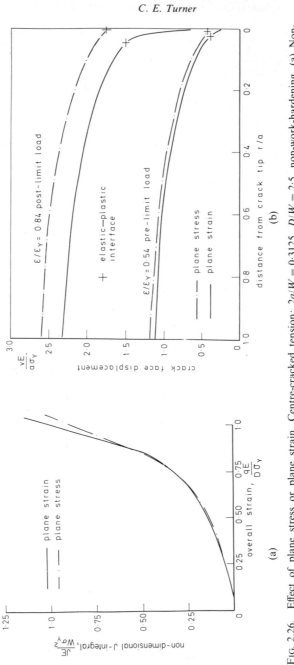

Fig. 2.26. Effect of plane stress or plane strain. Centre-cracked tension; $2a/W = 0.3125$, $D/W = 2.5$, non-work-hardening. (a) Non-dimensionalised J versus deflection; (b) crack profiles and tip opening [73].

structural mild steel, is also shown in these figures. The effect of hardening is significant in terms of load (Fig. 2.21), but rather small in terms of deflection (Fig. 2.24). These remarks are based on deep-notch test piece data and this point will be referred to again in Section 2.2.2.1.4. The general similarity between Fig. 2.24 for J versus deflection and Fig. 2.4 for COD versus deflection may be noted.

The effect of the degree of constraint imposed is illustrated by the results of plane stress and plane strain computations shown in Fig. 2.26a. Up to the limit shown, the results of J versus displacement are very similar but it must be noted that if crack tip details are studied the results are quite different. This is typified by the results shown in Fig. 2.26b. Displacements and strains are much higher and maximum stress lower for plane stress than for plane strain. This is not, of course, unexpected but emphasises the difficulty of characterising a structural (as distinct from formalised test piece) situation which will so often fall between the plane stress and plane strain cases.

(*iii*) *An energetic meaning for J in plasticity?* It has already been pointed out that for incremental plasticity materials J does not appear to have the physical meaning of potential energy release rate, i.e. the energy rate available to drive an increment of crack. Sumpter & Turner [73] computed a value of energy that could be explicitly defined for plasticity, to which the J contour integral values could be compared.[4] The term $\bar{J}$ was defined as

$$\bar{J} = \lim_{\Delta a \to 0} \left[\int_S F \Delta q \, \mathrm{d}S - \Delta U_{\mathrm{el}} - \Delta U_{\mathrm{pl}} \right] \qquad (2.27\mathrm{a})$$

where $F \Delta q$ is the external work integrated over the surface S to which the forces are applied. For fixed displacement

$$\bar{J} = - \left. \frac{\partial U_{\mathrm{el}}}{\partial a} \right|_q - \left. \frac{\partial U_{\mathrm{pl}}}{\partial a} \right|_q \qquad (2.27\mathrm{b})$$

that is, the difference (with due regard to sign) between the elastic and plastic energy rates each evaluated at fixed displacement.[5] Equation (2.27)

[4] The results come from Ref. [23] where the notation J^* was used, but is here changed to $\bar{J}$ to distinguish it from J^* as discussed in Section 2.2.2.2. The $\bar{J}$ discussion was first presented to a meeting on numerical methods at the University of Stuttgart in April 1973, but not published until 1976 [73].
[5] It should be noted that although $U_{\mathrm{T}} = U_{\mathrm{el}} + U_{\mathrm{pl}}$ where U_{T} is the area under the load displacement curve this expression cannot be interpreted as a statement of physical energy *rates*. When the rates $\partial/\partial a$ are formed, $\partial U_{\mathrm{T}}/\partial a = \partial U_{\mathrm{el}}/\partial a + \partial U_{\mathrm{pl}}/\partial a + \gamma$ (at fixed displacement), where γ is the surface energy.

is the energy rate available to overcome the Irwin–Orowan surface energy (i.e. including local surface plasticity) γ_e, for fracture in an elastic–plastic material under fixed grips. $\bar{J}$ is not, however, single-valued for an incremental plasticity material, but depends upon the loading path. The program available at the time did not incorporate the facility for growing a

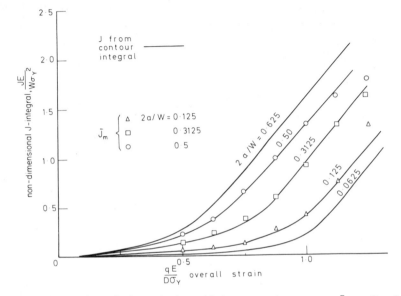

FIG. 2.27.   Comparison of J-integral values with the monotonic energy rate $\bar{J}_m$ as a function of displacement, q. Centre-cracked tension; various $2a/W$, $D/W = 2{\cdot}5$, plane strain, non-work-hardening [73].

crack whilst the load or displacement conditions remained fixed, i.e. for determining the energetically meaningful magnitude $\bar{J}_i$ (i for crack growth increment at fixed grip). The term $\bar{J}_m$ (m for monotonic loading) was therefore evaluated by computing U_{el} and U_{pl} for five different crack lengths in centre-cracked plates with $2a/W$ from 0·0625 to 0·625 using non-work-hardening incremental plasticity. For the three intermediate crack lengths $\Delta U_{el}/\Delta a$ and $\Delta U_{pl}/\Delta a$ were evaluated, where Δa is the difference in crack length between successive computations for the monotonic loading problem. $\bar{J}_m$ was then formulated according to eqn. (2.27) as the difference between the two terms. The results are shown in Fig. 2.27, where it is seen that this evaluation of $\bar{J}_m$ is within computational error of the value of J derived from the contour integral. Thus, for this one set of results,

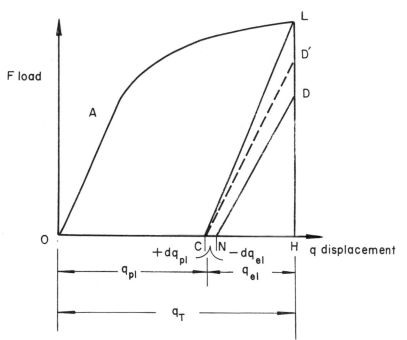

FIG. 2.28. Energy release rate available in elastic–plastic crack growth.

$J_{\text{contour integral}}$ is numerically equal (to computational accuracy) to $\bar{J}_m$, which is a definable energy difference for monotonically loaded cracks cut initially to different lengths to which no physically useful meaning has been attached.

An estimate of the elastic energy available on crack growth from a load deformation curve with extensive plasticity can be made by a direct extension of the well-known method for calculating G in LEFM from change in compliance. Referring to Fig. 2.28 for unloading with no crack growth, LC would be parallel to the loading line OA, ignoring effects of hysteresis and reversed plasticity. If crack growth occurs at constant displacement in a purely elastic situation, the load falls from L to D′. The final response, dF_{el}, comprises two terms, one due to change in net section area (corresponding to LD) and one due to change (increase) in net section stress (corresponding to DD′) such that together with the reduction in gross section stress, the overall displacement remains unchanged. The second term is not possible in plasticity if the net section stress, σ_n, is limited to σ_Y. After extensive yielding, a good estimate of the decrease in load, LD, is the change with crack length of the plastic limit load F_L. The additional drop in load D'D is thus $(dF_L - dF_{el})$. The elastic compliance for unloading from D, now DN, has a smaller slope (greater compliance) than LC and is equal to D'C just as for the LEFM case. The difference is that the unloading line from D does not pass through the transposed origin C as would have been the case for a purely elastic event unloading from D', so that there is a change in length $CN = +dq_{pl} = -dq_{el}$. The area corresponding to $GB\,da$ in the LEFM case is LD'C. An extra energy release D'DNC is available in the elastic–plastic case. This extra area may be written approximately as either $CN\cdot HD' \simeq -dq_{el}F_L$, or as $D'D\cdot HC \simeq -(dF_L - dF_{el})q_{el}$. The total energy rate available is here called I, thus

$$BI = -\partial U_{el}/\partial a|_q = B\bar{J} + \partial U_{pl}/da|_q \qquad (2.28\text{a})$$

$$= BG - F_L\,\partial q_{el}/\partial a \qquad (2.28\text{b})$$

$$= BG - q_{el}(\partial F_L/\partial a - \partial F_{el}/\partial a) \qquad (2.28\text{c})$$

Note dq_{el}, dF_L, dF_{el} are inherently negative so that $I > G$; q_{el} is a known elastic term. $dq_{el}/q_{el} = (dF_L - dF_{el})/F_L$; dF_L can be evaluated if the limit load F_L is known and dF_{el} can be found from elastic compliance values as for LEFM. Clearly a mild degree of work-hardening can be incorporated in the argument by use of a flow stress, but the expressions may not be adequate for a high degree of hardening or at a regime near the 'elbow' of the load deflection curve well before the net section stress reaches yield.

These equations are applied in Section 2.4.1.

The simple derivation just given is based on conventional plasticity, but according to the arguments of Rice [60] an evaluation of the energy rate available for crack growth in plasticity tends to zero as $\Delta a \to 0$, although giving the well-known Griffiths balance for elastic behaviour. Until recently this argument does not seem to have been discussed widely in the literature. It implies that the difference between monotonic loading and incremental crack growth is the essence of the problem. Several computations subsequent to Ref. [73] have taken these differences into account by modelling the changes in local crack tip plasticity as a crack spreads [69, 74–76], all of which find a non-zero energy release rate for the growing crack. Kfouri & Miller [77] suggested that this energy rate *must* be evaluated at a finite value of Δa, corresponding to some fracture process zone. Their term, called G^Δ, is conceptually similar to $\bar{J}_i$ described above, with the essential difference that, consistent with Ref. [60] $\Delta a \nrightarrow 0$ and the method of evaluation is from the local work of separation not from the global energy balance. It is described further in Appendix 2. By implication other finite element evaluations for the growing crack are fortuitously non-zero, for the reason that in numerical solutions the crack growth increment is perforce finite. The particular derivation leading to the conclusion that the external and internal (elastic + plastic) work rates balance with no surplus for crack growth (surface energy) in plasticity as $\Delta a \to 0$ has been challenged [78]. The whole argument has also been restated by Rice [79]. Yet further contradictory numerical evidence is outlined in Appendix 2 so there is urgent need for clarification of the issue.

It seems to the writer that there is agreement that, if a strong singularity in energy density exists in the crack tip model such that the energy rate is finite as $\Delta a \to 0$, then perforce the balance of external and internal energy rates *does* provide a specific term which corresponds to the effective surface energy. In LEFM terms, where both the stress and strain are proportional to $K/(2\pi r)^{1/2}$ the energy is singular according to $K^2/2\pi E r$ so that, integrated over the volume $2\pi r \, dr$ as $r \to 0$, a non-zero energy rate, G, is available that requires a corresponding surface energy rate G_c. A less strong singularity does not provide this non-zero term as $r \to 0$ and the several tentative solutions for the growing crack suggest singularities (such as the $\ln(1/r)$ term in Ref. [60]) that do not have the required strength. Thus, there is no term *in the model* that, as $\Delta a \to 0$, will provide the energy rate to offset surface energy, the normal requirements of equilibrium and compatibility in the continuum formulations themselves implying an exact balance of energy. Even the *onset* of crack growth (i.e. initiation as normally envisaged) implies a departure from the strong singularity of the monotonic solution that is carried over, at least by implication, from the HRR power law non-linear elastic solutions to the plasticity field. If an energy balance is to be used for these plasticity cases where the strong singularity is not found, then it seems necessary to invoke another mechanism such as finite COD, which here implies not just the opening of the crack profile but the dissipation of work 'at' the crack tip (for which 'crack-opening stretch' may be a more evocative term) as the crack growth occurs, thus again forcing a non-zero energy term into the continuum model. Without either the strong singularity or a device such as crack-opening stretch, it does not seem possible to uncouple the 'remote' plasticity from the 'local surface plasticity' as $\Delta a \to 0$ in the continuum formulation. These points are further discussed in Section 2.4.1.

(iv) *J estimates from measurements of compliance or work done.* While the foregoing discussion has focused on the interpretation of *J* in terms of an *available* energy rate, determination of *J* from compliance measurements is based upon its meaning as a *dissipative* energy rate, as evidenced by the general compliance relationship

$$J = \frac{-1}{B}\left(\frac{\partial U}{\partial a}\right)\Big|_q \tag{2.29}$$

where $U =$ total absorbed energy (or area under the load/load-point deflection curve) at load-point deflection q.

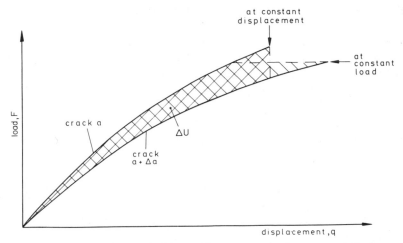

FIG. 2.29. J evaluated from $-\Delta U/\Delta a$ at either constant load or constant displacement. (Based on Ref. [61].)

The various measuring techniques for J described in Chapter 3 all derive from this relationship. The earliest of these, involving the assessment of J from the change in area under the aforementioned curve with increase in crack length, as shown in Fig. 2.29, was verified by both Boyle [69] and Sumpter [23] by comparison of J values from computed areas under the load deflection curves for successively longer initial cracks with corresponding J values from contour integrals. To within computational accuracy (of some 5%) the results agreed.

A method was subsequently proposed by Rice *et al.* [80] for determining J from the load/load-point deflection curve of a single deeply notched specimen in which plasticity is restricted to the vicinity of the ligament ahead of the crack. Verification was complicated by different views of the proper value to be inserted for the term U in the resulting formula

$$J = \eta U/B(W - a) \tag{2.30a}$$

Early applications were either restricted to cases in which the plastic component of work dominated, such as deep-notch ($a/W \geq 0.7$) bending and compact tension, or the total work was 'corrected' by subtraction of a term for the elastic displacement with no crack. Some test data supported this usage [81]. Meanwhile, following Turner's re-analysis of LEFM data [82] into the form

$$G = \eta_{el} U_{el}/B(W - a) \qquad (2.30b)$$

(where U_{el} is by definition the elastic term $Fq/2$ up to the point of interest, probably the (maximum) load at fracture in a brittle test) it was realised that relationships of the form of eqns. (2.30a) and (2.30b) could be established, at least for some configurations, based on total (elastic + plastic) work[6] and a wider interpretation of the η factor has since emerged.

Following here its usage in the J-work relationship, the immediate point of interest is that in Ref. [80] $\eta = 2.0$ for the deep-notch cases discussed and in Ref. [82] $\eta_{el} \approx 2.0$ for $S/W = 4, 0.4 \lesssim a/W \lesssim 0.8$, suggesting that for the deep-notch bend configuration total work could be used with $\eta \approx 2$. This usage was pursued by Sumpter [23] who suggested writing eqn. (2.30a) in the form of an addition of elastic and plastic components:

$$J = \frac{\eta_{el} U_{el}}{B(W - a)} + \frac{\eta_{pl} U_{pl}}{B(W - a)} \qquad (2.30c)$$

η_{pl} was derived from

$$J_{pl} = -\frac{1}{B}\frac{\partial U_{pl}}{\partial a} = \frac{-1}{B}\frac{\partial F_L q}{\partial a}\bigg|_q \qquad (2.31)$$

where F_L is the limit load. In bending $F_L \propto (W - a)^2$ whence $\eta_{pl} = 2$ when $\partial/\partial a$ is formed.

In single-edge notch tension $F_L \propto (W - a)$ whence $\eta_{pl} = 1$. η_{el} is calculated from known LEFM solutions as shown in Ref. [82].[7]

---

[6] 'Total work' implies no correction for components due to elasticity or presence of the crack. It must not of course contain extraneous errors due to testing procedures and must be determined up to the point of interest, probably initiation of crack growth, since J is not directly relevant after slow crack growth. Thus in practice the determination of 'total work' is likely to require careful measurement, as discussed in Chapter 3.

[7] Reference [82] defined η_{el} as the reciprocal of the term used here and omitted the contribution of shear deflection in evaluating it. Another definition of η_{el} is given in Appendix 4.

The relation is

$$\eta_{el} = \frac{G}{U_{el}/B(W-a)} = \frac{(W-a)}{C}\frac{\partial C}{\partial a} = (W-a)(Y^2 a/\int Y^2 a\, da) \quad (2.32)$$

where Y is the LEFM shape factor (eqn. 1.6). The compliance C (eqn. (1.2b)) is evaluated by integrating eqn. (1.2a) (with G expressed as K^2/E and hence in terms of Y) and using the unnotched compliance, C_o to evaluate the constant of integration. For the deep-notch three-point bend cases, as remarked earlier, $\eta_{el} \approx 2$ so that $\eta_{el} \approx \eta_{pl}$ and eqn. (2.30c) reduces to eqn. (2.30a) with U interpreted as total work and η given the value 2. Subsequent computations supported this (Fig. 2.30), even with a degree of work-hardening, for the deep-notch cases. Srawley [83] has come to the same conclusions by showing, yet more generally, that

$$\eta = \frac{J}{U/B(W-a)} = \frac{\partial \ln U}{\partial \ln (W-a)} \quad (2.30d)$$

where U is the total work, again pointing out that the elastic and plastic components can be combined into one term for the deep-notch bend cases just referred to with $\eta = 2$. This now seems generally agreed to be the most fruitful basis for the use of eqn. (2.30).

Merkle & Corten [84] have derived an expression for the compact tension piece, and a proposal for a test procedure based on eqn. (2.30c) has been made [85]. The usage of J in terms of work done for standard deep-notch test purposes has been detailed in Ref. [86] and is discussed further in Chapter 3. Robinson [12] made use of both Rice's [80] and Sumpter's [23] estimates for J in bend and centre-cracked pieces, finding, as already noted, good agreement between the J values at initiation for the two configurations.

At a fairly early stage in these developments Wou [87] had computed data for three-point bending (using non-hardening incremental plasticity; $S/W = 4$, $a/W = 0.5, 0.3, 0.1$), and found that for all three cases η was constant to within 10% from elasticity through to extensive plasticity. The values were 1.9, 1.8, 1.1, respectively, based on U taken as total work up to the deflection in question. The implication of Wou's finding that η did not vary appreciably with the extent of plasticity for the one rather shallow notch-depth studied was not pursued. Only the deep-notch cases were noted, where both η_{el} (from Ref. [82]) and η as computed by Wou were nearly 2. This tended to focus attention on the interpretation of U in eqn. (2.30a) as total work done, without correction, as previously remarked.

It was also noted [23], though not pursued at the time, that the slope of $J - q$ in the plastic region, for which conventional estimates then available from Ref. [80] for bending, give $J/q_{pl} = 2L(W - a)\sigma_Y/S$, was in accord with the cases $a/W = 0.5$ and 0.3 but not 0.1. It is now clear that these effects stem from the same cause and that, if it can be estimated satisfactorily, the η_{pl} factor has a usefulness not only in J- work relations (as in eqn. (2.30)), but also for J-displacement or strain relationships, as discussed in Section 2.2.1.4, and instability concepts, as discussed in Section 2.4.1. Let the definition of η_{pl} be

$$\eta_{pl} = - \left(\frac{W - a}{W}\right) \frac{1}{F_L} \frac{\partial F_L}{\partial(a/W)} \tag{2.33a}$$

where F_L is the limit load conveniently written for tension or bending

$$F_L = L\sigma_Y B(W - a)^N/S^{N-1} \tag{2.34}$$

$N = 1$ for tension and $N = 2$ for bending; S is interpreted either as span or as gauge length D as appropriate. L is the constraint factor, here taken as a geometric variable though also containing the plane strain effect,[8] if von Mises criterion is used. Since, for rigid plastic material,

$$J_{pl} = \frac{-1}{B} \frac{\partial U_{pl}}{\partial a}\bigg|_q = \frac{-1}{B} \frac{\partial F_L q}{\partial a}\bigg|_q = \frac{-q}{B} \frac{\partial F_L}{\partial a} \tag{2.31 bis}$$

and, from eqns. (2.33a) and (2.34),

$$\frac{\partial F_L}{\partial a} = - \frac{\eta_{pl} F_L}{W - a} \tag{2.33b}$$

then

$$J_{pl} = \frac{\eta_{pl} U_{pl}}{B(W - a)} \tag{2.30e}$$

and the slope

$$dJ/dq_{pl} = \eta_{pl} L(W - a)^{N-1}\sigma_Y/S^{N-1} \tag{2.35}$$

and η_{pl} can be rewritten as

$$\eta_{pl} = N - \left(\frac{W - a}{W}\right) \frac{1}{L} \frac{\partial L}{\partial(a/W)} \tag{2.33c}$$

It is immediately clear that η_{pl} departs from 2 in bending and 1 in tension when, in eqn. (2.33c), $\partial L/\partial(a/W)$ is not negligible. For deep notches, where

---

[8] That is, a further multiplicative factor of $2/3^{1/2}$ ($= 1.155$).

constraint is fully established, it may vary little with crack depth but in other cases, notably shallow notches, the second term in eqn. (2.33c) dominates over the first. Further discussion here is restricted to the usage in eqn. (2.30e) with the other topics followed up in Sections 2.2.1.4 and 2.4.1.

In using eqn. (2.30e) an additional feature emerges, consistent with Ref. [80], in the restriction of plasticity to the vicinity of the ligament ahead of the crack (i.e. the notch slip line fan or spiral region). Integrating eqn. (2.31) gives

$$U_{pl} = -B\!\int\! J_{pl}\, da + f \qquad (2.36)$$

where f is a function of variables other than a. In short $\int J_{pl}\, da$ relates only to the plastic work of the notched region. If there is yield remote from the notch and considering only

$$_o U_{pl} = {}_n U_{pl} + {}_r U_{pl} \qquad (2.37a)$$

where $_o U_{pl}$ is overall plastic work, $_n U_{pl}$ is plastic work in the notched region, and $_r U_{pl}$ is plastic work remote from the notch. For deep notches $_r U_{pl} = 0$ since there is no remote plasticity. For very shallow notches in tension once yield is uncontained U_{pl} is approximately proportional to gauge length since the load deflection diagram is barely affected by the shallow notch *if there is adequate work-hardening*, so that

$$_o U_{pl} \propto D; \; _n U_{pl} \propto D_n; \; _r U_{pl} \propto (D - D_n) \qquad (2.37b)$$

therefore

$$_n U_{pl} \simeq {}_o U_{pl}(D_n/D) \qquad (2.37c)$$

where D is overall gauge length and D_n is the effective gauge length of the notched region, estimated, for example, from the extent of the slip line region.

In using eqn. (2.30e) the term required is $_n U_{pl}$. For deep notches or no work-hardening, this is equal to $_o U_{pl}$, but for shallow-notch work-hardening cases eqn. (2.37c) applies. For shallow single-edge notch tension with slip at $45°$, $D_n \approx 2W$, for centre crack plate $D_n \approx W$. For *no work-hardening* in tension, plasticity is still restricted to the notched region so that a discontinuous relation appears (see Section 2.2.2.1.4) between the mild work-hardening and no work-hardening cases. In bending the situation is less clear. The central plastic hinge is of a diameter comparable to W but because $F_L = 1\cdot5\, F_Y$ (for bars of rectangular cross-section) plasticity spreads along the span on the upper and lower surfaces of the bar even with no hardening so that the various effects merge with no apparent

TABLE 2.1
REPRESENTATIVE VALUES OF η FACTOR, INCLUDING SHALLOW NOTCHES

a/W	0·025	0·05	0·1	0·3	0·5	Remarks
Bending						
$S/W = 4$						
any^a η_{el}	0·30	0·53	0·98	1·71	1·98	Eqn. (2.32) Theoretical
η_{pl}	0·25	0·50	1·0	2·0	2·0	Appendix 4} estimate
4 $\eta_{imax}{}^b$	0·37	0·65	0·98	2·1	1·96	Computed; mild work-hardening
8 $r_{imax}{}^b$	0·25	0·44	0·69	$>1\!\cdot\!45^c$	$1\!\cdot\!90^c$	Computed; mild work-hardening
SEN tension						
$D/W = 4$						
any^a η_{el}	0·048	0·096	0·20		1·0	Eqn. (2.32) Theoretical
η_{pl}	0·075	0·15	0·30			Appendix 4} estimate
4 $\eta_{max}{}^b$	0·070	0·15	0·30			Computed; mild work-hardening

DEN tension η_{pl} for $2a/W = 0.625$: computed, 0·44; estimated, eqn. (A.4.9): 0·57

^{a} Beyond gross yield the effective value would diminish for shallow-notch work-hardening cases: see text.
^{b} At about $q/q_Y \simeq 1\cdot0$ (see text and Fig. 2.30).
^{c} Computations not carried to large enough displacement.

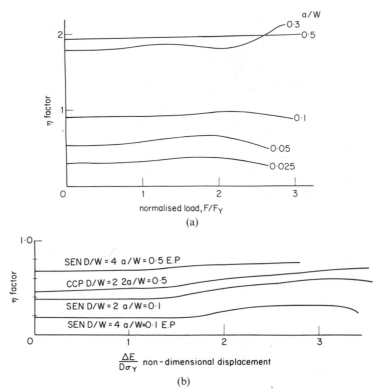

FIG. 2.30.   Variation of the η factor in the expression $J = \eta U/B(W - a)$ (eqn. (2.30a)): (a) with extent of plasticity, F/F_Y computed for three-point bending; $S/W = 4$, various a/W; $F_Y = 4B(W - a)^2\sigma_Y/6S$, the nominal load at first yield of the ligament; (b) with extent of plastic deformation $\Delta E/D\sigma_Y$; computed for various tension cases. EP = ends kept parallel.

discontinuity and depend on the span/width ratio including the limiting case of pure bending.

Some intermediate notch-depth ratio cases have been tested by Dawes [88] using eqn. (2.30c) with appropriate values of η_{el} and with $\eta_{pl} = 2$ in tests on weldments where moderate notch-depth ratios ($a/W \approx 0.2$–0.3) were required to locate the notch tip in the desired metallurgical zone of the weldment. Chipperfield [89] has also used eqn. (2.30c) in a similar way for impact tests on pieces modified to restrict the angle of bend, so that by testing successive pieces to different angles, initiation could be detected.

Such notch-depth ratios appear to be on the borderline of where the use of $\eta_{pl} = 2$ is acceptable. Some estimates of η_{pl} for shallow notches are made in Appendix 4 with a few values given here in Table 2.1 so that the general

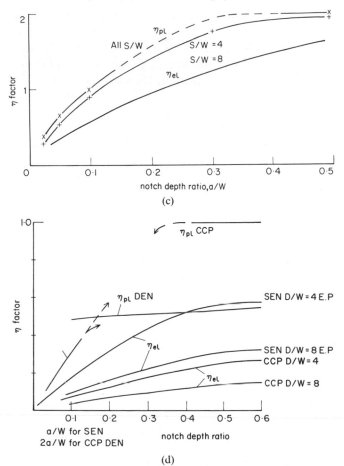

FIG. 2.30—*contd.* (c) Estimates of η_{el} and η_{pl} in three-point bending as a function of a/W for two span ratios S/W; (d) estimates of η_{el} and η_{pl} in various tension cases as a function of a/W.

trend can be seen. Further computed and estimated data are given in Fig. 2.30. It is seen that the rough agreement between η_{el} and η_{pl} is less close (on a fractional basis) than for the deep notches. Within this variation there is a trend for η_{pl} to be greater than η_{el} for contained yielding but the *effective value* of η_{pl} then tends to diminish. This effect is the gradual reduction from η_{pl} to some lower value as yield spreads away from the ligament. In tension this lower value would be $\eta_{pl}D_n/D$. There is thus a regime around $q/q_Y \approx 1.0$ where η_{pl} is more relevant than η_{el}, but the effect of work-hardening, if present, has not yet created the extensive remote plasticity that later brings about the further effects of gauge length or span. As will be seen in Section 2.2.2.1.4 these trends have their counterparts in terms of J–displacement graphs and will be further discussed there. Usage of the η factor to relate J to work done (eqns. (2.30a) to (2.30e)), must clearly be made with caution outside the rather deep notch cases and is probably best avoided if alternative methods are available.

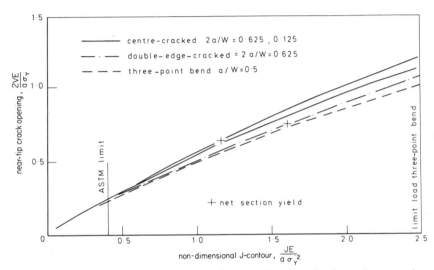

FIG. 2.31. Uniqueness of near-tip deformation as a function of J for various test-piece configurations. Plane strain, non-work-hardening, displacement measured at $a/160$ behind the tip [73].

(*v*) *A characterising meaning for* J. Since, according to eqn. (2.20), J characterises the crack tip singularity for a power law hardening non-linear elastic material, it may be asked whether this is also true for an incremental plasticity material. Recalling the difficulties of correctly modelling the crack tip behaviour, particularly in programs with constant strain elements as used for the great majority of elastic–plastic computations, the question cannot be answered categorically. Sumpter & Turner [73], however, attempted to study the question by examining some near tip features of various calculations. They argued that displacement is computed more accurately in a finite element study than stress or strain so that a useful guide to the characterising role of J would be gained by comparing near tip displacements in different geometries as functions of J, for a given mesh and computational routine. If the near tip displacements were indeed exactly characterised by J, then the results for the several geometries would be identical. In fact, a crack tip system would exist, unique for a given value of J. Their results, in Fig. 2.31, show that for the three configurations noted, the near tip displacements (at $x = -a/160$) are closely similar but not identical functions of J. Larsson & Carlsson (Chapter 1, Ref. [5]) had of course already shown that such different geometries gave different plastic zone sizes and they, as well as Rice, [90] had suggested that this was caused by the different in-plane stress, σ_x, parallel to the crack, near but not at its tip, corresponding to different values of the higher order terms in the series expansion for the complete stress fields. Rigid–plastic slip-line field solutions also suggest that the crack tip conditions will depend upon the geometry of the component.[9] The direct carry-over of the characterising role of J to real plasticity materials would imply that the several effects of plasticity in modifying the linear elastic field (and thus the aforementioned results of Larsson & Carlsson), and also of elasticity and perhaps work-hardening in modifying the rigid–plastic slip-line field solutions, all tend to diminish rather than accentuate any differences that may exist between the various configurations.

Figure 2.32 shows that the near tip displacement, v, as a function of J is linear on a logarithmic plot. This is indeed as would be expected from eqns. (2.20) and suggests that the use of incremental plasticity, with or without hardening, does not affect the role of J as a measure of the severity of crack tip deformation, at least for variations in the extent of yielding in a given geometry with its own implied degree of in-plane biaxiality. The significance attached to this observation in Ref. [73] was that the use of small test pieces fracturing with extensive yield is thus justified for the

---

[9] The hydrostatic stress, σ_H, increases from a low value at a free surface (where σ_x perpendicular to the surface is zero, σ_Y is the applied stress and σ_z that induced to maintain plane strain) according to the angle of rotation of the slip line. Thus there is no elevation of hydrostatic stress for a collapse system with straight slip lines, such as commonly envisaged for centre-cracked plates, but a significant increase along a curved slip line such as envisaged for plastic hinges in bending. The maximum elevation of stress from a free surface occurs for the deep edge crack for which the Prandtl solution gives $\sigma_{max} = (2 + \pi)k$ where k is the yield stress in shear. Using the Tresca shear stress criterion of yielding $\sigma_{max} \simeq 2 \cdot 57 \sigma_Y$; using the von Mises criterion $\sigma_{max} \simeq 2 \cdot 98 \sigma_Y$. The greater the hydrostatic tensile stress, the greater the in-plane tensile stress σ_x parallel to the crack close to its tip will be. Such solutions are discussed in some detail in connection with fracture in Ref. [54]. Some plate solutions are given in Refs. [91 and 92] and some bending problems are discussed in Refs. [93 and 94].

C. E. Turner

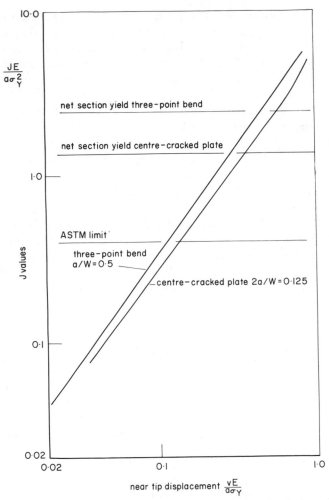

FIG. 2.32. Near linear log–log relationship between J and near-tip displacement. Three-point bending; $a/W = 0.5$; $S/W = 4$ and centre-cracked plate, $2a/W = 0.125$, plane strain, mildly work-hardening. (Based on Ref. [73].)

prediction of the fracture behaviour of large pieces of the same geometry that would fracture below yield. This may be so in particular for configurations in which a high constraint is maintained even in the gross yield regime, e.g. three-point bend. For centre-cracked plates, a configuration in which constraint is lost as gross plasticity spreads, it was shown [73] that the linearity between J and tip displacement (on a logarithmic scale) was not quite as closely maintained once net section yield had occurred.

If accepted as it stands, the use of J for small-scale testing purposes, from which a value of J_i is derived and hence K_{Ic} inferred for use in LEFM situations, is validated, subject to some uncertainties over the proportional amount of slow crack growth found in J and K testing, already mentioned in the preceding text and to be discussed further in Section 2.4.1. The extension of J, or indeed COD, to design in the plastic region requires that these concepts be yet more generally valid to include all conditions of loading, yielding and geometry. There seem to be various opinions as to how closely this is in fact so.

Hilton & Sih [95] have described some computations in which a remote transverse stress $_r\sigma_x$ is applied to a centre-cracked plate in addition to the opening stress $_r\sigma_Y$. They made finite element computations using a 'total' theory of plasticity, which would lead to results identical to eqns. (2.20), but used a stress–strain law linear up to σ_Y and power-hardening beyond. The crack tip was modelled by a special singularity element, and calculations made in plane stress. The results are expressed in terms of the ratio of the plastic to elastic strain concentration factors, which are related to J as outlined (following eqn. (2.21)) in connection with Ref. 56. The crack tip strains depend on biaxiality, the effect of negative transverse stress and low hardening being produced, but J contour values are not quoted, nor is the incremental case discussed.

Miller & Kfouri [96] performed incremental plasticity computations in plane strain with a mild degree of linear work-hardening using rectangular isoparametric elements including the effects of finite geometry changes. A transverse stress $_r\sigma_x$ is actively applied to the lateral edges of a centre-cracked plate, of value either zero or equal in magnitude—with either positive or negative sign—to the applied stress $_r\sigma_Y$ perpendicular to the crack. The results shown for an elastic–plastic computation are expressed in terms of the local σ_Y as a ratio to applied $_r\sigma_Y$, which is itself just under $\sigma_Y/2$. The 'ear' of plasticity is then of an extent around twice the crack length but still only reaching one quarter of the way to the remote boundary. The degree of plasticity is not very extensive, while well beyond LEFM. The results are shown in Fig. 2.33. The maximum stress appears at about $a/10$ (the first nodal point) where the values differ significantly for the three cases. However, Miller & Kfouri argue that, at the tip, there is a free surface both in reality and in a large-geometry-change model of the type they use, at which the transverse stress must be zero, so that in plane strain the axial component is $1\cdot155\,\sigma_Y$ in each case. Although no direct reference is made to

J in this last discussion, the conclusion that the stress at the crack tip is the same for all three stress ratios is just what a characterising argument would imply. However, the stress values adjacent to the tip (at $r = a/10$ for the elements used) vary from case to case. In the context of a characterising parameter, this would imply that, at that distance, the higher order terms in the series expansion of the stresses are already overriding the leading term.

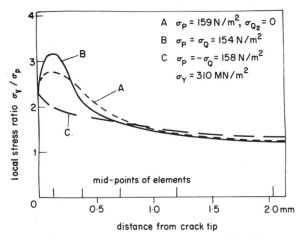

Fig. 2.33.   Effect of biaxial loading on stresses ahead of the crack tip. Centre-cracked plate; $2a/W = 0.125$; $D = W$; linear work-hardening slope $E/210$; finite geometry change [96].

Results of further biaxial computations made with the conventional assumptions of small geometry change (SGC) are show in Fig. 2.34. Plasticity extends to the free boundary and the smallest element is $a/160$.

Clearly, *J* is a function of the biaxiality in terms of applied stress or displacement, as in the work described previously [95, 96]. Figure 2.35 shows the near-tip deformations. These results tend to show that although at distances such as $a/40$ the crack displacements vary from case to case, near the tip these differences tend to disappear. For example [97], at $a/20$, two particular cases differ by 15 %, but at $a/160$ by only 5 %. These results cannot be accepted without questioning the role of large geometry change (LGC). Until recently the argument was not clear cut. From the results of Rice & Johnson [26] for the several values of E/σ_Y quoted, the extent of the zone in which the LGC results differ substantially for the SGC model is about $r = K^2/E\sigma_Y$. This distance is marked on Fig. 2.35 as J/σ_Y and within this zone the SGC computation is clearly not realistic. However, the LGC results are for uniaxial loading with contained plasticity, and the question whether $K^2/E\sigma_Y$ can indeed be translated as J/σ_Y for the more general conditions of loading as well as for small-scale yielding, is the very point at issue in the discussion of the characterising role of J. It therefore appears that, within the SGC formulation, crack tip deformation is tending to a value unique for a given value of $JE/\sigma_Y^2 a$ and that LGC formulations are too restricted either in generality or in degree of detail to show whether or

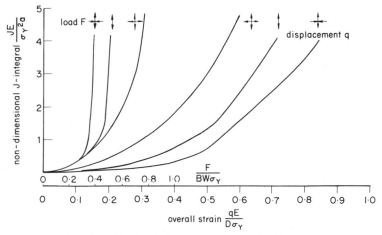

FIG. 2.34. J as a function of load and displacement for applied biaxial stress fields. Centre-cracked plate, $2a/W = 0.625$; plane strain; $D/W = 5$; mildly work-hardening [97].

not a unique zone exists at the tip in which a given microstructural process of fracture could be embedded. As pointed out by analytical arguments [26] and experiment [11, 12] such a continuum zone is of an extent of the order of δ (or J/σ_Y) and thus considerably larger than the crystallographic scale, though perhaps comparable to microstructural features such as inclusion spacing [26].

Additional information on stresses near the crack tip is given in Fig. 2.36a [23] for plane strain non-hardening incremental plasticity finite element computations. The corresponding extent of yielding is shown in Fig. 2.36c. In these particular results conventional constant-strain small-displacement elements were used. The slight drop in stress at the crack tip is not necessarily genuine and may well be a consequence of the unrealities of the constant strain element when used at a singularity. It will be noted that the highest stress values (for non-work-hardening) are found for small yield zones, and that, as plasticity spreads, the local maximum stress decreases slightly despite the maintenance of plane strain (Figs. 2.36b and d).

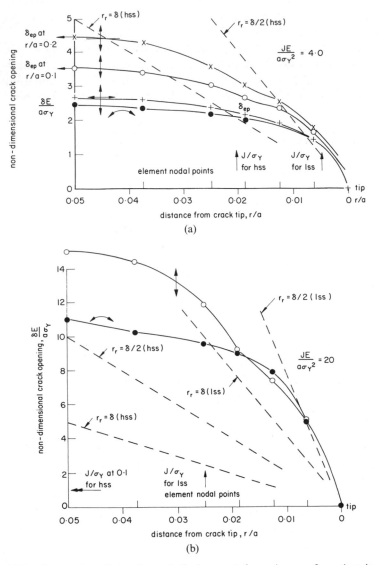

FIG. 2.35. Comparison of near-tip crack displacements for various configurations in plane strain. (a) Centre-cracked plate, $2a/W = 0.625$ with biaxial loadings and three-point bending, $a/W = 0.5$; mildly work-hardening, at $JE/a\sigma_Y^2 = 4.0$; (b) plate with shallow centre crack, $2a/W = 0.0625$, uniaxial tension, and three-point bending, $a/W = 0.5$, non-hardening, at $JE/a\sigma_Y^2 = 20.0$. The positions of possible measuring points for COD at $J/\sigma_Y\delta$, or $\delta/2$ from the tip are indicated for $E/\sigma_Y = 800$ (low-strength steel, lss) and $E/\sigma_Y = 200$ (high-strength steel, hss) [97].

For a work-hardening computation (Figs. 2.36b and d) the local stress $_1\sigma_Y$ tended to increase with deformation to a value of about $5 \cdot 2\sigma_Y$ in one particular plane strain computation in bending. (The degree of hardening can be described by $\sigma/\sigma_Y = 1 \cdot 16$ at $\varepsilon/\varepsilon_Y = 10$; $\sigma/\sigma_Y = 1 \cdot 25$ at $\varepsilon/\varepsilon_Y = 30$.)

For both the work-hardening and the non-hardening cases, as plasticity becomes well established, the stress pattern passes from the near singular (i.e. decreasing rapidly away from the tip) to a near uniform value over a region of several elements. This is perhaps in accord with the HRR analytical model in the non-hardening limit but is possibly accentuated by the various differences between the analytical and computational models such as the plasticity laws used and the representation of work-hardening by other than a simple power law.

In the computations described, the local crack tip strain continued to increase throughout. The values reached are perhaps arbitrary, depending on element size. In particular cases, at an average strain overall of $0 \cdot 9\,\varepsilon_Y$, local values of $_1\varepsilon_Y/\varepsilon_Y = 160$ were found for plane stress and $_1\varepsilon_Y/\varepsilon_Y = 50$ in plane strain, at distances $a/160$ and $a/320$ respectively from the crack tip. These values, derived from a small-geometry-change computation, are of course very much open to question, but the general trend of rapidly increasing strain as the tip is approached is in keeping with the HRR model.

The extent of the plateau region, in which the stress $_1\sigma_Y$ is roughly constant, is about $a/20$ for the first of the cases described here and about $a/40$ for the second. It is not clear whether this difference is due to the change from tension to bending or from non-hardening. The extent does not vary greatly with continued deformation, and in both cases the plateau is first established at about $J/\sigma_Y^2 a = 0 \cdot 6$ (the data from Ref. [23] for the first case establish this value only as between $0 \cdot 5$ and $0 \cdot 8$) when the plastic zones are comparable to a in extent, not yet reaching the far boundary though possibly influenced by it. Thus (for low-strength steels) the plateau is of an extent about $10 J/\sigma_Y$ when first established. This is well beyond the distance J/σ_Y (marked in Figs. 2.36a and c) taken to be indicative of LGC effects [97]. Stresses evaluated at the edge of the plateau should therefore be reliable in both computational and physical respects, although as general yield is reached, the extent of the plateau is more nearly comparable to the probable extent of LGC effects and the stress values are thus questionable.

In broad terms, for the non-hardening calculations, once the plateau is established the local stress has a value $m\sigma_Y$ over a region near the tip (but outside the LGC zone) of about $m = 3$ for three-point bend, 2–2·4 for double-edge notch and 1·4–1·8 for centre-cracked tension, all in plane strain. For a centre-cracked plate non-hardening, in plane stress, the

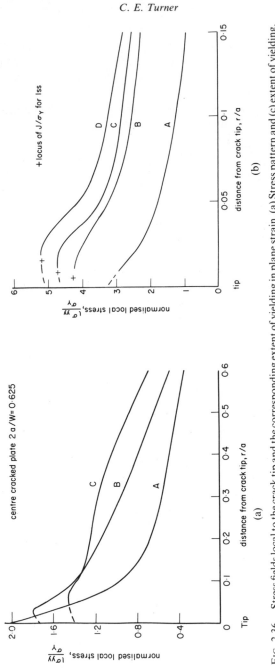

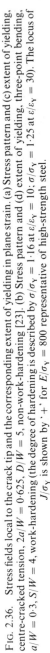

FIG. 2.36. Stress fields local to the crack tip and the corresponding extent of yielding in plane strain. (a) Stress pattern and (c) extent of yielding, centre-cracked tension, $2a/W = 0.625$, $D/W = 5$, non-work-hardening [23]. (b) Stress pattern and (d) extent of yielding, three-point bending, $a/W = 0.3$, $S/W = 4$, work-hardening (the degree of hardening is described by $\sigma/\sigma_Y = 1.16$ at $\varepsilon/\varepsilon_Y = 10$; $\sigma/\sigma_Y = 1.25$ at $\varepsilon/\varepsilon_Y = 30$). The locus of J/σ_Y is shown by '+' for $E/\sigma_Y = 800$ representative of high-strength steel.

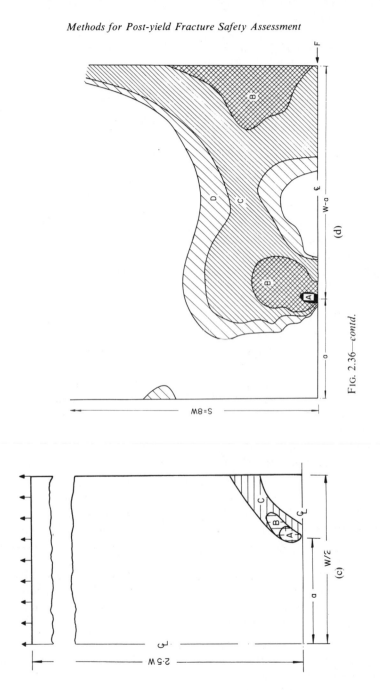

FIG. 2.36—*contd.*

computed value of $_1\sigma_Y$ was limited to $1\cdot15\,\sigma_Y$ with $\sigma_{net} = 1\cdot02\,\sigma_Y$. In the work-hardening case the plateau stress $_1\sigma_Y$ rises with deformation. When the plateau is first established ($JE/\sigma_Y^2 a \simeq 0\cdot6$):

$$_1\sigma_Y \simeq 3\cdot4\,\sigma_Y;\ _1\sigma_H \simeq 2\cdot7\,\sigma_Y;\ \varepsilon_v \simeq 3\cdot2\,\varepsilon_Y;\ \bar{\varepsilon}_{pl} \simeq 1\cdot0\,\varepsilon_Y;\ \bar{\sigma} \simeq 1\cdot0\,\sigma_Y$$

In more extensive plasticity ($JE/\sigma_Y^2 a \simeq 4\cdot2$):

$$_1\sigma_Y \simeq 5\cdot2\,\sigma_Y;\ _1\sigma_H \simeq 4\cdot6\,\sigma_Y;\ \varepsilon_v \simeq 6\cdot0\,\varepsilon_Y;\ \bar{\varepsilon}_{pl} = 11\,\varepsilon_Y;\ \bar{\sigma} \simeq 1\cdot2\,\sigma_Y$$

where ε_v is the volumetric (elastic) strain and $\bar{\sigma}$, $\bar{\varepsilon}$ are the von Mises equivalent stress and strain.[10] The maximum and hydrostatic stresses, $_1\sigma_Y\,\sigma_H$ have increased with extensive yield more than the work-hardening effect and are greater than a Prandtl slip-line field estimated for deep notches, augmented by plane strain and hardening (cf. $(2(3)^{1/2})(1\cdot2)(1 + \pi/2) \simeq 3\cdot56$ for both).

The volumetric strain is not predicted by slip-line models, but could, arguably, be approximated by use of the bulk modulus with any estimate of σ_H augmented by an allowance, if known, for work-hardening. In reality the full rigour of complete (elastic plus plastic) plane strain would clearly not be maintained in gross yielding as it is in the computations.

Very recently, the question of uniqueness of the field described by J in incremental plasticity has been studied by McMeeking & Parks [98] in a large-geometry-change formulation with light work-hardening starting from an initial crack with tip radius $b_0/2$. The value of b_0 is shown not to be important for extensive yielding. The authors' specific interest is in minimum ligament size sufficient to maintain a J-dominated regime but comparison is made of δ/b_0 versus $J/\sigma_Y b_0$ in a manner similar to Fig. 2.31 for deep-notch pure bend and centre-cracked plate. The results are indeed similar to Fig. 2.31, with perhaps rather wider spread. The bend data remain linear into widespread plastic deformation whereas the CCP data show some departure from linearity in gross yielding.

For the specific interest in minimum ligament size it is concluded that $W - a > 25$ or $50\,J/\sigma_Y$ is indeed adequate in deep-notch bending or the comparable compact piece but probably not adequate for CCP. Biaxial effects are not specifically discussed but the difference of endurance of the J-dominated field into gross plasticity is ascribed to the easy or constrained slip pattern that arises in the various configurations, and thus to the transverse constraint developed.

---

[10] $\bar{\varepsilon}_{pl}$ increases rapidly above the values just quoted as the tip is approached, whereas $\bar{\sigma}$ does so only slightly because the rate of hardening is small, but the tip data are not considered sufficiently realistic to be of use.

In these very deeply notched pieces the remote regions are entirely elastic and it is not clear to what extent the results are fully representative of structural configurations in which there is extensive yield of the whole component. As in Ref. [73] it is inferred that the bend piece retains a similar crack tip deformation system throughout its history, whereas the CCP pattern tends to alter as yield spreads. The results are here supported by local stress patterns that were not thought meaningful in Ref. [73]. The plateau and decrease yet nearer the tip, described in Fig. 2.36(a) for small geometry change, are indeed reflected in this more exact model. Thus the weight of present evidence is that the characterising role of *J* is maintained into the elastic–plastic regime, but as plasticity becomes sufficiently extensive for the remote boundaries to dominate the deformation field, the tip details tend to differ. Opening deformation very near the tip appears to remain linked to *J* more closely than the maximum local stresses, but the precise extent of these differences is not yet adequately quantified, nor the effect on fracture behaviour rationalised.

2.2.2.1.4 *Design application*

In the present generation of work most elastic–plastic computations are made in two dimensions for either plane stress or plane strain. In reality the situation is usually between the two. Figure 2.37 demonstrates this for a load extension diagram computed by Riccardella & Swedlow [70]. In Fig. 2.38 Harrison [99] shows how much of the fracture data on weldments fall between the plane stress and plane strain cases. Harrison [99] shows computed load versus clip gauge records in three-point bending (from Refs.

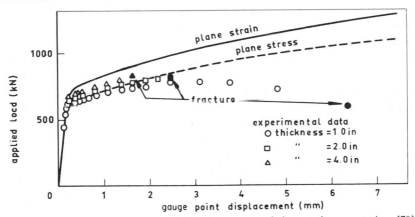

F ɪ ɢ. 2.37.   Tensile test data compared with plane stress and plane strain computations [70].

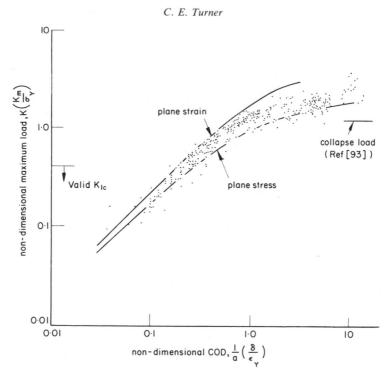

FIG. 2.38.   Fracture data from weldment tests compared with plane stress and plane strain
computations in terms of COD and maximum load K [99].

[22 and 23]) for plane strain and plane stress, for work-hardening typical of structural steel. He then uses experimental data mainly from weld-metal tests on 63-mm-thick material, to plot maximum load, expressed as K_{max} versus COD (Fig. 2.38). The test results are not themselves valid fracture data, since K_{max} is obviously not valid and initiation was not detected for COD, but Fig. 2.38 illustrates that the experimental relationship between load (to which K_{max} is proportional) and displacement (of clip gauge, to which COD is proportional) scatters across the whole spectrum between plane stress and plane strain, with some tendency towards not maintaining full plane strain for the higher ductilities. Thus, care must be taken in translating load or deflection data from a test piece, if analysed in plane strain conditions to model the crack, to a structural configuration that overall may be nearer plane stress. The relationship between load and displacement in the plastic region depends markedly on the assumption of plane stress or plane strain, and if the constraint factor is large, a load substantially below collapse in plane strain may be above collapse in plane stress.

The recommended restrictions on size for a degree of plane strain required to give a thickness-independent value of J at crack initiation (mentioned briefly in connection with Ref. [65] as $B > 50J/\sigma_Y$) are discussed more fully in Section 2.4.2 and Chapter 3, but if J is to be inferred from load or displacement data, or if loads at fracture are to be predicated for a real component, then clearly the degree of plane stress or plane strain must somehow be quantified.

Three-dimensional crack studies and a possible way of interpolating between plane strain and plane stress in two-dimensional configurations are outlined briefly in Section 2.4.3, though not specifically in relation to J.

Users of J have, of course, examined its applicability to problems of design. A serious limitation in the formulation of J is the exclusion of residual body-force and thermal stress problems and also of pressure applied on the crack face, within the boundary of the contour. A modification of J to allow usage with thermal or residual terms is outlined in Section 2.2.2.2 and in Chapter 4. In the following, only data on monotonic mechanical loading of homogeneous material have been used. For regular design applications, more general and readily accessible methods for the prediction of J must be found than the finite element computations discussed so far, well suited though they may be for certain problems.

One such method is extrapolation from LEFM. In general it is found, (see, for example, Fig. 2.21) that the value of J rises but little above the LEFM value (i.e. $J/G \simeq 1$), while the extent of yielding is contained within an outer elastic field. Expressed in terms of eqn. (2.25), $Y^* \simeq Y$, based on LEFM. If the plastic zone correction factor is included in the estimate and attributed to Y rather than to crack length, i.e. eqn. (2.26) is now written

$$K = Y^*_{el}\sigma\sqrt{a} \qquad (2.38)$$

then $Y^* \simeq Y^*_{el}$ up to about $\sigma/\sigma_Y = 0.6$ to within perhaps 5 %, and in eqn. (2.25) $R \simeq K$, when K is evaluated with the plastic zone correction factor. Such estimates are marked on Fig. 2.21 and on several of the figures to be discussed in the paragraphs which follow.

The significance of this is that without further calculation J can be estimated for any configuration for which K is known. While the restriction of this statement to contained yielding rules out most test piece applications, where the situation may be the fracture of a small piece after

C. E. Turner

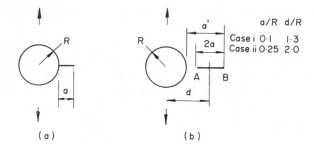

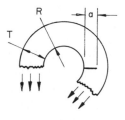

(c)

FIG. 2.39. Some two-dimensional design configurations for which *J* data have been computed. (a) Crack emanating from a hole in a plate in tension; (b) crack buried near a hole in a plate in tension; (c) radial crack in a thick-walled cylinder.

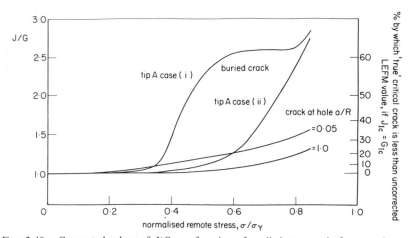

FIG. 2.40. Computed values of *J/G* as a function of applied stress ratio for a crack at, or buried near to, a hole in a plate in tension (Figs. 2.39a and b). Plane strain, work-hardening with yield plateau up to $\varepsilon/\varepsilon_Y = 3\cdot0$ followed by hardening, $\sigma/\sigma_Y = 0\cdot85\ (\varepsilon/\varepsilon_Y)^{0\cdot15}$ [101].

extensive yield, it leaves open application to many design problems other than those involving overload.

A typical geometry of interest is a crack emanating from the edge of an unreinforced hole (Fig. 2.39a) where the nominal applied stress is, say, $2\sigma_Y/3$ so that with a stress concentration of 3, there is local yield before the existence of a crack. As shown in Fig. 2.40 (from Ref. [100]) the elevation of J over G is quite modest (i.e. $J/G < 1·5$), while the plastic regime is contained within the outer elastic field, in this case up to a remote stress of about $0·86\,\sigma_Y$, at which point $\varepsilon/\varepsilon_Y \simeq 12$ for the uncracked body computation. Similar results for a crack buried beneath a hole (Fig. 2.39b) are also shown in Fig. 2.40 (Ref. [101]). The increase in J/G above unity is more marked (i.e. $J/G < 2·5$) than for the edge crack.

When the ligament between the crack and the hole is very small, J can be arbitrarily high but the practical implication of the failure of such a thin ligament does not seem important, the tip remote from the hole creating the condition that might lead to complete fracture. Compared on a basis of physical crack length ($2a$ for the buried crack, a for the edge crack) Ref. [101] shows that the results of the two cases are comparable. A further feature of the crack-at-a-hole problem is seen in Fig. 2.41 by plotting

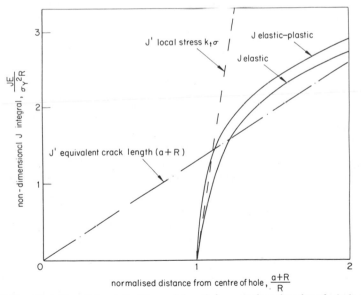

FIG. 2.41. Variation of J and G with crack length for a crack at the edge of a hole under a remote stress $\sigma/\sigma_Y = 0·67$. Plane strain, work-hardening as for Fig. 2.39 (k_t is the elastic stress concentration factor in the absence of the crack) [100].

$JE/\sigma_Y^2 R$ as a function of $(a + R)/R$. The resulting graph is nearly bi-linear, corresponding to the well known LEFM treatment of the problem either as a short crack in the local elevated stress field, 3σ, or as a long crack of length $(a + R)$ in the remote stress field, σ. The crack in the local elevated stress field concept is valid up to about $a \leq 0.2 R$ beyond which the longer equivalent crack concept is relevant. This corresponds to an equivalent crack $R' = a + R$ defined by

$$R' = 0.265 \, J_c E'/\sigma^2 \qquad (2.39)$$

If $R > R'$ the critical crack size will be relatively small, and small increases in crack length cause large increases in J (or K), whereas if $R < R'$ the system is far less sensitive to small increases in crack length so that, if possible, toughness, hole size and applied stress should be related to maintain the condition $R < R'$. Prediction of the load to cause initiation of crack extension in this geometry, based on the equality between computed J contour values and J_i data obtained from small three-point bend specimens, agreed to within $\pm 10\%$ for aluminium alloy 2024-T351 [102].

Another geometry investigated is the two-dimensional representation of an axial crack in a thick-walled cylinder (Fig. 2.39c). Results [101] presented in Fig. 2.42 for a shallow crack in terms of the J/G ratio again indicate that deviations from the elastic case are minor ($\leq 20\%$) until finally more or less uncontained yield of the wall would be met whereon J/G increases rapidly.

It does not seem possible to fit a precise limit for the value of load where J/G increases rapidly. Clearly, since this is where yield ceases to be contained, it is somewhat before collapse, though perhaps but little, in terms of load. It is suggested that, with some uncertainty, the limit below which J/G does not greatly exceed unity is about 'net section yield' as described in Section 2.2.2.1.3 from slip-line limit loads. Where there is first an increase in J as yield spreads to a local surface (as in the hole-in-a-plate case (Fig. 2.40) and the shallow crack in a cylinder (Fig. 2.42)) followed by a much greater increase as yield spreads to the remote boundary, it is this latter condition that would be predicted. Care must also be taken in two-dimensional models of three-dimensional problems, i.e. the semi-elliptical part-through crack, since gross yield is likely to occur in the back face ligament, rather than by loss of in-plane containment.

Where LEFM-based approximations for J are thought inadequate, various approximations may be considered for obtaining an alternative estimate.

Methods for estimating J in terms of a load–deformation curve of the

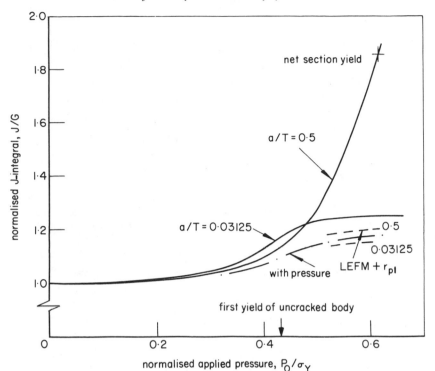

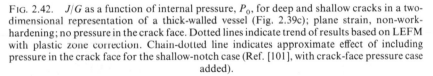

FIG. 2.42. J/G as a function of internal pressure, P_0, for deep and shallow cracks in a two-dimensional representation of a thick-walled vessel (Fig. 2.39c); plane strain, non-work-hardening; no pressure in the crack face. Dotted lines indicate trend of results based on LEFM with plastic zone correction. Chain-dotted line indicates approximate effect of including pressure in the crack face for the shallow-notch case (Ref. [101], with crack-face pressure case added).

structure, analogous to those of Refs. [103 and 104], have been applied by Shih [105] to several anti-plane strain cases based on linear plus power law behaviour; the results compared favourably with finite element computations using the embedded singularity element method. One particularly simple formulation is given here. The load–deflection curve is assumed to be represented by a linear portion $q = CF$, where C is the compliance, plus one or more sections of a power-hardening curve of the form $q = \psi r^n$ springing from the same origin (Fig. 2.43). Both C and ψ will be functions of crack length, a, but not of load, F. Values for ψ and n are found by fitting to an experimental curve or by any approximate method that is available. There is a discontinuity of slope at the junction, which can

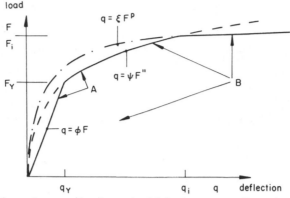

FIG. 2.43. Linear plus power-law fits to a load deflection curve to allow a simple estimate for *J*. Curve A linear plus one power-law section, curve B linear plus two power-law sections.

be minimised by taking several short lengths of different curves if so desired. It is necessary to express the variation of yield load F_Y, with crack length. For a point in the plastic region an estimate might be made by non-hardening collapse load methods, or of course experimental data might exist. For the two-curve fit (Fig. 2.43, curve A)

$$B_n J = \left[1 - \frac{n}{n+1} \right][X - (n-1)Y]F_F q_F + \left[\frac{n}{n+1} - \frac{1}{2} \right][X + 2Y]F_Y q_Y$$

(2.40a)

where $X = C'/C$ and $Y = F_Y'/F_Y$ and the prime denotes differentiation with respect to crack length. B_n is the net thickness (usually B, but less than B if side-grooved test pieces are in question). The suffix Y refers to the yield point, and suffix F to the point, perhaps at fracture, at which J is to be evaluated. The elastic compliance C is often known experimentally as a power of a, in the form $C = C_0 + C_1 a^t$ so that C' is easily formed. It must also be noted that F_Y may be a function of a geometric dimension such as a ligament $(W - a)$ (e.g. in a cracked plate or beam) or possibly of some other dimension of the component (such as the arm width of a Double Cantilever Bond (DCB) test piece), which may itself be a function of crack length, and this must be allowed for in the formation of F_Y'. For the three-curve fit (Fig. 2.43, curve B)

$$B_n J = \left[1 - \frac{p}{p+1} \right][X - (p-1)Y]F_F q_F + \left[\frac{p}{p+1} - \frac{n}{n+1} \right][X + 2Y]F_i q_i$$

$$+ \left[\frac{n}{n+1} - \frac{1}{2} \right][X + 2Y]F_Y q_Y \qquad (2.40b)$$

The suffix i refers to the intermediate junction of the n power and p power parts of curve B, which is a point chosen arbitrarily to fit the data.

Figures 2.44a and b show the non-dimensionalised results of computations for various beams in three-point bending, including moderate work-hardening, with shallow notches, where plasticity first breaks forward to the notched surface and then spreads along the span, despite the decrease in moment, while the central moment increases towards its limit value, augmented by the work-hardening. The conventional deep-notch estimate for the slope $\partial J_{\text{pl}}/\partial q = 2 \cdot 912\,\sigma_{\text{u}}(W - a)/S$ was shown to work well [72] for the deep-notch case when based on a constraint factor of $1 \cdot 456$ and σ_{u} rather than σ_{Y} to allow for hardening. As already discussed (Sections 2.2.2.1.3 (iv)) in terms of $J = \eta U/B(W - a)$ this deep-notch estimate is inappropriate for the shallow-notch cases. The slope $\partial J_{\text{pl}}/\partial q$ is related to η_{pl} and an estimate is made for these terms in Appendix 4. Figures 2.45a and b show similar results for shallow-notch tension. The conventional estimate of $\partial J_{\text{pl}}/\partial a = L\sigma_{\text{Y}}$ appears adequate for non-hardening cases with deep or shallow notches (Fig. 2.25) but, as already discussed (Section 2.2.2.1.3 (iv)) is inadequate for shallow notches with hardening when yield of the remote shank of the component can occur before collapse of the notched ligament is reached. Some estimates of this effect are again made in Appendix 4. The significance in the present context is that shallow-notch circumstances seem more relevant to design problems than do the deep-notch cases used for material test purposes, and further discussion of these results is therefore amalgamated with the more general question of a simple J design curve.

The conventional tensile estimate $\partial J_{\text{pl}}/\partial q = L\sigma_{\text{Y}}$ is satisfactory for deep-notch pieces or for non-hardening shallow-notch pieces where the remote ends do not yield. The effect of non-hardening for a shallow-notched centre-cracked plate is shown by the dashed line in Fig. 2.45b (taken from Fig. 2.25). This may be contrasted to the hardening case for similar (though not identical) geometry in which the chain-dotted curve agrees with the hardening shallow-edge notch results.

For shallow notches Muscati & Turner [106] remarked that the toughness seemed greater than expected from deep-notch data predictions. Chell & Spink [107] and Milne & Worthington [108] have also remarked on this trend. (It should be noted that some of the data in Refs. [106 and 107] refer to the same tests.) These results seem worthy of further study since a larger margin of safety is implied for shallow-notch cases over and above fracture mechanics predictions.

A design curve approach very similar to that for COD has been proposed [109] although not elaborated to include rules for residual stress,

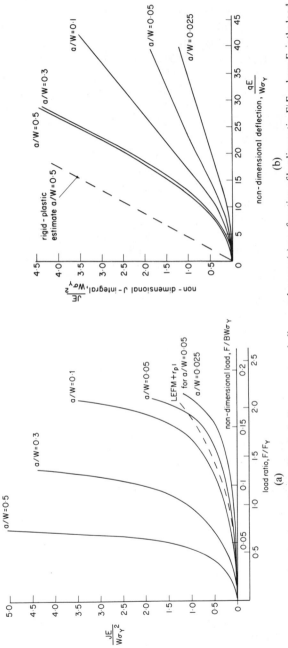

Fig. 2.44. *J* values for three-point bending, including some very shallow notch cases: (a) as a function of loading ratio F/F_Y where F_Y is the load to cause first yield in an unnotched beam, LEFM plus plastic zone correction shown dotted for one case; (b) as a function of deflection q. The dotted line is the estimate of slope for $a/W = 0.5$, based on Ref. [72]. $S/W = 8$; plane strain; work-hardening as in Fig. 2.10 [106].

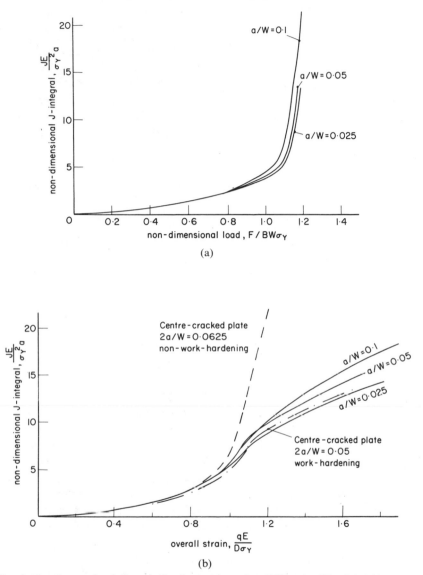

FIG. 2.45. *J* curves for shallow single-edge notch tension, $D/W = 4$, $a/W = 0.1, 0.05, 0.025$; plane strain, work-hardening as for Fig. 2.10: (a) as a function of load; (b) as a function of displacement, q. Also shown in (b), shallow centre-cracked tension, $D/W = 2.5$; $2a/W = 0.0625$, non-work-hardening (dotted, from Fig. 2.24) and $D/W = 4$, $2a/W = 0.05$, work-hardening (chain-dotted). Note that the tension is applied by displacements that cause the ends to remain parallel, not by pure axial loading.

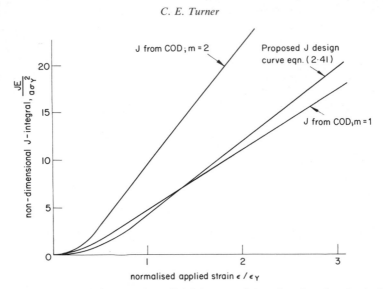

FIG. 2.46.   Proposed design curve (eqn. (2.41)) in terms of J as a function of strain $\varepsilon/\varepsilon_Y$ [109] with COD design curve (eqn. (2.14)) transformed with $m = 1$ and $m = 2$ (as in eqn. (2.50)).

weldments, equivalent crack sizes, etc., as for COD. The equations proposed can be written in non-dimensional form as

$$JE/\sigma_Y^2 a = 1.254\,\pi(\varepsilon/\varepsilon_Y)^2 \qquad\qquad \varepsilon/\varepsilon_Y < 1.0 \qquad (2.41a)$$

$$JE/\sigma_Y^2 a = 1.254\,\pi((2\varepsilon/\varepsilon_Y) - 1) \qquad \varepsilon/\varepsilon_Y \geq 1.0 \qquad (2.41b)$$

The factor 1.254 is the well known LEFM surface flaw factor of 1.12 in terms of K (here squared). The diagram in Fig. 2.46 is entered at some known strain ratio. In Ref. [109], it is suggested that $\varepsilon/\varepsilon_Y$ be estimated for a stress concentration problem from the stress–strain curve by picking out the stress and strain such that the Neuber relationship $k_t^2 = k_\sigma k_\varepsilon$ is satisfied where k_t is the elastic concentration factor and k_σ and k_ε the plastic stress and plastic strain concentration factors respectively. The two additional lines plotted in Fig. 2.46 refer to the COD design curve given by eqns. (2.14) and will be discussed in Section 2.2.3.

The present computational results for notches with $a/W \leq 0.1$, all in plane strain, are shown in Fig. 2.47 in terms of $JE/\sigma_Y^2 a$ as a function of displacement ratio q/q_Y or *nominal elastic strain* $\varepsilon/\varepsilon_Y$. The ordinate is in accord with eqns. (2.41) and equivalent to that for the COD design curve. The abscissa is chosen for simplicity since, as seen in Figs. 2.24 and 2.25 the curves of J versus deflection already have the parabolic plus linear form

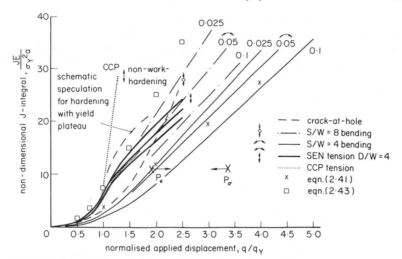

FIG. 2.47. Collected values of $JE/\sigma_Y^2 a$ for shallow-notch and other design-type computed cases as a function of deflection ratio, q/q_Y. All are for plane strain and all with mild work-hardening except the one case marked. P_σ shows the point to which a shallow-notch bending case P_ε would be transformed if treated in plane stress instead of in plane strain, for the unnotched body only.

being sought. Moreover, calculation of q or nominal elastic ε is simple, an acceptable design overload requirement possibly being expressed as attainment of a deflection that is some small multiple of the yield deflection. In unnotched tension q/q_Y is of course the actual strain ratio $\varepsilon/\varepsilon_Y$ whether ε is elastic or plastic, but in bending this is so only for the elastic case. In Fig. 2.47 the deflection ratio q/q_Y for bending is related to the non-dimensionalised deflection $qE/\sigma_Y W$ (as used in previous figures) by

$$\frac{qE}{\sigma_Y W} = \frac{\alpha}{6} \frac{q}{q_Y} \frac{S^2}{W^2} \qquad (2.42)$$

Shear deflection is not allowed for so that $\alpha = 1$. If shear were allowed for[11] by elastic methods, then for $S/W = 4, \alpha \simeq 1\cdot24; S/W = 8, \alpha \simeq 1\cdot06$, and the bending data would be moved closer to the origin by these proportions, thus bringing the two cases somewhat nearer to agreement. There seems no reason to suppose that data for beams of yet larger span/width ratio would not fall further towards the top left of the figure.

[11] For beams of rectangular cross-section in three-point loading the deflection including shear is $q \simeq q_b (1 + (4 W^2/S^2))$ where q_b is the simple bending term $q_b = FS^3/48 EI$.

The data for a crack at a hole in a plate are included by interpreting the abscissa as $(k_t\varepsilon/\varepsilon_Y)$, where k_t is the elastic SCF. This implies that if in the LEFM region K is written $K = k_t\sigma\sqrt{\pi a}$ (as is commonly the case for short cracks) the term $k_t\sqrt{\pi}$ is *not* interpreted as the LEFM shape factor Y. The $\sqrt{\pi}$ (and a free surface factor of $1\cdot12$ if desired) is taken as Y and $(k_t\sigma)$ as the applied stress and hence strain, on the grounds that this gives a much better grouping of the SCF and non-SCF results. In the plastic region the abscissa is taken as $k_t q/q_Y$, with $k_t = 3$ and q/q_Y computed for the uncracked body (but including the hole) for $D = 12R$, $W = 20R$.

The upper edge of the data so far discussed can be represented by

$$JE/\sigma_Y^2 a < 2\pi\,(q/q_Y)^2 \qquad\text{for}\qquad q/q_Y < 2\cdot0 \qquad (2.43a)$$

$$JE/\sigma_Y^2 a < 20\,[(q/q_Y) - 0\cdot75] \qquad\text{for}\qquad q/q_Y > 2\cdot0 \qquad (2.43b)$$

There is a slight discontinuity at the junction of these equations that is accepted for the sake of simplicity. Equation (2.43a) is more conservative for many problems than the use of individual LEFM solutions for which $Y^2 < 2\pi$. The spread of data around $q/q_Y \simeq 1\cdot0$ or $1\cdot5$ is such that no one curve can represent it (on the basis of Fig. 2.47) to within a factor of about four-fold, and it is supposed that data could be found to fall appreciably above eqn. (2.43). Indeed some of the present results do so very marginally.

Before discussing an alternative basis for plotting the data, two points seem of great importance. The first relates to non-work-hardening cases. The shallow centre-cracked plate (CCP) data from Fig. 2.25 are shown as marked in Fig. 2.47. It falls substantially 'above to the left' of the main data for the moderate work-hardening cases. Inclusion of the same degree of work-hardening translates the CCP case to fall within the main set of results, emphasising that the discrepancy is not due to configuration but to the presence or absence of work-hardening that, as discussed in Section 2.2.2.1.3, determines whether or not yield spreads to the remote ends of a shallow-crack piece. Deep-notch data will not often be directly relevant to design cases, but will follow a simple trend in that, on the basis of load (Fig. 2.21) or displacement (Fig. 2.24) the work-hardening cases fall between the elastic and the non-hardening results. For the shallow-notch data of most relevance to Fig. 2.47, where, with work-hardening, the remote ends of the test piece yield, this is not so unless plotted on a basis of J versus U (where $U = f(Fq)$). On the basis of J versus q or J versus ε, a heavily work-hardening case (in the limit giving a load-displacement diagram that is merely a continuation of the elastic line) will fall near an extrapolation of the linear elastic parabolic regime. The non-hardening cases also rise steeply, whereas the mild or moderate hardening shallow-notch cases, because of yield of the remote ends, provide the main group of results well 'down to the right' of either the non-hardening or high-hardening cases. The diagram should therefore be used with caution for either extreme of non-hardening or very high work-hardening. It is open to speculation whether discontinuous hardening (as in the conventional representation of mild steel) would give a curve at first, following the rapid rise of the non-hardening case followed by a 'swing down towards the right', as sketched schematically in Fig. 2.47.

The second important point is the choice of plane stress or plane strain. As already noted [70], [99], real cases tend to fall between the two and it was remarked [106] that the experimental load deflection diagram agreed very closely with the unnotched computation in plane stress. This is not at all unexpected for a bar of cross-section $W = 2B$. In such a bar the notched tip may well be in plane strain, or substantially so, according to the criterion $B > 25$ or $50 \, J/\sigma_Y$ and the evidence of COD data such as Ref. [11].

Thus, should J be evaluated from a plane strain computation at the same load or same deflection, as found in the uncracked body calculations? All the data for Fig. 2.47 are for plane strain but for one of the bending cases relevant to Ref. [106] the effect of this uncertainty is shown by replotting one point P_ε, arbitrarily selected near the elbow of the load–deflection curve, assuming that the value of load found in the plane strain computation of J for the notched case implied a corresponding deflection for the plane stress computation of the unnotched case, P_σ. A load near collapse in plane stress may be barely beyond the substantially linear region of plane strain. In some cases it may be possible to work in terms of the ratio to the collapse load, yield load or yield deflection, implying plane stress values for the unnotched component and plane strain for the notched computation without translating numerical values from one case to the other. The problem is further discussed in Section 2.4.3. If the notch were itself in plane stress for a given load the value of J might be considerably increased (Fig. 2.26), but the critical toughness J_i or J_c would probably also be substantially increased and general experience of fracture problems seems to show that the plane stress case is less demanding in practice.

Reverting to the basis on which the results are plotted it is seen that if the ordinate is altered to $JE/Y^2\sigma_Y^2 a$ (in effect J/G_Y where G_Y is the LEFM G evaluated at $\sigma = \sigma_Y$) then not only is the divergence between the LEFM cases greatly reduced, but some of the effect of a/W is eliminated from the plastic regime. As already noted, choice of q/q_Y or $\varepsilon/\varepsilon_Y$ for tension is immaterial, but a very simple model of bending suggests that the latter would nearly eliminate the effect of span seen in Fig. 2.47. If plasticity is established predominantly as a central hinge circle with total rotation 2θ, then $q_{\text{pl}} \simeq \theta S/2$ and the shear strain $\gamma_{\text{pl}} \approx \theta/\pi$. Assuming a uniaxial system, $\bar{\varepsilon} = \gamma$ and from elastic beam theory for an unnotched beam $q_Y = (\sigma_Y/E)$ $(W/4)(S^2/W^2)$ neglecting shear. For the notched beam a correction of C/C_o can be introduced with the compliances evaluated as outlined below (eqn. (2.32)). Thus, finally, the abscissa of Fig. 2.47 can be transformed, since $q/q_Y = 2\pi(\varepsilon/\varepsilon_Y)(W/S)(C_o/C)$ so that, combining with the linear slope J/q_{pl} (eqn. (2.36))

$$\frac{JE}{\sigma_Y^2 a} \simeq \frac{\eta \pi L}{3} \frac{W - a}{a} \frac{C_o}{C} \frac{\varepsilon}{\varepsilon_Y} \tag{2.44a}$$

From eqn. (2.32) $C_o/C = \eta_{\text{el}} [W/(W - a)][(W/a)(S/W)(1/18\,Y^2)$ so that, as an alternative to eqn. (2.44a)

$$\frac{JE}{Y^2 \sigma_Y^2 a} = \frac{J}{G_Y} = \eta_{\text{pl}} \eta_{\text{el}} \frac{\pi L}{54} \frac{S}{W} \left(\frac{W}{a}\right)^2 \frac{1}{Y^4} \frac{\varepsilon}{\varepsilon_Y} \tag{2.44b}$$

For shallow notches $\eta_{\text{pl}} \simeq 10 a/W$ (Appendix 4) $\eta_{\text{el}} \simeq 72 a/W$ $[1 - (a/W)]/S/W$, $C_o/C \simeq 1 - 2(a/W)$; $Y \simeq 2[1 - (a/W)]$. Thus

$$\frac{JE}{\sigma_Y^2 a} \simeq 10 \left(\frac{W - a}{W}\right) \frac{\varepsilon}{\varepsilon_Y} \tag{2.44c}$$

$$\frac{JE}{Y^2 \sigma_Y^2 a} = \frac{J}{G_Y} \simeq 2 \cdot 5 \left(\frac{W + a}{W}\right) \frac{\varepsilon}{\varepsilon_Y} \tag{2.44d}$$

These are approximate linear relations in the plastic regime that have to be added to the LEFM values. The virtue of J/G_Y is that the LEFM regime is reduced to one curve although the spread of computed results seems marginally less for the plastic regime when plotted in the form without Y^2. If these equations are accepted at face value there is little effect of span and only a small effect of a/W. It appears that the trend with a/W should reverse from eqn. (2.44c) to eqn. (2.44d), whereas the computed data follow the same trend, decreasing as a/W increases (up to $a/W = 0 \cdot 1$) for both graphs. This is partly due to the many approximations made in the above analysis and partly perhaps to an inadequate representation of the Y factor for small a/W by only one or two terms since the usual expressions for Y in powers of a/W are not convergent series. However, the above analysis presents the general trends of a J design curve in a rational manner. The computed data are plotted against strain in the uncracked body in Fig. 2.48. The basis for the hole-in-plate case is correspondingly altered. For $\sigma_{nom}/\sigma_Y > \frac{1}{3}$ (where σ_{nom} is the remote stress, σ) the notional local stress at the hole, $_1\sigma = k_t\,\sigma_{nom} > \sigma_Y$. In such cases $_1\sigma$ is restricted to σ_Y or $n\sigma_Y$ where n is some

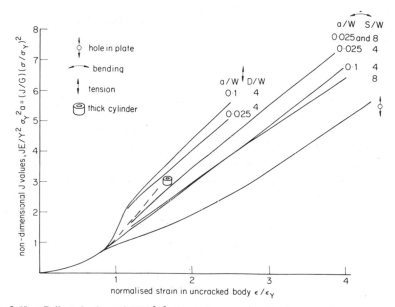

FIG. 2.48. Collected values of $JE/Y^2\sigma_Y^2 a$ for shallow-notch and other design-type computed cases as a function of strain, $\varepsilon/\varepsilon_Y$, in the uncracked body.

allowance for work-hardening, so that $k_\sigma < k_t$ and in accordance with the spirit of Ref. [109] becomes $k_\sigma = n\sigma_Y/\sigma_{nom}$, whence $k_\varepsilon > k_t$ and becomes $k_\varepsilon = k_t^2\sigma_{nom}/n\sigma_Y$. The basis for the strain axis in Fig. 2.48 is then $k_\varepsilon(\varepsilon/\varepsilon_Y)$ rather than $k_t(\varepsilon/\varepsilon_Y)$, with a factor n rising to $1\cdot2$ as an approximation to the computed results [100]. Results for the two-dimensional shallow axial–radial crack in a thick cylinder [101] ($a/B = 0\cdot03125$; radius ratio 2) are also included, based on the hoop strain at the bore, uncracked. In the J computation [23] pressure on the crack face is not included. Here the elastic term for crack face pressure, $K = 1\cdot12p\sqrt{\pi a} = Y_p p\sqrt{a}$ has been added to the elastic–plastic solution by addition of the coefficient Y_p to the computed value of Y^* (eqn. (2.25)) contributing, in this case, some 35–40 % of the final value of J.

To cover the upper edge of the (tensile) data, three equations are needed because of the 'knee' around $\varepsilon/\varepsilon_Y \simeq 1 \cdot 2$ (Fig. 2.48).

$$\frac{J}{G_Y} = \left(\frac{\varepsilon}{\varepsilon_Y}\right)^2 \quad \text{(i.e. LEFM)} \quad \text{for} \quad \frac{\varepsilon}{\varepsilon_Y} \leq 0 \cdot 85 \qquad\qquad (2.45a)$$

$$\frac{J}{G_Y} \leq 5\left(\frac{\varepsilon}{\varepsilon_Y} - 0 \cdot 7\right) \quad \text{for} \quad 0 \cdot 85 < \frac{\varepsilon}{\varepsilon_Y} \leq 1 \cdot 2 \text{ (contained yield)} \qquad (2.45b)$$

$$\frac{J}{G_Y} \leq 2 \cdot 5\left(\frac{\varepsilon}{\varepsilon_Y} - 0 \cdot 2\right) \quad \text{for} \quad \frac{\varepsilon}{\varepsilon_Y} > 1 \cdot 2 \text{ (uncontained yield)} \qquad (2.45c)$$

The first equation might be extended a little further by the use of the LEFM plastic zone correction factor. The second and third equations are empirical expressions to fit the upper edge of Fig. 2.48 with no intentional margin.

The biaxial data used in Fig. 2.34 are for deep notches, and may not be representative of the shallow-notch behaviour. They are therefore not included in Figs. 2.47 and 2.48. The unreality of the rigorous plane strain computation is emphasised in that the shank remote from the notch hydrostatic tension exists so yield cannot occur for biaxial loads of equal magnitude and sign. In the present study, even for the shear loading case, collapse of the mildly hardening ligament occurs before first yield of the whole section. In this situation the effect of biaxiality can be allowed for roughly (to within, say, 20%) by the normal arguments for multiaxial stresses.

In the near-elastic region the slope of the load–displacement curves scale according to elastic strain in the remote ends, since, at least for gauge lengths such as here ($D = 5W$), the displacement of the notched region makes but a small contribution to the overall value. Thus, using the LEFM compliance for the uniaxial case, the effective compliance for the biaxial can be estimated. This picture extends up to a load F_{Yn} estimated from yield of the *ligament* in which the net stress $\sigma_n = \sigma/[1 - (2a/W)]$ (here $\sigma_n = 2 \cdot 66\,\sigma_Y$ for $2a/W = 0 \cdot 625$) is combined with the transverse stress, $+\sigma$, 0, or $-\sigma$, under plane strain conditions.

The plastic regime can also be estimated roughly by differentiating eqn. (2.32) for the slope $\partial J_{pl}/\partial_q$, with σ_Y replaced by $\bar{\sigma}_y$, the axial component calculated from the equivalent stress in the ligament. Clearly, $\bar{\sigma}_y > \sigma_Y$ for the biaxial tensile loading and $\bar{\sigma}_y < \sigma_Y$ for the shear loading. As remarked earlier, with shallow notches $\bar{\sigma}_y \simeq \sigma_Y$ and yield would not occur for the equi-biaxial case in plane strain. In the present deep-notch case, up to the extent shown Fig. 2.25, LEFM is a good representation of the biaxial tensile data, the degree of yielding, although stretching to the free boundary, not being extensive enough to cause J to rise appreciably above G. Estimates of J as a function of load beyond the LEFM regime are very crude, and indeed have not been attempted for other configurations even for the uniaxial case, although use of the Dugdale model is possible, as described in Section 2.3.2. If, in the uniaxial case, yield of the ligament (and hence the notional end of the LEFM regime) occurs at F_{Y1} and a value $J/G(>1)$ is reached at a load $F_1|$, then, since G is not affected by biaxiality, the same value of J/G is reached in the biaxial case at a load F_2 where $(F_2 - F_{Y2}) = (F_1 - F_{Y1})(\bar{\sigma}_{y1}/\bar{\sigma}_{y2})^2$ where $\bar{\sigma}_{y1}$ and $\bar{\sigma}_{y2}$ are the axial components of the equivalent stresses in the ligament for the uniaxial and biaxial cases. How general these heuristic estimation factors may be cannot be stated without more data for other biaxial configurations, notably with shallow notches.

The crucial question of what strain is reasonably required for adequately ductile behaviour is not clear. Figure 2.48 includes a treatment of strain concentration but not of residual stress. Following the COD design curve practice, the requirement for load-induced strain perhaps between $1 < \varepsilon/\varepsilon_Y < 2$, based on experience, would be supplemented by a strain $\varepsilon/\varepsilon_Y \to 1$ for residual stress $\to$ yield stress level, making a total required strain of perhaps 2 or $3\varepsilon/\varepsilon_Y$ and hence a given value of toughness, J_i, for some postulated or known crack size a.

Three-dimensional crack studies and a possible way of interpolating between plane strain and plane stress in two-dimensional configurations are outlined briefly in Section 2.2.3, though not specifically in relation to J.

2.2.2.1.5 *Critique of the method*

Much of what has been said about COD in Section 2.2.1.5 also holds for J. One desirable feature of a J method would be the offer of continuity with well-established LEFM procedures. Although this is notionally so for COD, in fact the uncertainties over the value of m in the relationship $J = m\sigma_Y \delta$ (cf. Section 2.2.3) imply that the use of K and δ will not normally give contiguous results. The virtue that, for linear behaviour, J degenerates to G may, however, also be lost unless the test methods are identical. The practice of finding J from U (eqn. (2.30)) will not ensure an identical value for J_c as for G_c (and hence K_c) found from the standard LEFM method. Indeed, the value of J at initiation, J_i, may not agree with G_{Ic} (from K_{Ic}), since the latter may allow a certain stable crack growth, small on a fractional scale (i.e. less than 5 % change in slope) for a large test piece, but large in absolute value (1 or 2 mm for test pieces 50–200 mm wide) and hence significant in terms of the 'no growth' (or perhaps the crack opening stretch) allowed in the J test method. Thus J_i might underestimate K_{Ic} for materials that show any stable growth in LEFM testing. Indeed there seems to be a certain fortuitousness that J_i equals G_{Ic} when the former relates to the start of cracking at mid-section of a minimal degree of plane strain, whereas the latter relates to the instability or some specified non-linearity of a much more rigorously defined degree of plane strain. A method has been proposed [85] (see Chapter 3), in which the J test is conducted using clip gauges so that if a near elastic result is obtained the method of analysis ensures continuity with LEFM, and some support for the results of that method have been published [110]. This test method is also capable of allowing for some degree of slow stable growth if, by agreement of all concerned, the initiation of cracking is thought to be unreasonably restrictive. However, as discussed in Ref. [85], the test should then be conducted on full thickness pieces to model the real material properties, and the effect of geometry, in modifying the amount of slow growth found in various configurations, is still unclear. This point of slow stable growth, also raised in connection with the COD method, will be discussed in Section 2.4.1.

In summary, the J method offers a good two-dimensional treatment for all configurations, that is compatible with LEFM, but restricted to mechanical stressing with no thermal or residual terms, unless the J arguments are extended to include modifications (Section 2.2.2.2) of which further examination seems desirable. The concept relates to crack initiation. There is a general agreement between both COD and J but far less detailed work seems to have been attempted on stress concentrations, residual stress and so on in terms of J, nor has the collection of J data for weldments matched that which has occurred for COD. The reason for the different degree of emphasis on weldments in the various fracture toughness methods appears to lie in the nature of the driving force for the bulk of the research, notably general structural steels with many welding variables in the case of the COD method and the reactor safety problem with closely specified weldments that have a resulting toughness comparable to the parent plate in the case of J [111]. In the writer's view, the spread of any design method to a broader range of circumstances is likely to call for a more extensive treatment of weld toughness problems, so that procedures containing both test techniques and design proposals that can be used for weldments seem highly desirable.

There is some evidence to support the contention that a critical value of either J or COD characterises the onset of crack growth, but the effects of biaxiality and loss of constraint through the thickness are still far from resolved and tests over a wide range of absolute size are still desirable. The main appeal of the J concept is perhaps its notional breadth of coverage that offers a framework for a coherent scheme of analysis, testing and design, whereas a number of the empirical approaches to be discussed in Section 2.3 offer a more restricted coverage, although possibly providing a better fit to reality within those narrower bounds.

2.2.2.2 The J* and Other Extensions to the J Integral
A modified concept called the J^* integral was introduced by Blackburn [112] and used in extensive computational studies [113–116]. It is defined [113] by:

$$J^* = \lim_{\rho \to 0} \int_{\Gamma^1} \left\{ \frac{1}{2} \sigma_{ij} \frac{\partial u_i}{\partial x_j} \mathrm{d}x_2 - T_i \frac{\partial u_i}{\partial x_1} \mathrm{d}s \right\}$$

$$= \int_{\Gamma^3} \left\{ \frac{1}{2} \sigma_{ij} \frac{\partial u_i}{\partial x_j} \mathrm{d}x_2 - T_i \frac{\partial u_i}{\partial x_1} \mathrm{d}s \right\}$$

$$+ \lim_{\rho \to 0} \int \int \left\{ \frac{1}{2} \sigma_{ij} \frac{\partial^2 u_i}{\partial x_1 \partial x_j} - \frac{1}{2} \frac{\partial \sigma_{ij}}{\partial x_1} \frac{\partial u_i}{\partial x_j} - \frac{\partial}{\partial x_3} \left(\sigma_{i3} \frac{\partial u_i}{\partial x_1} \right) \right\} \mathrm{d}S \quad (2.46)$$

where Γ^1 is a circle of radius ρ around the crack tip and Γ^3 is another contour beyond Γ^1 also surrounding the crack tip, traversed anti-clockwise; s is the path length along Γ^3 and S is the area enclosed by Γ^3 and Γ^1. Both contours are in the x–y plane, or x_1–x_2 plane in the notation of eqn. (2.46), not containing any other singularities with the crack extending in the x_1 direction.

It will be seen that the density of stress working term, Z, in eqn. (2.18), that was itself an integral of the form $\int \sigma \, \mathrm{d}\varepsilon$, has been replaced by $\frac{1}{2}\sigma\varepsilon$. The strain ε now refers to the final state of strain so that the term is independent of the history by which the state was arrived at. Rice [117] comments that such a transformation will give a path-independent integral. However, there seems to be no assurance that such a term is necessarily meaningful in either energetic or characterising senses, except for the elastic case, and it is notable that the discussions of energy in Refs. [113 and 116] are not fully explicit on this point.

The first merit claimed for J^* is that, to within computational accuracy, Hellen has shown that the values remain path independent for incremental plasticity calculations including some with thermal stress terms. The immediate point is that if it is accepted that, for plasticity, both J and J^* are sensibly path independent in practice, J^* has the additional merit of being

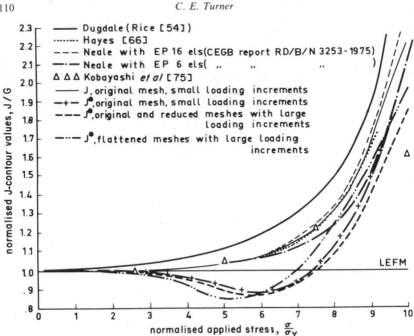

FIG. 2.49. Comparison of J^* integral with other plasticity computations for a centre-cracked plate, $2a/W = 0.25$, and with the Dugdale infinite plate model [117].

able to accommodate the thermal and indeed residual stress term that cannot be included in the J formulation without loss of path independence. This point is further elaborated in Appendix 6. This appendix also brings out a third, most important claim for J^*, namely its applicability to three-dimensional geometries. The variation of J^* with extent of plasticity is shown in Fig. 2.49, together with the value for J. More examples are given in Chapter 4. It is seen that there is a generally similar trend, with J^* falling below J and indeed, for a small degree of yielding, below G.

It is not clear to the writer to which point (initial growth from a pre-existing defect, or final rapid propagation of the slow growth) J^* is claimed to be relevant. Reference [112] merely suggests that 'J^* should be used as a parameter to determine whether fracture will occur from a comparison with its known critical value'. Reference [113] at one point suggests 'J^* is proposed as a general criterion for onset of crack growth ...' but at another point discusses fast fracture as if J^* were to be applied to that event. In broad terms, it is suggested that most experimental data supporting J could also be interpreted as supporting J^*.

Another modification to the *J* integral has been proposed by Bergkvist & Huong [55] for application to cases of axial symmetry. Again a surface integral is added. A cylindrical element is considered about the x_3 axis with an axially symmetric crack in the $x_1 x_2$ plane (Fig. 2.50) (using the notation of Ref. [55]). The equivalent to the *J* integral is given as

$$J_{\text{R}} = \int_{\Gamma} (W n_1 - n_j \sigma_{ij} u_i, 1) \, \mathrm{d}l + \int_A \left(\frac{\sigma_{22} u_1}{x_1^2} - \frac{\sigma_{11} u_1, 1}{x_1} - \frac{\sigma_{13} u_3, 1}{x_1} \right) \mathrm{d}A \tag{2.47}$$

where u, σ are the standard notation for displacement stress; W is 'strain energy per unit volume', $\mathrm{d}l$ is crack extension, Γ is the closed curve that

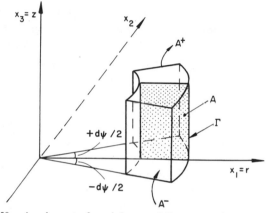

FIG. 2.50. An element of crack in an axially symmetric body [55].

generates the piece, and *A* the cross-section of Γ in the $x_1 x_3$ plane (Fig. 2.50). In the derivation no reference is made to the stress–strain properties of the material to which this integral is relevant. Since the phrases 'energy release' and 'energy flow to the crack edge' are used, elastic material seems implied. By implication this need not be restricted to linear, although in the illustrative example, it is. In this restricted case the definition of 'strain energy' is not in question, but if the integral were to be applied to plasticity problems presumably the same arguments over the most appropriate definition of strain energy and the interpretation of J_{R} as a characterising rather than energetic term would arise, as has been discussed for the *J* integral.

Yet another modification to the *J* integral has been proposed by Ainsworth *et al.* [118] for use in two-dimensional thermal cases in the $x_1 x_2$

plane. Again an area integral is added. For a crack extending in the x_1 direction:

$$J_\theta = \int_\Gamma [W\,dx_2 - \sigma_{ij}(\partial u_i/\partial x_i)\,dS_j] + \int_A \int \sigma_{ij}(\partial\theta_{ij}/\partial x_1)\,dA \quad (2.48)$$

where the total strain ε has been decomposed into a mechanical component ε' and a thermal component θ. The first term has the usual meaning for the J integral and the second is taken over the area A within the contour to correct for the contribution of the thermal effects of which only the incompatible component contributes to the energy. For no thermal term the result is identical to the conventional J integral, unlike J^*. The thermal problem is of course restricted to cases where the elastic constants do not vary with position (due to variation of temperature). Use in plasticity is envisaged with the implied restriction on the energetic meaning of J_θ. It would clearly be of great interest if it could be demonstrated that for a given value of J_θ in a thermal problem the crack tip deformation were comparable to those for the same value of J in a mechanical problem but as far as the writer is aware, this has not been studied.

These modifications promise a number of advantages over the conventional formulation of J, whilst departing from the simple contour form. This may not be too important in computational uses. More work is required to clarify the meaning of the various proposals in both the energetic and crack tip characterising senses, and approximate formulae developed to aid exploratory use.

Bui [119] has derived an integral identical in concept to the J integral but expressed in terms of the complementary form, i.e. in eqn. (2.18) Z is now $\int\varepsilon\,d\sigma$ instead of $\int\sigma\,d\varepsilon$ and the second term $u_i\partial\sigma_i/\partial x$ instead of $T_i\,\partial u_i/\partial x$. A possible use of this formulation would seem to be in problems where uniform pressure is applied to the face of the crack. This is not permitted within the ends of the contour in the normal formulation of J but in Bui's formulation it would seem to be a permissible feature, since there would be no contribution along the crack face between the tip and ends of the contour with $\partial\sigma/\partial x = $ zero. The writer is not aware of any reported usage of this feature in the literature.

2.2.3 Relation between COD and J

The quantitative relation between COD and J may be stated by writing eqn. (2.12b) in the generalised form:

$$J = m\sigma_Y\delta \quad (2.49a)$$

It should be noted that eqn. (2.12) relates to G (from LEFM) and is a formal relationship only for the BCS–Dugdale model. If now, with extensive plasticity, $J > G$, then there are four factors—constraint (as in eqn. (2.17)), work-hardening, the increase of J and G, and the lack of uniqueness in the definition of δ in most elastic–plastic models—that interact to give some particular value to m. All four effects also depend on the degree of yielding, although some tend to cancel so that for virtually all studies $1 \leq m \leq 3$. Thus, eqn. (2.49a) is simply an arithmetical relationship giving a particular value of m when all the variables are fixed.

J can be related schematically to COD in the plastic region for three-point bending or tension, using the relationship developed for strain in the text before eqn. (2.44a). With a rotation of 2θ then approximately $\delta = n(W - a)2\theta$ where n is the 'rotational' factor that defines the apparent centre of rotation [8]. Thus, using eqn. (2.36) with $q_{pl} = \theta S/2$

$$J \simeq (\eta L/4n)\sigma_Y \delta \qquad (2.49b)$$

For deep notches $\eta \simeq 2$, $L \simeq 1 \cdot 5$, $n = 0 \cdot 4$ or $0 \cdot 45$ so

$$J \simeq (1 \cdot 67 \text{ to } 1 \cdot 87)\sigma_Y \delta \qquad (2.49c)$$

in fair agreement with the values of m in eqn. (2.49a). In a similar approximation for centre-cracked tension eqn. (2.36) $N = 1$, $L \simeq 1$, $\eta \simeq 1$, $q_{pl} = \delta$ (over a gauge length corresponding to the slip-line fan) so that in this case

$$J \simeq \sigma_Y \delta \qquad (2.49d)$$

An example of this relationship for bend data given by Harrison [99] is in essence similar to the evidence of Fig. 2.19b. The experimental data are the same as those in Fig. 2.38, which were seen to fall between the results computed for plane strain and plane stress. Further comparisons can therefore be made using the computed curves as bounds on the experimental data. Thus computed clip gauge mouth opening data (from Refs. [22 and 23]) are turned into COD values by the formula from Ref. [8] and then compared with computed J contour values (Fig. 2.51). This eliminates the differences in definition of COD discussed in connection with Fig. 2.8, but can only be applied to the bend pieces for which DD 19 holds. Expressed in terms of $J = m\sigma_Y \delta$, $m \simeq 1 \cdot 3$ for plane stress and $1 \cdot 7$ for plane strain. Harrison also shows that K from equivalent energy and K with a plastic zone correction factor, based not on eqn. (1.10) but on an effective crack length such that the compliance matches that of the actual test piece, correlate equally well with COD for these particular data.

The relationship between J and COD has been examined by Chell & Spink [107]. They use a finite width solution for the BCS–COD model and take the restraining stress in the plastic zone (f in eqn. (2.7)) and σ_u rather than σ_Y or $m\sigma_Y$ and thus write $J = \sigma_u\delta$. They then compare J evaluated by the contour integral, using the data for Ref. [22] shown in Fig. 2.21, with their calculations of J from COD. One such comparison is shown in Fig.

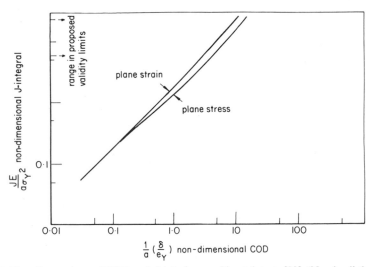

Fig. 2.51.   Comparison of COD and J data from weld notch tests [99]. (Nearly all the data fall between the two bounds shown, as in Fig. 2.38.)

2.52. The non-dimensional abscissa is $F/BW\sigma_u$ (instead of $F/BW\sigma_Y$ in Fig. 2.21) and the ordinate $(JE/\sigma_u^2 W)^{1/2}$ instead of the group $(JE/\sigma_Y^2 W)^{1/2}$ used in the J studies. For the non-hardening computation $\sigma_u = \sigma_Y$. For the hardening case the translation is less clear, since the degree of hardening recorded in Ref. [22] is considerably more than the rather small amount recorded in Ref. [107]. The curve falls above and to the left of the non-hardening case because of the change in abscissa. As remarked by Chell & Spink, the results are very sensitive to stress when near collapse. Comparison on a deflection basis would be of interest. It is not clear to what extent the finite size correction used would overcome the unsatisfactory features inherent in Fig. 2.2: perhaps indeed quite realistically for deep-notch problems, where the remote ends of the component are elastic in both model and reality, but perhaps not so satisfactorily for shallow-notch work-hardening cases.

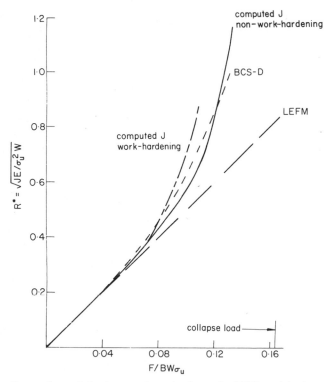

FIG. 2.52. Comparison of plastic stress intensity factor by HSW model adapted to finite width and J computations. Computed data the same as used for Fig. 2.22. $S/W = 4$, $a/W = 0.5$; note that terms are normalised with respect to σ_u not σ_Y [107].

By combining eqn. (2.49) with eqn. (2.11), used to define the non-dimensional COD, ϕ,

$$\frac{JE}{\sigma_Y^2 a} = 2\pi m \phi \tag{2.50}$$

enabling the COD design curve given by eqns. (2.14) to be plotted on Fig. 2.46 in order to compare it with the J design curve for various values of m. For $m = 2$ the COD curve is seen to yield far higher J values than the J design curve given by eqns. (2.41), i.e. the latter is clearly less conservative. This is not surprising, since the COD curve contains a degree of conservatism by intention whereas eqns. (2.41) do not.

At elastic levels, LEFM predicts (cf. eqn. (2.9)):

$$GE = \sigma^2 \pi \bar{a} \tag{2.51}$$

where here the $\bar{a}$ notation is used to imply the crack equivalent to a thorough-crack of length $2a$. Writing $G = m\sigma_Y\delta$ and re-arranging, using eqn. (2.11):

$$\phi = (\varepsilon/\varepsilon_Y)^2/2m \qquad (2.52)$$

There is thus conservatism built into eqn. (2.14a) amounting to a safety factor of $2m$. For centre-cracked tension m is probably near unity since there is little constraint. For high-constraint cases m may be taken as $\sqrt{3}$ (from LEFM plane strain) or even as high as 3 (from Prandtl slip-line field solutions). A precise value is not argued here beyond the notion that there is an in-built conservatism of 2 or perhaps more.

An excellent exposition of the COD philosophy as at present used and a brief comparison of the COD and original J design curve (Fig. 2.46) is given by Harrison *et al.* [120].

The quasi-elastic Dugdale model underlying the COD design curve does not give a good estimate of overall displacements, since stress and hence remote strain are always increasing elastically as δ increases. For little work-hardening a closer picture [4] is that the whole remote strain is provided by the crack opening at virtually constant remote stress. Thus, beyond the small-scale yielding regime of eqn. (2.14a), $\varepsilon \sim \delta$ and a linear regime is fitted on to the elastic part to give eqn. (2.14b). This picture is not always correct, as seen in the foregoing discussions of J for shallow-notch work-hardening cases. The picture may in part be attributable to the existence of a yield plateau in the structural steels used for many of the early COD studies in which fracture occurred, or the test was terminated, at gross strains little if any beyond the plateau.

Another interpretation can be found:
From eqn. (2.8), using the expansion of ln sec $(\pi\sigma/2\sigma_Y)$ and eqn. 2.11

$$\phi = \frac{4}{\pi^2}\left[\frac{1}{2}\left(\frac{\pi\sigma}{2\sigma_Y}\right)^2 + \cdots\right] = \frac{1}{2}\left(\frac{\sigma}{\sigma_Y}\right)^2 + \cdots \qquad (2.53)$$

This can be written $\phi = \phi_{el}$ + correction terms. Then, from LEFM, again using $\bar{a}$ as the equivalent crack,

$$\bar{a} = K^2/\pi\sigma^2 = r_{pl}(2\sigma_Y^2/\sigma^2) = r_{pl}/\phi_{el} \qquad (2.54)$$

and from eqn. (2.16a)

$$\bar{a} = K^2/2\pi m\sigma_Y^2\phi = r_{pl}/\phi \qquad (2.55)$$

to within some uncertainty over the value of m and the precise value of r_{pl} thereby defined. This again shows the whole process as a correction factor to LEFM, dependent only on stress level and independent of geometry. The current use of $\bar{a}$ rather than a, based on the LEFM shape factor Y, allows variations in the evaluation of COD for different configurations, which is a considerable advance on the original infinite plate model, but the plasticity and geometry factors remain independent, whereas in the J method they are interacting.

As far as the writer is aware, the BCS-based finite size correction factors [50, 107] have not been applied to δ from the COD design curve as such, though strictly more applicable than the LEFM-based factors. They have been used in related COD studies (Section 2.3.2).

An explanation for the higher degree of conservatism in the COD design proposals may be found by relating the origin of the two procedures to the two basic purposes of post-yield fracture mechanics, laboratory testing or structural design, defined in Section 1.7. In principle, either of the leading parameters J or δ can be used for either purpose. In reality, most J studies are centred on the first purpose (although Begley & Landes in their early formulations envisaged the latter) and COD studies are mainly aimed at the second purpose (although Wells in his early formulations envisaged the former). It is essential to recognise these differences in purpose if the differences in recommendations of the J and COD schools are to be reconciled. In the writer's opinion, it is these differences in purpose rather than the technical differences between J and COD that lead to the different usages proposed. Early COD tests showed both a size-dependence, to be discussed in a general manner in Section 2.4, and also an effect of in-plane constraint [6] which led to recommendation of the more severe three-point-bend test over the notch tension test. Use of fatigue notches altered the temperature at which these events occurred but did not remove them. At a slightly later stage [14] the great effect of slow stable tearing was realised so that a distinction was made between δ_i for COD at crack initiation and δ_m for COD at maximum load. The choice of which to use is partly at the discretion of the user and is still not fully resolved, as discussed in Section 2.2.1.2, but the four measures (full thickness, high constraint, sharp notches and use of δ_i or δ_m) allow the user to make an assessment of fracture toughness with a degree of conservatism realistic for his purposes, and allowances for residual stress, though somewhat arbitrary, are incorporated, with the whole procedure fairly extensively validated both by the original experimental data on which the curve is based and by many subsequent in-service applications. The writer is not aware of extensive use of COD in thermal strain cases, although the concept is obviously as physically reasonable as for mechanical strains. Again, related work using the BCS model has been conducted [121] as outlined in Section 2.3.2.

The practical problem of over- or under-matching weld strength, wherein the deformation is non-uniform along the test piece, has not been studied in depth by either technique. Conventional analyses do not apply to this non-uniform pattern without modification to the local–remote parameter relationship and as noted in Appendix 3 the use of J would be restricted to variations in yield stress perpendicular to the crack plane (which is of course the main variation in question). Some preliminary attempt has been made to recognise the problem in terms of COD [110]. An experimental approach is outlined in Section 2.3.9.

2.3 ADDITIONAL METHODS FOR DESIGN APPLICATIONS

2.3.1 Introductory Remarks

In outlining the several fracture mechanics concepts that have been developed outside the main stream of COD and J work, it is convenient to group together five directed predominantly at the elastic–plastic and contained yield situation (Sections 2.3.2 to 2.3.6) and three others that very clearly encompass uncontained and general yield (Sections 2.3.7 to 2.3.9). In most cases the proposals are not directed exclusively at the one regime rather than at the other but were perhaps first developed with the one application in mind and may then have been extended to the other. Some are clearly pressure-vessel oriented and contain particular efforts to embrace thermal stresses, whilst others are directed more towards general welded structural steelwork. These trends are obvious from the example data and references quoted, but it is doubtful whether there has been a wide enough application of most of the methods for distinct conclusions to be drawn that one is superior to the other for a particular class of problem. The presentation here emphasises what the present writer sees as the assumptions, strengths and limitations of the methods, where possible using J or COD concepts as a unifying theme.

2.3.2 Extensions to the BCS–Dugdale Model

Heald, Spink & Worthington (HSW) [122] used the BCS–Dugdale model to define an *apparent toughness* in the post-yield region, incorporating the LEFM shape factor

$$K_A = Y\sigma_f\sqrt{a} \tag{2.56}$$

where σ_f represents the nominal remote stress at fracture. The BCS–Dugdale model is used to relate the remote stress σ_f to the crack tip

COD by eqn. (2.7) but for fracture δ_c is written in terms of K_{Ic} (eqn. (2.4)) and the restraining stress (f in eqn. (2.7)) taken as σ_u. The LEFM shape factor Y is introduced to account for geometry, so that the plasticity and geometry terms do not interact. Thus

$$\sigma_f = \frac{2}{\pi}\sigma_u \cos^{-1}\left[\exp\left(\frac{-\pi K_{Ic}^2}{8\sigma_u^2 a Y^2}\right)\right] \qquad (2.57)$$

If $a \to 0$, $\sigma_f \to \sigma_u$; if the exponent is small, $\sigma_f \to K_{Ic}/Y\sqrt{a}$ as in LEFM, and it is argued that a solution that tends to the correct answers for these two limits must be a reasonable interpolation in-between.

Early results, such as in Ref. [122], show very good agreement between experiment and prediction for a variety of configurations and materials. No positive attempt is made to explain the complexities of in-plane constraint, the effect of gauge length on strain in the COD model and the likelihood of slow stable crack growth in ductile rather than LEFM fractures. Indeed, slow crack growth is not discussed so that it is not clear whether the good agreement between experiment and prediction is because of or despite slow growth, or fortuitously, in its absence. The writer has not seen weldment data analysed by this method.

The general relation between the HSW proposals and J has already been seen for deep-notch bending (Fig. 2.52). Some shallow notch bend results from a rotor forging are shown in Fig. 2.53 from Ref. [106]. This reveals a weakness of the method as originally formulated. The stress σ_u is essentially tensile. In bending the outer fibre stress can be equated to σ_u and the load derived by elastic or plastic analyses. It is also possible to relate σ_u to a strain ε_u, if the stress–strain curve is known, and relate this strain to a load by, for example, a finite element analysis. The two curves labelled elastic $K_{Ic} = 66$ and plastic $K_{Ic} = 66$, using the HSW theory (Fig. 2.53), were obtained one by conventional elastic bending theory and the other from an elastic–plastic finite element analysis of the uncracked body. This latter predicts a failure load for the shallowest notch some 50 % higher than the elastic methods. Unfortunately uncertainties concerning the experimental data of Fig. 2.53 do not permit an assessment of these computational alternatives. The deep-notch test $a/W = 0.5$ is valid to ASTM LEFM requirements and gives $K_{Ic} = 66 \text{ MN/m}^{-3/2}$. The tests at $a/W = 0.2$, $a/W = 0.3$ are not valid in respect of notch depth, and give $K_Q = 84.5 \text{ MN/m}^{-3/2}$.[12] Previous tests had shown no scatter greater than $= 5 \%$ for a given radial position in the

---

[12] $K_Q = K_{Ic}$ value obtained from tests not satisfying validity requirements of Chapter 1.

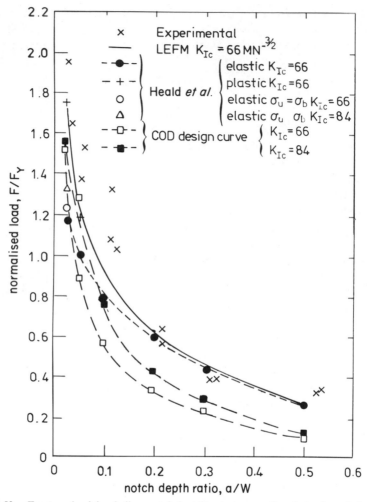

FIG. 2.53.   Fracture load in shallow-notch bending tests as a function of crack length according to COD design curve; HSW model and J analysis in plane strain. $S/W = 8$, work-hardening described as for Fig. 2.11. F_Y is calculated load to first yield of unnotched beam [106].

original forging. It therefore appears either that the material was not uniform, or that undetected sources of error were present. Another difficulty is whether the shallow notch tests representing a structure should be treated by plane strain or plane stress. In short, these uncertainties, quite typical of practice, more than mask the variation in the several plasticity theories used, all of which perforce predict results below the LEFM line whereas the test data perversely tend to fall above it.

As noted earlier, Chell *et al.* [50] have extended the BCS–Dugdale model to finite plates by constructing Green's or weight functions for the problems. Good agreement is found with the computed solutions of Hayes & Williams [49]. In LEFM the stress intensity is related to the Green's function [50] by

$$K(a) = \int_0^a G(a, x)\sigma(x)\,dx \qquad (2.58)$$

for an edge crack of length a in the x-direction subject to remote stress σ (which may be a function of x).

Displacements of the crack face are

$$\delta(x) = 2/E' \int_x^a G(x', x)K(x')\,dx' \qquad (2.59)$$

Expressions for the Green's function $G(x', x)$ are constructed [123] from the LEFM shape factor Y and certain inverse trigonometric functions of x/x' and x/W (where W is plate width). Plasticity solutions of the BCS–Dugdale type are constructed [50], just as in the original BCS–Dugdale model, by superposition of the appropriate elastic solutions for the remote applied stress and the crack face load components that sum to f (Fig. 2.2) along the line of extent $(a_1 - a)$ ahead of the real crack which represents the plastic zone in this model (eqn. (2.6)).

This usage of the BCS model for various geometries has been extended to finite-width centre-cracked and single-edge tension [50], three-point bending [107] the CKS pieces [124] and to three-dimensional penny-shaped and semi-circular surface cracks [125] (see Section 2.4.3). Although in a formal sense these solutions are not part of the COD design method as generally understood, nor indeed of the original HSW model, they form a valuable addition to the methods available for post-yield fracture analysis. Indeed, whether these recent papers should be regarded as 'COD', from which concept they clearly stem, or as 'J', is almost a matter of semantics, since in the BCS model the two terms are formally (as distinct from arithmetically) related by the $J = f\delta$ relationship. An application of these analyses to single-edge notched tension pieces is made by Chell & Davidson [126], who found that the toughness expressed as K_{Ic}, derived from J or δ, increases as crack length decreases in the post-yield regime for a 1%Cr, 1%Mo, $\frac{3}{4}\%$V steel of $580\,\text{MN/m}^2$ proof stress. Despite reference to considerable non-linearity the records do not appear to show the very extensive plasticity discussed elsewhere in this chapter. Slow crack growth is thought not to occur on the evidence of close inspection, but in the absence of a micro-mode change it is doubtful whether such evidence is conclusive. They conclude that K_{1c} is not a material constant in the post-yield regime. The test technique uses a J or COD analysis but is not one of the preferred standard methods (Chapter 3) so the argument on post-yield toughness again remains open to question.

Chell has also studied some pressure-vessel and thermal-transient nozzle problems of great interest, in terms of estimates for J and δ from his model [127]. The application of the modified BCS–Dugdale model to thermal and residual stress cases is outlined in Ref. [121]. An important part of the argument is that although the limit load of a whole structure is not affected by thermal, residual or other self-equilibratory stresses, the thermal and residual problem must be studied for the *local* region of interest for which the fracture may be influenced by the thermal and residual stresses. They point out that in some cases thermal systems behave like load-controlled systems and in other cases like deformation-controlled systems. Chell [127] quotes J for the BCS model as

$$J = \frac{8}{\pi^2 E} a\sigma_L^2 Y^2 \ln \sec \frac{\pi\sigma}{2\sigma_L} \qquad (2.60)$$

which follows at once from the $J = f\delta$ relationship just quoted, together with eqn. (2.7). However, he here sets the restraining stress f equal to σ_L the limit load stress for collapse of a component assumed to be in plane stress. This usage of limit load reflects a suggestion that has recently come to the fore in the UK, as described in Section 2.3.7, and is tending to replace the usage $f = \sigma_u$ in the original HSW model.

2.3.3 Newman's Two-parameter Method

Newman [128] has suggested an elastic–plastic analysis for surface or through cracks based on the Neuber [129] relation

$$k_t^2 = k_\sigma k_\varepsilon \qquad (2.61)$$

where k_t is the elastic stress concentration factor (SCF), k_σ the plastic SCF and k_ε the elastic strain concentration factor. This relationship will be discussed later. The elastic SCF is written for an elliptical hole in a plate

$$k_t = \left(1 + 2\sqrt{\frac{a}{\rho}\frac{\alpha}{\beta}}\right) \qquad (2.62)$$

where ρ is the notch root radius, and α and β the LEFM correction factors for external boundaries and ellipticity respectively.[13] k_σ and k_ε are written

$k_\sigma = {}_1\sigma/\sigma_n$ (i.e. local stress over net section stress)
$k_\varepsilon = {}_1\varepsilon/\varepsilon_n$ (i.e. local strain over net section strain)

$\varepsilon_n = \sigma_n/E_n$ (where E_n is the secant modulus for the net section stress and strain). Combining all these relationships

$$\sqrt{{}_1\sigma_1\varepsilon E} = \sigma_n \sqrt{\frac{E}{E_n}} \left(1 + 2\sqrt{\frac{a}{\rho}\frac{\alpha}{\beta}}\right) \qquad (2.63)$$

For fracture ${}_1\sigma = \sigma_f$; ${}_1\varepsilon = \varepsilon_f$; $\rho = \rho^*$ a so-called effective crack tip root radius, all taken to be material properties at fracture for given temperature, strain rate, etc. This relationship is then rearranged (by keeping the term in $\sqrt{a/\rho}$ on the right-hand side and grouping the other two on the left-hand side) and then multiplying by $\sqrt{\pi}$ to give

$$(\rho^*\sigma_f\varepsilon_f E\pi/4)^{1/2} - (\rho^*\sigma_n\pi E/4E_n)^{1/2} = \sigma_n\sqrt{\pi a}\frac{\alpha}{\beta} = K_{le} \qquad (2.64)$$

where K_{le} means the K value according to a simple LEFM determination. If the test were valid by LEFM requirements for plane strain then $K_{le} = K_{lc}$, of course. The first term on the left-hand side is written K_f and designated a fracture toughness 'property'; the second term is written as $K_f n\sigma_n/\sigma_u$, where σ_u is the tensile strength, whence $n = \sigma_u/(\sigma_f/\varepsilon_f E)^{1/2}$ is designated a second toughness 'property'.[14]

The relationship is usually written

$$K_f = \frac{K_{le}}{\left(\dfrac{E_n}{E}\right)^{1/2} - n\dfrac{\sigma_n}{\sigma_u}} = \frac{K_{le}}{\phi} \qquad (2.65)$$

ϕ is shown in Fig. 2.54, the slope of the linear regime being $\phi = 1 - n(\sigma_n/\sigma_u)$. It is not clear in Ref. [128] just how the particular curve shown was derived or whether this is a schematic relationship only. For $\sigma_n < \sigma_u$, E_n is taken as E, since it is argued that net section yield does not occur until somewhat after $\sigma_n = \sigma_Y$. For brittle materials, with $E_n = E$ and $n = 1$, LEFM is recovered. At least two tests must be conducted for

[13] That is, $\sqrt{\pi(\alpha/\beta)}$, as appears in eqn. (2.64) is the LEFM shape factor, Y, as used hitherto, but based on net section stress, σ_n, not gross section stress, σ.
[14] Newman uses m for this term, but n is used here to avoid confusion with the term m in the $J = m\sigma_Y\delta$ relationship.

different crack sizes but the same thickness material, to find the two values K_f and n for that thickness.

It seems to the writer that in eqn. (2.63), $\sqrt{a/\rho} \gg 1$, so that in eqn. (2.65) the corresponding term in n must be much smaller than $\sqrt{E_n/E}$ that arises from the term 1 in eqn. (2.63). n is also essentially positive. In data given in Ref. [128] n varies from -0.35 to $+1.0$, seeming to imply either that the rationale is inadequate or that eqn. (2.65) must be regarded as empirical. In

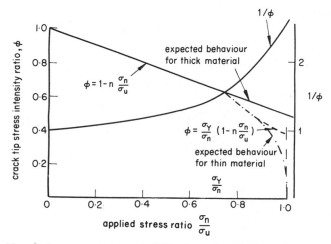

FIG. 2.54. Newman's two-parameter method. The factor $\phi = K_{Ie}/K_f$ as a function of applied stress ratio σ_n/σ_u. (After Ref. [130] with $1/\phi$ superimposed.)

a later paper [130] Newman writes, 'If n is equal to unity the equation represents behaviour of high toughness materials (plane stress fracture)'. This may be so, but it implies $\rho^* = 0(a)$ and thus not physically identifiable in the origins of eqn. (2.63). Negative n values do not seem to be reported in the more recent studies. The ϕ diagram in Fig. 2.54 is somewhat modified [130] by cutting off the square corner at $\sigma_n/\sigma_u = 1$ as shown, so that the linear region extends only to $\sigma_n \leq \sigma_Y$. For $\sigma_Y < \sigma_n < \sigma_u$ the slope is increased to give $F = (\sigma_Y/\sigma_n)(1 - n(\sigma_n/\sigma_u))$. Expressions for ϕ in the linear regime are given, based on the LEFM Y factors, and σ_u is now the nominal stress to cause collapse of the net section (based on the tensile strength rather than yield or some flow stress and not apparently including a geometric constraint factor).

A wide range of fracture data on certain steels, aluminium and titanium alloys is examined for a variety of test configurations with a very

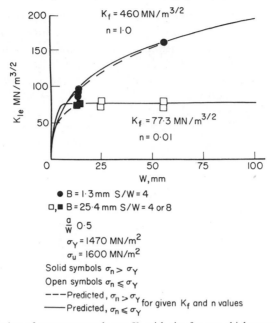

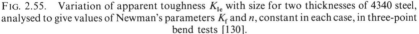

FIG. 2.55. Variation of apparent toughness K_{Ie} with size for two thicknesses of 4340 steel, analysed to give values of Newman's parameters K_f and n, constant in each case, in three-point bend tests [130].

satisfactory consistency of the results. Some data are shown in Fig. 2.55 for two thicknesses of steel. K_{Ie} is a function of width, but K_f is constant (for each thickness) for all widths with $n = 1$ for thin sheet and $n = 0.01$ for thick plate, for which K_f is in fact also constant though by implication it would not be so for yet smaller test pieces. A general picture of the toughness–stress level–thickness relationships for an aluminium alloy is shown in Fig. 2.56.

In a broad-brush way, the concept can be related to COD or J, at least for plane strain. From the definition attached to eqn. (2.64)

$$K_f = \left(\rho^* \sigma_f \varepsilon_f E \frac{\pi}{4} \right)^{1/2} \tag{2.66}$$

$$\therefore \quad K_f^2 / E = \left(\frac{\pi}{\delta} \frac{\rho^*}{\rho} \frac{\sigma_f}{\sigma_Y} \right) \sigma_Y 2 \rho \varepsilon_f \tag{2.67a}$$

$$= m \sigma_Y \delta \tag{2.67b}$$

$$= J \tag{2.67c}$$

where the COD, δ, is identified as the strain ε_f over the gauge length 2ρ (i.e. the original crack width) and $(\pi\rho^*\sigma_f/\delta\rho\sigma_Y)$ is identified notionally as a constant, m (as in eqn. (2.12)). For plane strain with moderate work hardening, the computations discussed in Section 2.2.2.1.3 suggest that the maximum stress at the crack is about $4\sigma_Y$; if ρ^* is taken as the radius of the opened crack tip it is about $\rho(1 + \varepsilon_f)$ and ε_f is of the order of 0·5 on the micro scale so that $m \simeq (\pi/\delta)(1·5)(4) \simeq 2·3$, in good agreement, probably fortuitously so, with values found in Section 2.2.2. The use of a second parameter n is reminiscent of the HSW model (Section 2.3.2), in that it allows results to tend to LEFM at the one extreme and to the tensile strength at the other. If, as it appears in plane stress, $\rho^* = 0(a)$, so that $n \rightarrow 1$, some process zone or even plastic zone seems implied. As noted, this may improve the practical usefulness of the method, but seems to undermine its derivation from the Neuber equation where ρ is an actual tip radius. The Neuber equation is only proven for shear loadings [129] and for notches, not cracks. The writer is not aware of strict support for its usage in plane stress or plane strain and for cracks. A rationalisation can be made if it is supposed that, just as in LEFM,

$$K = \lim_{\rho \to 0} k_1(\sigma/2)(\pi\rho)^{1/2} \tag{2.68a}$$

so in power-law material $\varepsilon/\varepsilon_Y = (\sigma/\sigma_Y)^N$

$$K_\sigma = \lim_{\rho \to 0} k_\sigma\sigma(\pi/I)^{1/N+1}\rho^{1/N+1} \tag{2.68b}$$

$$K_\varepsilon = \lim_{\rho \to 0} k_\varepsilon\varepsilon(\pi/I)^{N/N+1}\rho^{N/N+1} \tag{2.68c}$$

where K_σ, K_ε are plastic stress and strain intensity factors and I is a constant as in the HRR solutions (Section 2.2.2) [56, 57]. k_σ, k_ε are the plastic stress and strain concentration factors. Clearly, eqns. (2.68b) and (2.68c) must be valid for $N = 1$, since eqn. (2.68a) is then recovered. Assuming, with no other justification, that eqns. (2.68b) and (2.68c) are valid, and since in the HRR solutions $J = I K_\sigma K_\varepsilon$, where $JE \rightarrow K^2$,

$$k_1^2 = k_\sigma k_\varepsilon(\sigma/\sigma_Y)^{N-1} \tag{2.69a}$$

assuming the limits are well behaved. However, if for a crack at a stress concentration, the K value is related to the local stress[15] in the absence of a

[15] It is clear from the formulation used by Buekner [132] that the meaning of σ in the formula for K is the value of the stress in the uncracked body where the crack will subsequently be, although it is often referred to loosely as the 'remote stress', following the original derivation for a crack in an infinite plate where the two terms are, of course, the same.

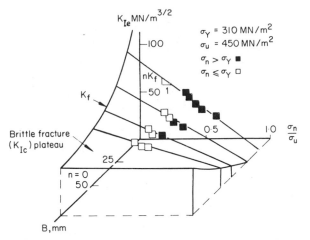

FIG. 2.56. Apparent toughness K_{Ie} at failure for compact tension specimens of 2219-T851 aluminium alloy as a function of nominal failure stress ratio and thickness [130].

crack, $k_t\sigma$ for LEFM (as it indeed usually is for small cracks in the region of stress concentration), and by analogy, to $k_\sigma\sigma$ in the non-linear case rather than to the 'remote' stress, σ, then

$$k_t^2 = k_\sigma k_\varepsilon \qquad (2.69b)$$

Thus this expression might be expected to provide a useful basis for crack studies whilst $JE \simeq K^2$, i.e. for contained yielding. Some experimental evidence has suggested [131] that for uncontained yield the relationship (2.69b) is not valid. On either of these arguments relating the method schematically to J (or COD) $(J/G)^{1/2} = 1/\phi$ (from eqn. (2.65)). The factor $1/\phi$ is superimposed on Fig. 2.54 from the original data and clearly follows the trend of $(J/G)^{1/2}$ or (Y^*/Y) as in Fig. 2.21 (if the ordinate were normalised so that the graph started at unity instead of at Y) or indeed of the ln sec plasticity term of the HSW and related models (Fig. 2.52).

2.3.4 Tangent Modulus (Gross Strain) Method

This concept was proposed in terms of gross strain [133], following detailed analysis of surface-flawed (part-through crack) and bend pieces tested in connection with the HSST programme conducted in the USA [134–136]. The object of the approach is 'to measure directly the plastic strain capacity of a nuclear pressure vessel steel, in the presence of sharp flaws' and to answer such questions as: 'Will a flaw of given size and shape cause fracture at a given location in a structure if a specified amount of plastic strain is

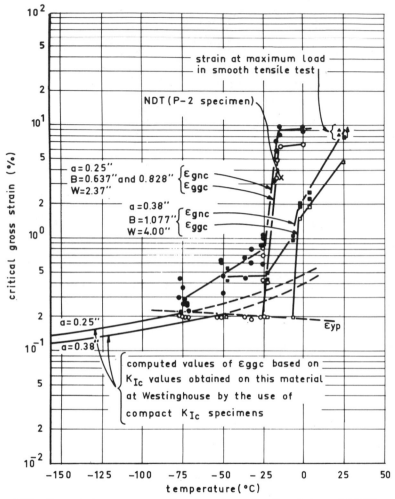

FIG. 2.57.   Gross strain as a measure of toughness for surface-flawed plates of A533B. ε_{gnc} is gross strain at the net section but remote from the notch; ε_{ggc} is gross section strain remote from the notch plane [133].

expected to occur at the location in the absence of the flaw? The feature that gave rise to the analysis was the abrupt increase in gross section strain remote from the notch as a function of temperature in surface-cracked tension tests on A533B (Fig. 2.57). This behaviour is similar to the well-known ductility transition in the structural steels [137] and the effect of size and the tension or bending nature of the stresses on that transition [138].

It was remarked [133] that the sharp transition occurred only with

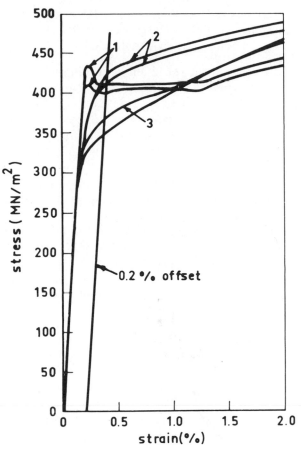

F IG. 2.58.   Variations in stress–strain curves relevant to the test data of Fig. 2.57 [133].
Stress–strain curves taken at room temperature (+ 1·5 °C for the upper curve of group 3) for
the following locations: (1) mid-plate material, 635 mm from bottom end of plate; (2) bottom-
block material, 635 mm from bottom end; (3) bottom-block material, 864 mm from bottom
end.

stress–strain curves with a yield plateau, found in mid-plate-thickness
material of the plate of A533B used. In near-surface layers the stress–strain
curves were more rounded (Fig. 2.58) and the transition in ductility was
spread over a wider range in temperature.

In these and other tests, three strain measurements are referred to:

ε_{gnc} the strain in the net (notched) section but remote from the notch
itself (e.g. on the side face);

ε_{ggc} the strain in the gross section remote from the notch plane;

COD   apparently the mouth-opening COD, not the crack tip opening
as defined in all the COD procedures discussed in Section 2.2.1;
hereafter referred to as 'mouth COD'.

The precise pattern of the transition in ductility is shown to vary with
choice of ε_{gnc}, ε_{ggc} or mouth COD as the ductility parameter, and with size of
defect, plate width and thickness.

At this stage no particular method is proposed to incorporate these
geometrical factors, although the effect of loss of constraint by yielding and
that of net to gross area are discussed. It seems significant that the tests are
of small equivalent a/W ratio (less than 10% based on the area of the
section) so that, contrary to many deep-notch test configurations, the
possibility of the net section work-hardening sufficiently to allow gross
section yield is evident.

Through a series of intermediate reports [139–141] presenting additional
experimental data on test pieces in pure bending and in bending with
backing bars, the concept is taken from what Ref. [139] describes as the
'gross strain concept [that] is so simple that it is easily dismissed as being
obvious,' to a method [142] of some complexity. Four features are
included:

   (i)   a relationship between local and gross strain, outlined below;
   (ii)   a relationship between K_c and K_{Ic} as a function of constraint;
   (iii) a description of K_{Ic} as a known experimental function of
          temperature, whence, using (ii), K_c is also known as a function of
          temperature;
   (iv)   an assessment of the effective value of the tangent modulus in the
          work-hardening range for various tension and bending loads.

A 'notch ductility factor', described below, is defined. Its usage appears
closely analogous to fracture toughness in that a critical value at fracture is
found from one test (in fact from K_{Ic} here) and taken as a material
toughness property, and an applied value is found for the loading
configuration of interest. For a component that has yielded, this applied
value is here summed over elastic, yielding and work-hardening regimes
and therein lies its difference from K. The process is however reminiscent of
the approximate analyses for J (cf. Section 2.2.2.1.3) in which a plasticity
term is added on the elastic value.

Step (i): The stress–strain curve with yield plateau is approximated to in a
stepwise manner, with elastic modulus E, a yield plateau, and then a linear-
hardening regime of tangent modulus E_s. The local notch root strain is

related to the work-hardening modulus by the equation (in incremental form)

$$d_1\varepsilon\sqrt{\rho} = (2Y\sqrt{a}\sqrt{E_g/_1E})\,d\varepsilon_g \qquad (2.70)$$

where in terms of Section 2.3.3, $_1\varepsilon$ is the notch root strain; $_1E$ is the modulus at the notch; while ε_g is the strain remote from the notch; E_g is the modulus remote from the notch; Y and a are the standard LEFM shape factor and crack length terms; and ρ is the notch root radius. Note that a sharp crack is not apparently implied at this stage. The term $d_1\varepsilon\sqrt{\rho}$ is the 'applied' notch ductility factor. The writer has not seen the derivation of eqn. 2.70. The result can be obtained in terms of total strain from the Neuber expression for stress and strain concentration factors by reasoning similar to that discussed in Section 2.3.3.

To evaluate the critical value at fracture, eqn. (2.70) is then re-written, taking $E_g = E$ and $E = E_s$, the value of the tangent modulus in the strain-hardening range. This equation is applied to the fracture problem in terms of total rather than incremental strains, with $_1\varepsilon = \varepsilon_f$; $\varepsilon_g = \lambda$, where ε_f is

$$\varepsilon_f\sqrt{\rho} = (2Y\sqrt{a}\sqrt{E/E_s})\lambda \qquad (2.71)$$

Re-expressing the normal LEFM relationship for K in terms of strain

$$\lambda = (K_c/\sigma_Y)\varepsilon_Y/Y\sqrt{\pi a} \qquad (2.72)$$

and equating λ between eqns. (2.71) and (2.72) gives

$$\varepsilon_f\sqrt{\rho} = (2K_c\sqrt{E}/\sigma_Y\sqrt{\pi E_s})\varepsilon_Y \qquad (2.73a)$$

$$= C(K_c/\sigma_Y) \qquad (2.73b)$$

for given values of E, E_s and ε_Y. The term $\varepsilon_f\sqrt{\rho}$ is called the 'notch ductility factor at fracture' and eqn. (2.73) is assumed to hold for elastic or plastic situations.

Step (ii): In Ref. [141] the Irwin relationship [143] is used

$$K_c/K_{Ic} = \sqrt{1 + 1\cdot4\beta_{Ic}^2} \qquad (2.74a)$$

where

$$\beta_{Ic} = (K_{Ic}/\sigma_Y)^2/B \qquad (2.74b)$$

B is here replaced by $2a$, where a is the depth in the thickness direction of the part-through crack. The rationale for this is that the distance from the point of maximum constraint to the nearest free surface is the factor of interest. For a through crack this is $B/2$; for a part-through crack with $a < B/2$ it is a. It is also suggested that eqn. (2.74) is of most use for finding K_c if K_{Ic} is known, but where K_c is known and K_{Ic} required, Ref. [141] suggests the use of a graph of

$$1/\beta_c = \sqrt{1 \cdot 4/(K_c/K_{Ic})^2 [(K_c/K_{Ic})^2 - 1]^{1/2}} \qquad (2.75a)$$

where

$$\beta_c = (K_c/\sigma_Y)/2a \qquad (2.75b)$$

from which K_c/K_{Ic} can be read off from a known value of β_c.

Step (iii): The temperature-dependence of K_{Ic} or K_c is taken for A533B to be

$$K_{Ic}/\sigma_Y = A_T/(T_\infty - T) \qquad (2.76a)$$

where T is temperature (°F) and T_∞ is the temperature above which linear mechanics does not hold. This is taken as about 125 °F and correlates roughly with the onset of upper shelf behaviour in the Charpy V test or as the point of inflection in $\frac{5}{8}$ in (16 mm) DT tests. A_T is an empirical coefficient, $A_T = 125\,°F$ here. If K_c is involved eqn. (2.76) is re-written

$$(K_c/\sigma_Y)(K_{Ic}/K_c) = A_T/(T_\infty - T) \qquad (2.76b)$$

where K_{Ic}/K_c is derived from Step (ii) in terms of β_c.

Step (iv): In evaluating eqn. (2.70) for any problem, the tangent modulus at the yielding notch root is required. It is proposed [142] that an effective tangent modulus can be found for each different configuration. Thus, it is suggested that the effective modulus is

$$E_g = \int_0^B (d\sigma/d\varepsilon)\,dz\sqrt{B} \qquad (2.77)$$

where z is distance in the thickness direction, and the strain ε is taken as

$$\varepsilon = \varepsilon_0 + (\varepsilon_1 - \varepsilon_0)z\sqrt{B} \qquad (2.78)$$

where ε_0 and ε_1 are the front and back face strains at the cracked section, whence

$$E_g = (\sigma_1 - \sigma_0)/\varepsilon_1 - \varepsilon_0) \qquad (2.79)$$

that is, the effective tangent modulus is the stress difference between the faces divided by the strain difference. For tension, $d\sigma/d\varepsilon = E_g$, so that the effective modulus is the tensile stress–strain tangent hardening modulus, $E_g = E$ in the elastic range; $E_g = 0$ on the yield plateau; $E_g = E_s$ in the (linear) strain-hardening range. The notch ductility factor (as eqn. (2.71) but with the appropriate values of modulus) is summed over the three terms (thus implying incremental form) to give:

$$\sum \Delta\varepsilon\sqrt{\rho} = 2Y\sqrt{a}\sqrt{E/E_s}(\varepsilon_Y) + 0 + 2Y\sqrt{a}(\varepsilon_f - \varepsilon_s) \qquad (2.80)$$

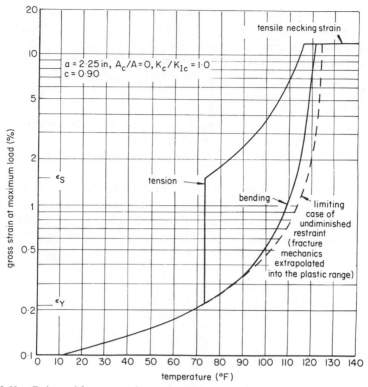

FIG. 2.59. Estimated fracture strain as a function of temperature for pure tension and pure bending with a part-through thickness defect [142]. (ε_s is the strain at the end of the yield plateau, a and c define a semi-elliptical crack. $A_c/A = 0$ denotes that the ratio of cracked total area is negligible; $K_c/K_{Ic} = 1\cdot0$ denotes that departure from plane strain is neglected.)

where ε_f is the final strain, ε_Y the yield strain and ε_s the end of the plateau or start of the hardening regime.

For bending: $\sigma_1 = -\sigma_0$; $\lambda_i = -\lambda_0$ and $E_g = \sigma_0/\varepsilon_0$ which is the secant modulus, E_{sec}, at the front or back face. Thus, $E_g = E$ in the elastic range; $E_g = \sigma_Y/\varepsilon$ in the yield plateau range, where $\varepsilon_Y \leq \varepsilon \leq \varepsilon_s$ and $E_g = E_s(1 + (\varepsilon_d/\varepsilon))$ where $\varepsilon_d = (\sigma_Y/E_s) - \varepsilon_Y$ and $\varepsilon \geq \varepsilon_s$. The applied notch ductility factor is found by summing the three terms of which the latter two have to be integrated between the appropriate strain limits. The critical notch ductility factor at fracture is in principle a function of constraint (Step (ii)), that requires some judgement, and temperature (Step (iii)). Figure 2.59 shows the results of the above analysis, assuming $K_c = K_{Ic}$ in eqn. (2.73); $C = 0\cdot90$, for the particular case of a surface crack 56 mm deep in a piece of large cross-section in which the difference between net and gross cross-sectional area is neglected. Clearly the tension and bending curves are quite different because the effective tangent modulus is never zero with a strain gradient as found in bending. The different effects on gross and net section strains, ε_{ggc}, ε_{gnc} (Fig. 2.57) are not found, since the effect of net and gross area has deliberately been ignored in the example. The next step is to include this area effect in a further extension of Step (iv).

It is argued that the reason why the vertical rise in gross strain from the tension case (Fig. 2.58) is not found in practice (Fig. 2.57) is that some bending occurs across the net section due to the presence of the defect, so that $E_g \neq 0$ in the yield plateau regime. As an approximate analysis the ratio of the reduction in stress at the back face, $\Delta\sigma_b/\sigma$ is found by assuming that a linear variation in stress from σ on the front face to $\sigma - \Delta\sigma_b$ on the back will compensate for the moment created by the loss of stress over the semi-elliptical cracked area so that the line of action of the load still passes through the mid-point of the gross section. When the front face just yields, the magnitude of the bending effect between back and front face is thus $(\Delta\sigma_b/\sigma)\sigma_Y$ and the effective tangent modulus from eqn. (2.79) is this value divided by the difference in strain across the cracked section. The strain on the yielded face will rise to ε_s and that on the back face will remain at some still-elastic value given by the bending effect as $\varepsilon_Y(1 - \Delta\sigma_b/\sigma)$ so that the effective modulus is given as

$$E_g = \frac{(\Delta\sigma_b/\sigma)\sigma_Y}{\varepsilon_s - \varepsilon_Y(1 - (\Delta\sigma_b/\sigma))} \tag{2.81}$$

The term $\Delta\sigma_b/\sigma$, evaluated as previously explained, depends of course on the ratio of the cracked area to the gross area. The final picture to emerge is shown in Fig. 2.60. The various strain points $\varepsilon_1, \varepsilon_3, \varepsilon_4$ are found by equilibrium from the requirement

$$\sigma A = \sigma_{\text{net}}(A - A_c) \tag{2.82}$$

where σ is the gross section stress, σ_{net} the net section stress, A the gross cross-sectional area and A_c the cracked area. At point 1 where the net section just yields, $\sigma_{\text{net}} = \sigma_Y$ so that $\sigma_1 = \sigma_Y(A - A_c)/A$, whence $\varepsilon_1 = \varepsilon_Y(A - A_c)/A$. Similar arguments obtained at point 3, where $\sigma = \sigma_Y$ and at point 4 where $\sigma_{\text{net}} = \sigma_u$.

The various patterns of gross and net section strains, such as Fig. 2.57, are thus accounted for in terms of bending or tension, and gross or net section effects. Further detailed examples of application are given in Ref. [142].

In examining the method it seems desirable to distinguish between the empirical relationships for a particular material or event and broad points of principle, which it should be possible to relate to fundamentals of yielding fracture mechanics, however the detail is expressed. Much of the detail of the method is in fitting effective toughness data to temperature or degree of constraint and known K_{1c} or K_c data. The methods used here are

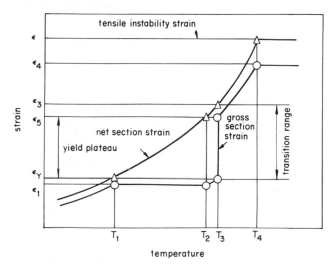

FIG. 2.60. Schematic diagram of net and gross section fracture strain as a function of temperature for a material with an ideally plastic yield plateau prior to strain hardening [142].

empirical and could be grafted on to any fracture concept. The writer doubts whether the simple formulae of Step (ii) such as eqns. (2.74) and (2.75) have wide generality and this point is recognised in Ref. [141].

The form of eqn. (2.73b) is immediately apparent from LEFM discussions of the plastic zone but the value of C to be assigned to a process zone is not then clear. The above argument in terms of a blunt notch allows a value to be assigned to C, although it could be argued, as in Neuber's theory, that such a term is inherently a material property, which must surely depend upon the micro-mode of fracture. Thus the present theory implies that it should be applicable only to a regime in which the ratio of yield strain and work-hardening to elastic modulus dominates the fracture process, since no other factor enters into the constant, C.

In the very particular application seen so far, namely A533B tensile, bending and vessel tests, the authors of the method are also able to call on a wealth of detail, e.g. yield strains for particular plates, actual crack sizes after slow growth and so on, that is denied to the user of more general methods for a wide variety of purposes. Thus the method seems to succeed in great detail where this is available, but has not, as far as the writer knows, been applied by others to a wider range of circumstances.

Reference [144] offers a review of the tangent modulus method making criticisms broadly similar to those now given, and suggests there are wide limits of uncertainty in the use of the method. It concludes trenchantly: 'Lacking theoretical justification and bedevilled by inexact computations the Tangent Modulus seems to be of little usefulness, indeed,' but that hardly does credit to the attempt to include features still not covered by the better known methods.

There remains the generality of approach. In essence, this seems to be that when notched/gross area ratio is comparable to yield/tensile strength ratio, the effects of work-hardening can dominate the strains remote from the notch whilst the notch tip itself is following its supposedly unique pattern of intense local deformation. To the extent that the part-through crack is a three-dimensional situation there is little analytical or computational ability against which to judge this statement. However, in relation to shallow-notch two-dimensional cases, the feature of whether or not hardening of the notch section allows yielding of the remote structure was discussed in the previous section in terms of J. The behaviour of J–q graphs for shallow notches with and without hardening (Fig. 2.45b) is quite different. Moreover, shallow-notch tension and bending, where work-hardening affects only the outer fibres at mid-span, differ somewhat in their response to hardening.

If the critical value of toughness is taken to be a function of temperature, it is clear that at low toughness the nominal fracture strains for the various tension and bending cases will fall in a different sequence to that for a higher value of toughness, and it is suggested that this behaviour reflects the major features of Fig. 2.59. It is accepted that yet further complexities of Fig. 2.60 for the part-through crack case cannot be reproduced in the computations discussed. It must also be recalled that the computations as shown allow only for simple continuous hardening with no yield plateau, although this feature could be incorporated.

In summary, the general features of the tangent modulus method are not at all contradictory to the predictions of J or COD technique, although the modulus method contains additional complexities and empirical relationships not as yet incorporated into those methods.

2.3.5 Equivalent Energy

This method was developed before the J contour approach became well known [145, 146]. In the writer's view, the method can be rationalised to a certain extent in terms of J and beyond that should be treated with caution as essentially empirical. A comparison of J and equivalent energy methods in these terms has been given in Refs. [147 and 148].

One application of the method is primarily directed at the testing problem where it is desirable to measure toughnesses on pieces smaller than required for valid LEFM testing, although if the general principles of the method were indeed established then they would also be applicable to other structures.

In essence the area under a non-linear load extension diagram for a post-yield toughness test is measured up to the first attainment of maximum load. This area is scaled to the triangular area defined by any arbitrary point in the linear part of the record, to give a ratio, β. The dimensions of the test piece are scaled by β and the load by β^2. The load so obtained, F_Q, is used to determine K_Q for the scaled-up piece. This value is called K_{Icd} (d for dimension;[16] i.e. K_{Ic1} where 1 refers to a 1-inch-thick test piece). This procedure can be rationalised very simply.

By definition for the scaled test piece

$$K_{Icd} = Y\sigma\sqrt{a} \tag{2.83}$$

where $\sigma = F/BW$ (or some function other than BW with dimensions of area such as BW^2/S for bending), $F = \beta^2 F_1$; $B = \beta B_1$; $W = \beta W_1$; $a = \beta a_1$, where suffix 1 refers to the actual test piece on which the ductile diagram was obtained.

Thus

$$K_{Icd} = \sqrt{\beta}(Y\sigma_1\sqrt{a}) \tag{2.84a}$$

This is a K value for point F_1, where the stress is σ_1 (see Fig. 2.61) scaled by $\sqrt{\beta}$.

Thus

$$K_{Icd} = K_{F_1}\sqrt{\beta} \tag{2.84b}$$

where K_{F_1} is the K value at point F_1. Noting $\beta = U\sqrt{U_1}$ by definition,

$$G_{Icd} = G_{F_1}U/U_1 \tag{2.85a}$$

$$= U\left(\frac{F_1^2}{2\beta}\frac{\partial C}{\partial a}\right)\Big/(F_1^2 C/2) \tag{2.85b}$$

[16] Not to be confused with K_{Id}, a notation often used for dynamic initiation toughness.

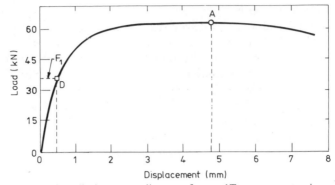

FIG. 2.61.   Load-point displacement diagram for a 1T compact tension specimen, illustrating how K_{Icl} is calculated. (After Ref. [146].)

from conventional LEFM, since F_1 is on the linear part of the diagram. In accordance with eqn. (2.30) this can be written

$$G_{\mathrm{Icd}} = \eta_{\mathrm{el}} U_{\mathrm{el}}/B(W - a) \qquad (2.85c)$$

where

$$\eta_{\mathrm{el}} = ((W - a)\,\partial C/\partial a)/C \qquad (2.86)$$

and is the same term as that introduced in eqn. (2.30) [82]. It has already been shown that, provided J_{c} is identified with G_{c} (recalling some uncertainties over the amount of slow crack growth in a plastic or elastic test for toughness) eqn. (2.85c) holds for either the required elastic situation (i.e. G_{c}) or the actual plastic situation (i.e. J_{c}) provided $\eta_{\mathrm{el}} = \eta_{\mathrm{pl}}$. As discussed in Section 2.2.2.1.3 this is widely recognised to be so for deepnotch three-point bend, $S/W = 4$, and nearly so for compact tension geometries. In so far as $\eta_{\mathrm{el}} \neq \eta_{\mathrm{pl}}$ for other configurations, then this rationalisation of equivalent energy methods in terms of J is limited, but the less well known evidence discussed in Section 2.2.2.1.3 and Appendix 4 that there is a rough equality between η_{el} and η_{pl} over a wider range of cases than hitherto recognised allows a corresponding extension of the regimes for which there is a similarity between the J and equivalent energy concepts.

It is of course possible that this rationalisation is incorrect or inadequate. Uncertainties over the physical meaning of J_{Ic} may be relevant. Clearly for crack initiation J_i is required in the J-based theory so that U would be U_i at initiation. In the foregoing, U is to maximum load and the degree of slow crack growth there entailed would in the writer's opinion be material- and temperature-dependent and indeed probably geometry-dependent. The writer is not aware what arguments are used over slow stable crack growth with equivalent energy methods. If, therefore, the equivalent energy method is more extensively validated by experiment than appears likely from the foregoing, there is an implication on the degree of slow crack growth incorporated in U and in the scaled test piece. This point has, of course, arisen in discussions of all the other testing procedures; the rationalisation of how much slow growth to tolerate, if any, and how to translate it for one test configuration to another remains to be resolved and thus seems to the writer to be an unknown variable in any application of equivalent energy based on other than crack initiation.

The foregoing refers only to the determination of toughness from a test invalid by conventional LEFM standards. For application to a structure, the following rationale is offered [145]: 'The equivalent energy method has its basis mainly with the laws of geometric similitudes.' '... a choice of which measurable quantitatives may be the most useful in predicting fracture for all modes of failure as applicable to water reactor pressure vessels' has to be made. In order to encompass both LEFM and yielding regimes, 'the energy to maximum load was chosen as the fundamental quantity governing the fracturing process'. No mention is made of slow crack growth. 'The major parameter necessary in the application of the equivalent energy method is the volumetric energy ratio. This is simply the ratio of corresponding normalised energy-absorbing capacities up to maximum load of two geometrically similar structures. This quantity is affected by both thickness and temperature and possibly geometry.' For determining energy ratio, 'the load deflection curve normal to (across) and at the location of the flaw, but with the flaw not present, was chosen as being relevant'. To apply the method the essential steps are: (i) a load–deflection curve for the unflawed structure must be found from either a model, computational or theoretical procedure, (ii) a model of the flawed structure must be tested to determine where on this unflawed load–deflection curve the maximum load to fracture is found, (iii) the volumetric energy ratio must also be found, or assumed from a given graph. For brittle fracture in the LEFM regime the volumetric energy ratio is simply the ratio of the dimensions, and the equivalent energy method reduces to LEFM. If the structure or model behaves with limited ductility at the position of the flaw but with the flaw not present, then the volumetric energy ratio is still taken as the ratio of the dimensions. If the model behaves with extensive ductility at the position of the flaw but with the flaw not present, then the ratio must be established by direct experiment or, in most examples quoted, Fig. 2.62 is used. This diagram is itself based on dynamic tear test data in the upper shelf toughness regime. Some evidence is given [145] that for several tension and compact tension

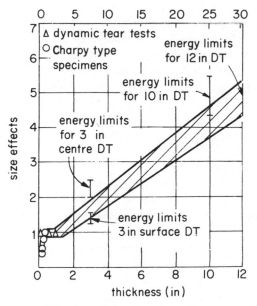

Fig. 2.62. Diagram used for determining the volumetric energy ratio in the equivalent energy method [145].

geometries 'data... tend to support the volumetric energy ratio as being independent of geometry', although the warning is added that, 'Should the volumetric energy ratios depend more strongly on geometry than indicated... then the type of laboratory specimen one may find applicable to pressure vessels may be limited'.

To apply the method to a structure the basic formula used is

$$K_{\mathrm{Icd}} = Y\sigma_{\mathrm{f}}\sqrt{a_{\mathrm{d}}S_{\mathrm{df}}} \qquad (2.87)$$

S_{df} is the volumetric energy ratio between a model of thickness d and the structure of thickness f. The volumetric energy ratios of the specimen from which K_{Icd} is obtained and for the structures of interest must be the same within 'acceptable limits'. Y is the usual LEFM shape factor; a_{d} is the defect depth in the model structure, thickness d. Still referring to eqn. (2.87), in Ref. [145] K_{Icd} is described as the 'fracture toughness property of the material in thickness d. This value is determined at the temperature of interest from any fracture toughness specimen from which valid K_{Ic} values may be obtained'. In the contemporary Ref. [146] K_{Icd} is defined as 'the fracture toughness obtained from a specimen of thickness d' and the 'method ... for obtaining the K_{Icd} value from standard fracture toughness tests' is that outlined in the foregoing, i.e. involving the equivalent energy concept. Clearly, if the method is correct, the value obtained from an LEFM valid test or from the method proposed must be identical but if the method is not correct then different predictions of structural behaviour will be predicted according to which toughness data are used. In eqn. (2.87), σ_f is the equivalent stress on the extrapolation of the linear part of the normalised load–deflection (i.e. stress–strain) curve such that the area defined equals that 'for the structure', taken up to 'maximum load'. Note that the stress–strain curve 'for the structure' is defined as 'the structure in the locality of the flaw but with the flaw not present'. The 'maximum load' (normalised) is, however, the maximum load (stress) of the *flawed* component. It appears that, apart from the different termination or maximum load point, the normalised diagrams are identical so that a small flaw is implied that does not significantly alter the detailed shape of the diagram. This would clearly not be so for 'structures' in the form of test pieces with deep notches unless the phrase 'flaw not present' is taken to imply only that a sharp crack is absent (in order to obtain as extensive a curve as possible), but a blunt notch to give the same nominal geometry must be included.

It is envisaged [145] that either model or structure, or both, may behave in a partly ductile manner in the region of the defect but with the flaw not present, but in a brittle manner when the flaw is present. No reference is made to the possibility that the model (small scale) may still behave in a ductile manner with a flaw present but the structure (large scale) may behave in a brittle manner. It will be recognised that this is the question of whether or not ductile–brittle transitions are size-dependent, to be discussed in Section 2.4.2.

In the writer's view, the method is in essence similar to the J elastic–plastic concept, and should be applicable provided (i) the micro-mode transition is either absent or is independent of size, or the method is applied so that all experiments and calculations are either wholly above or below (i.e. do not straddle) this transition, (ii) the method is restricted to cases where the elastic and plastic η factors (eqn. (2.30)) are reasonably similar, or the differences between them are allowed for by finding the volumetric energy ratios independently. As noted in Appendix 4 the configurations for which the η factor is not very sensitive to the extent of plasticity may not be as restricted as at one time it seemed. In short, consistencies between the methods can be seen, but the correctness of the one rather than the other can only be checked from the still awaited definitive experiments.

2.3.6 Non-linear Energy Rate $\tilde{G}$

As first propounded [103] this concept envisaged a small degree of yielding and/or slow crack growth notionally in plane stress. A term, $\tilde{G}$ called 'the total energy release rate, per unit thickness' was introduced and defined as

$$\tilde{G} = -\left.\frac{\partial U}{\partial a}\right|_q \tag{2.88}$$

where U is referred to initially in connection with 'the mechanical work (energy) per unit thickness corresponding to any displacement' and subsequently as 'the energy U'. The displacement (Fig. 2.63) is a non-linear function of the load F described by

$$q = \frac{F}{m} + k\left(\frac{F}{m}\right)^n \tag{2.89}$$

where m is the stiffness of the linear section of the record, and k and n are fitted constants. This relation is regarded as either a non-linear elastic description or a deformation plasticity description of a notch test with no unloading. The work done, $U = \int_c^q F\,dq$, is found and the differential with respect to crack length is taken at constant displacement. Regarded as the differential of a work-done term, with no implication on available energy

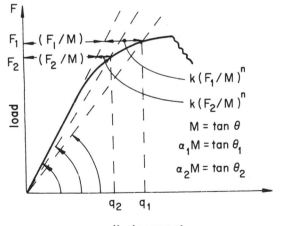

FIG. 2.63.   Non-linear load–deflection diagram described by eqn. (2.74) [103].

release, the formulation appears to be precisely analogous to the experimental method described in Chapter 3 and Ref. [61] to evaluate J. For non-linear elastic material it would also be identical to $\bar{J}$ (eqn. (2.27)). The term $\tilde{G}$ is written

$$\tilde{G} = \tilde{C}G \tag{2.90}$$

where G is the conventional linear elastic term and $\tilde{C}$ is a correction term for non-linearity given by

$$\tilde{C} = \left[1 + \left(\frac{2n}{n+1}\right)\left(\frac{1-\alpha_2}{\alpha_2}\right)\left(\frac{F_1}{F_2}\right)^{n-1}\right] \tag{2.91}$$

where α_2 is defined according to Fig. 2.63.

Application of the above approach to certain centre-cracked ($2a/W = 0.5$) Al-alloy data at a net section stress of less than $0.5\,\sigma_Y$ results in $\tilde{G}$ values exceeding the LEFM-based G by about 24%. This difference is surprisingly large compared with the remarks made in Section 2.2.2.1 that for most configurations J/G rises only some 10% for contained plasticity. The load–deformation records [103] though extending little beyond the linear (e.g. a deflection only perhaps 30% greater than the linearly extrapolated value) appear nearly horizontal. This implies to the writer that the non-linearity is probably due to stable tearing rather than to extensive plasticity, for which purpose the $\tilde{G}$ method then formulated seems inadequate.

In a further presentation [104] $\tilde{G}$ is defined as

$$\tilde{G} = \left\{\frac{d}{da}[U_{ext} - (U_{el} + U_{pl})]\right\} \tag{2.92}$$

where U_{ext} is work performed by external forces, and U_{el} is the elastic and U_{pl} the plastic components of the internal (strain) energy. Thermal activity is neglected here, as in most other discussions of fracture. It is then pointed out that for deformation plasticity and no slow crack growth or unloading, this term is formally identical to J. As described in Section 2.2.2.1.3 these terms are strictly definable in plasticity as J and eqn. (2.92) is identical to eqn. (2.27). In Section 2.2.2.1.3 it was further pointed out that $\bar{J}$ is not path-independent for incremental plasticity materials, so the above terms are $\bar{J}_m$, the monotonic loading case. We then face the dilemma that for non-linear elastic or notionally deformation plasticity with no unloading, $\bar{J}(\simeq J = G)$ is an energy available based on cracks cut to different initial length and not physically meaningful, and for incremental materials $J_m \neq J_i$ where $\bar{J}_i$ means the available energy release as the crack grows under fixed load or displacement. Thus, in the first part of Ref. [104] the meaning of G is the same as $\bar{J}_{\mathrm{m}}$. In the second part of Ref. [104] $\tilde{G}$ is at first evaluated for no slow crack growth. The formulation is different from Ref. [103], since it is now conducted at constant load. The work done (at constant load) is $\Delta U_{\mathrm{ext}} = F\Delta q$, so the change in internal (strain) energy is

$$\Delta U = \Delta U_{\mathrm{el}} + \Delta U_{\mathrm{pl}} \tag{2.93}$$

and

$$\Delta U_{\mathrm{el}} = \frac{F^2}{2} \Delta\left(\frac{1}{m}\right) \tag{2.94}$$

as for LEFM. Using eqn. (2.89) for q

$$\Delta U_{\mathrm{pl}} = \Delta\left(\frac{nk}{n+1}\right)\left(\frac{1}{m}\right)^n F^{n+1} \tag{2.95}$$

so that finally

$$\tilde{G} = \left\{1 + \frac{2nk}{n+1}\left(\frac{F}{m}\right)^{n-1}\right\} G \tag{2.96}$$

In the present writer's understanding, the correction factor in eqn. (2.96) arises from the formulation of ΔU_{pl} from the monotonic loading curve, so that it is $\bar{J}_{\mathrm{m}}$ (Section 2.2.2.1.3 (iii)) that is again being found.

In the last section of Ref. [104] the case of a growing crack is considered, i.e. not just the initiation of a crack under load, but the energy release rate available after significant stable growth. Since $\tilde{G}$ has been identified with $\bar{J}$ of Section 2.2.2.1.3 by the present writer, the attempt to evaluate $\tilde{G}$ for the growing crack is an attempt to formulate $\bar{J}_i$ which, in terms of eqns. (2.28), is $I - \partial U_{\mathrm{pl}}/\partial a$ where I was the total energy available by elastic unloading from the elastic–plastic situation. The difficulty of formulating a useful energy balance for the growing crack in plasticity was discussed in Section 2.2.2.1.3. To obtain an approximate estimate the assumption is made in Ref. [104] that n and k do not depend on crack length or growth and eqn. (2.89) is written twice, once in terms of the actual load F, reached by a load path including slow growth, and once in terms of $\bar{F}$ the hypothetical (higher) load that would have been found had there been no slow growth. On the assumption that n and k, which belong to the latter condition of no growth, are not changed, the two are related in [103] by the elastic stiffnesses.

$$\bar{F} = m_0 q_{\mathrm{el}} \qquad \text{and} \qquad F = m_{\mathrm{c}} q_{\mathrm{el}} \tag{2.97}$$

where m_0 is the initial stiffness (before growth); m_c is the stiffness at the critical situation after growth (which can be found by elastic methods if the final crack length is known); and q_{el} is the elastic displacement supposed equal in both cases.

Thus all terms in eqn. (2.96) which relate to $\bar{F}$ and the hypothetical curve are replaced by $F_c = \bar{F}m_c/m_0$, where F_c is the critical (final) load after crack growth. In the limit this would be the point of unstable growth. This substitution gives[17]

$$\tilde{G} = \left\{ 1 + \frac{2nk}{n+1} \left[\frac{F}{m_c} \right]^{n-1} \right\} \left(\frac{m_0}{m_c} \right)^2 \frac{F_c^2}{2B} \frac{d}{da} \left(\frac{1}{m_0} \right) \qquad (2.98)$$

The expression is admittedly approximate but is questioned here on two counts. Firstly, no attempt seems to have been made to formulate $\tilde{G}$ for the growing crack at the *onset* of growth. It seems tacitly assumed that $\tilde{G}$ as already evaluated in eqn. (2.96) (identified here as $\bar{J}_m$) is indeed the same as $\tilde{G}$ at initiation for the growing crack (identified in Section 2.2.2.1.3 as $\bar{J}_i$). This may indeed be so (see Section 2.4.1) but the point does not seem clearly established, although both energy available and remote dissipation rate (i.e. I and $\partial U_{pl}/\partial a$, eqn. (2.28)) may differ for the two cases. Also the modification itself appears fallacious in that the load F and the hypothetical load $\bar{F}$ are, for extensive yielding, related by plasticity effects as well as elasticity. In the limit of near collapse behaviour the relationship is of the form derived in connection with Fig. 2.28, where the plastic change is related to change in collapse load. Presumably for small-scale yielding this term is small in relation to the elastic compliance effect, but in that case it is accepted that $\tilde{G} = G$ and LEFM gives an acceptable estimate of behaviour although there may be an intermediate range where the plastic component of the change in load is neither negligible nor estimable from the collapse load.

Reference [149] discusses the effect of biaxial loads on $\tilde{G}$ and J, using a finite element method for linear plus power law hardening material following total theory plasticity. By way of example a particular centre cracked plate, $2a/W = 0.5$, $D/W = 2.5$, is discussed under axial stress σ and transverse stress $k\sigma (-3 \leq k \leq +3)$, in plane stress. G is defined as either $\partial U/\partial a|_q$, where U is strain energy, or $\partial V/\partial a|_Q$, where V is complementary energy. J is defined by the contour integral (eqn. (2.18)). It is found that for a given value of $\sigma/\sigma_Y \tilde{G}$ is a function of the biaxiality ratio k. Some results are

---

[17] Ref. [104] omits the squared power from the term (m_0/m_c) and writes d/dc. These are believed to be typographical errors.

shown in Fig. 2.64. It is stated that the 'numerical values of the J integral are similar to those of the non-linear energy release rate' (that is, $\tilde{G}$). Reference [149] continues, 'However, we do detect that the numerical difference between $\tilde{G}$ and J increases as applied stress increases, especially at larger k values (positive or negative)'. G is shown to be some 8 % lower than J for $k = -3$ and 4 % higher for $k = +3$ (at $\sigma/\sigma_Y = 0.4$), the variation being

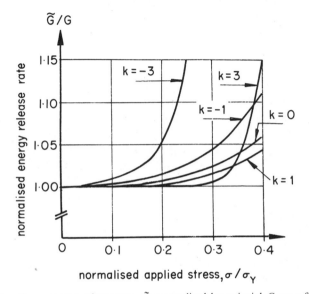

FIG. 2.64.   Non-linear energy release rate, $\tilde{G}$, normalised by uniaxial G, as a function of applied stress σ/σ_Y for various biaxiality ratios: $k = $ transverse stress/axial stress; centre-cracked tension; $D/W = 2.5$; $\sigma_Y = (0.002E/\alpha)^{1/n}$, $n = 13$, $\alpha = 0.02$; $\nu = 0.33$ [149].

systematic, not random. The origin of this difference is not discussed, nor is the accuracy of the computational data—for example, whether constancy of J with variation in path length is a function of k. The present writer does not understand why J as a measure of $\partial P/\partial a$ is not precisely the same as either $\partial U/\partial a|_q$ or $\partial V/\partial a|_Q$ as inferred from the previous discussion of Refs. [103] and [104].

Returning to the effect of biaxiality, it may be noted that Fig. 2.64 is consistent with Fig. 2.34 for $k = 0$ or ± 1. Note $\tilde{G}/G$ (and hence J/G) for $k = 0$ rises to only about 1.05, implying that plasticity is still well contained. For the biaxial ratios $k = \pm 1$, gross yield of the plate would not have occurred at $\sigma/\sigma_Y = 0.4$. The more extreme results for $k = +3$ have no counterpart in Fig. 2.34. For $k = +3$ the trend of $\tilde{G}/G$ departs from that for

$k = +1$ and rises rapidly as σ/σ_Y increases above about 0·3, at which point, for $k = \pm 3$ uncontained yield would have occurred (including the remote ends) under the influence of the transverse component.

If, as discussed in Section 2.2.2.1.4, the rapid upswing of the curves for J/G is attributable to yield of the ligament, and the same reasoning applies to Fig. 2.64, then, since the ligament stress is 2σ axially, with $k\sigma$ transversely, the values of σ/σ_Y at which the upswing should occur (for von Mises' criterion in plane stress) are

k	$+3$	$+3$	$-0\cdot1$	0	$+1$
σ/σ_Y	0·23	0·38	0·38	0·50	0·58

and this seems reasonably in accord with the data.

Clearly in the present writer's view the physical meaning of the term $\tilde{G}$ is not a rate of energy available if the computations are taken to refer to plasticity rather than non-linear elasticity and a characterising role is then ascribed to either $\tilde{G}$ or J. The question of whether a particular value of J, however obtained, describes a unique crack tip situation as argued in connection with Fig. 2.35 is not addressed.

2.3.7 Two-criteria Method

A two-criteria method has been proposed by Dowling & Townley [150]. This simply suggests that failure occurs when the applied load reaches the lower of either a load to cause brittle failure in accordance with LEFM or a collapse load. At the present stage of development the collapse load is estimated from whatever exact, bound or model data may be available, but related to the ultimate strength rather than the yield strength. It is also recognised that behaviour does not change abruptly from one regime to the other and a single transition between the two states is based on the Heald *et al.* [122] yielding fracture mechanics model. Finally, as a tentative design proposal, a defect acceptance curve is suggested, into which a factor of 2 is introduced for both LEFM and collapse estimates. In principle there is little more to be said. The validity of the method is shown by extensive comparison with experiment [150]. In many cases the original data on toughness are inadequate and best estimates perforce have to be accepted. One such set of test data, well authenticated, is shown in Fig. 2.65 for compact tension tests on pieces up to 250 mm thick. Only the LEFM and collapse estimates are shown, the transition region being omitted for clarity. In Fig. 2.65 the ordinate is p^ϕ which is a non-dimensional failure parameter defined as either

$$p_f^\phi = \sigma_{0f}/\sigma'_{0u} \qquad (2.99a)$$

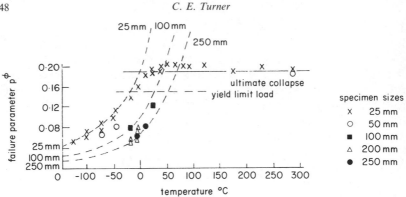

FIG. 2.65. Compact tension data plotted according to the two-criteria method. The failure parameter p^ϕ is the ratio of failure stress to collapse stress of the unflawed component (Ref. [150], based on data from Wessel *et al.* (1969). *Fracture* 1969, Chapman & Hall, London, and Ref. [146].

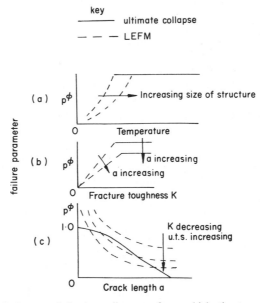

FIG. 2.66. Basic fracture behaviour diagrams from which the two-criteria method is synthesised, showing the variation of the failure parameter p^ϕ with temperature, toughness and crack length [150].

or

$$p_u^\phi = \sigma_{0u}/\sigma_{0u}' \qquad (2.99b)$$

where σ_{0f} is the failure stress (i.e. nominal remote stress) according to LEFM; σ_{0u} is the collapse stress for the flawed component based on ultimate strength; and σ_{0u}' is the collapse stress of the unflawed component based on ultimate strength. All these stresses are functions of geometry. σ_{0f} is of course a function of K_{Ic} and is therefore a function of temperature, whereas p_u^ϕ is independent of temperature. Change in absolute scale also affects p_f^ϕ but not p_u^ϕ. Figure 2.66 shows three plots—against temperature, toughness and crack size—useful for discussing fracture problems. One particular set of test data is shown in Fig. 2.67 for cracked aluminium

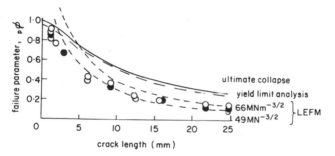

Fig. 2.67. Fracture data for aluminium alloy cylinders expressed in terms of the failure parameter p^ϕ (Ref. [150], based on data from Anderson, R. B. & Sullivan, T. L. (1966). NASA Tech. Note TND-3252, and Getz, D. F. *et al.* (1963). Paper 63, *Trans. ASME*, WA-187).

cylinders. The results are in good agreement, except near the transition between the two modes of failure. For this transition region Dowling & Townley proposed a generalisation of the HSW formula by writing

$$\frac{L_f}{L_u} = \frac{2}{\pi}\cos^{-1}\left[\exp -\left(\frac{\pi^2}{8}\frac{L_K^2}{L_u^2}\right)\right] \qquad (2.100)$$

where L_f = failure parameter (load, stress, pressure, etc.); L_u = limit parameter based on ultimate strength; and L_K = failure parameter based on LEFM.

All the test data discussed in Ref. [150] are shown non-dimensionalised in this way in Fig. 2.68 together with the tentative defect acceptance curve (dashed line) incorporating the factor of 2 for safety but omitting the transition region. No doubt some of the scatter in Fig. 2.68 could be reduced by use of more exact COD or J treatments in which the plasticity parameter is itself a function of geometry, but much of the scatter is due to inadequate toughness data that could be reflected in any method of analysis and there is no doubt that eqn. (2.100) gives a good near fit to the experimental results for a variety of actual engineering structures. The user is left with some doubt on whether the factors of constraint and slow crack growth are not important for fortuitous or fundamental reasons, and on how the variable toughness of weldments would be included in the method.

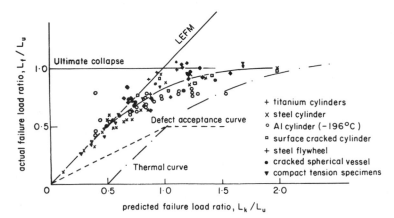

FIG. 2.68. Experimental data for failure load, L_f, as a function of load predicted for failure by LEFM, L_K, normalised by predicted collapse load, L_u, for the two-criteria method (Ref. [150], with thermal curve added from eqn. (2.102) [151]).

The method has been extended by Townley to cover thermal loadings [151]. The basic argument is that, in the elastic regime, thermal stress is as effective as mechanical stress in inducing a K value that, at K_{Ic}, will cause cracking. On the other hand, self-equilibrating thermal stresses do not contribute to plastic collapse. Specific reference is made to the Dowling–Townley curve (eqn. (2.100)) relating to 'failure initiation' and it is postulated that, under thermal stress alone, failure will not occur outside the LEFM regime until the elastically calculated K value reaches K_{Ic}. Equation (2.100) is then modified by the inclusion of a term l_{th} defined as 'the hypothetical external load of the same kind as L, which gives the same linear stress intensity factor at the crack tip as the actual thermal load applied to the structure'.

$$\frac{l_{th}}{L_K} = \frac{K_{I\text{ thermal}}}{K_{Ic}} \qquad (2.101)$$

to give an effective failure load (compare eqn. (2.100)).

$$\frac{L'_f}{L_u} = \frac{2}{\pi}\cos^{-1}\left[\exp-\left(\frac{\pi^2}{8}\frac{L_K^2 - l_{th}^2}{L_u^2}\right)\right] \qquad (2.102)$$

The general relationship of L'_f to L_f is shown in Fig. 2.68 by the chain-dotted line marked 'thermal curve'. The position where this line cuts the abscissa, L_K/L_u, depends of course on the value of l_{th}, and the type of loading chosen

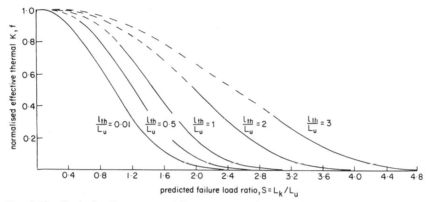

FIG. 2.69.   Ratio $f = K_{\text{thermal effective}}/K_{\text{thermal actual}}$ as a function of ratio S of predicted failure load to collapse load for various thermal loading ratios [151].

to represent l_{th} should, as far as possible, give a stress pattern similar to the thermal one. An example suggested is that at a nozzle corner intersection, a pressure load simulates thermal shock better than an applied force or moment. This seems to pre-suppose that the exercise is being conducted in terms of pressure loading. Indeed, if separate mechanical load systems were present, presumably they would have to be converted to some equivalent load, defined by adding K values for the LEFM regime and resulting in a modified collapse load L_u from a collapse interaction curve for the combined load system. Such difficulties are of course implicit in other yielding fracture analyses.

From the basic proposition that thermal stress affects fracture in the LEFM regime but not at collapse, the difference $(L_f - L'_f)$ is treated as the fraction of the K_{thermal} that is effective in causing failure

$$f = (L_f - L'_f)/l_{th} = K_{\text{thermal effective}}/K_{\text{thermal actual}} \qquad (2.103)$$

One presentation is then to plot f against L_K/L_u for various values of thermal loading l_{th}/L_u (Fig. 2.69). An alternative presentation, for thermal stress alone, is offered in Fig. 2.70. If L_K for thermal loading (i.e. $L_K = l_{th}$) and the mechanical collapse load L_u are estimated, then the diagram is entered at the appropriate point along the abscissa whence the ordinate R can be read off. R is defined as the 'ratio of the elastically calculated stress intensity which will cause failure to the critical stress intensity of the material' (which would of course be unity for a LEFM calculation) so that

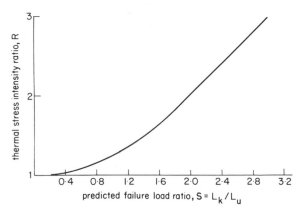

FIG. 2.70.   Ratio R of elastic stress intensity to cause failure by thermal stress alone to K_{Ic} as a function of ratio S of predicted failure load to collapse load [151].

$Rf = 1$ at failure by thermal stress alone. This statement says that $(K_{thermal\ effective}/K_{thermal\ actual})(K_{thermal\ failure}/K_{Ic}) = 1$ at failure, since, for thermal stress alone, $K_{thermal\ effective} = K_{Ic}$ when $K_{thermal\ actual}$ reaches $K_{thermal\ failure}$.

The concept seems a bold approach to extrapolate failure analysis into the thermal regime. Some remarks are made on parts of Fig. 2.69 where loads opposite in sign to L_K could exist, and it is noted that these might in some cases cause failure alone and in other cases not, though presumably they could always be outweighed by sufficiently strong thermal effects. The effect of eqn. (2.100) in rounding off the original linear LEFM and collapse regime plots (Fig. 2.65) are also discussed. The relevance to crack initiation or complete failure is not clear to the writer. Even if a crack is initiated in the LEFM regime at a value of K_{Ic}, it does not follow that propagation will occur unless K increases with crack length (as of course it often does for mechanical problems). Conversely, even though thermal stress will not cause collapse, it could presumably cause a crack to initiate at some high local value of stress or strain at a pre-existing defect at a value governed by total COD rather than just $K_{thermal\ effective}$. In short, the extent of 'failure' being guarded against itself seems to vary with the ductility (near LEFM or near collapse) of the structure being studied. This may indeed be quite rational but does not seem explicitly defined, and as the author points out [151] there is little if any experimental evidence to guide the formation of the model.

It will be recalled that, in Ref. [121], the independence of overall collapse load from thermal or residual stress was accepted, but an effect of these stresses on the risk of fracture in a region of local yield was specifically argued, contrary to the view expressed in the foregoing.

A further formalisation of the two-criteria method has been given in Refs. [152] and [153]. A failure diagram is here presented with abscissa S_r where

$$S_r = \sigma/\sigma_1 = L/L_u \qquad (2.104)$$

σ is the stress at the point of interest, in the absence of a defect; in general a membrane and bending component may have to be added. L is the corresponding load or pressure applied. σ_1 is the limit stress for the flawed component, based on a flow stress here taken as $(\sigma_Y + \sigma_u)/2$ (contrary to Ref. [2.150], where σ_u is used). L_u is the load or pressure to cause plastic collapse.

K_I is calculated and K_{Ic} found by the conventional methods of LEFM (although a lower bound procedure for finding K_{Ic} from invalid tests, based on a modification of the equivalent energy method (Section 2.3.5) is recommended if valid data are not available). Weldments are not specifically mentioned in Ref. [150] the toughness merely taking into account 'material variability . . . and any other variables which might affect the data'. The ordinate to the diagram is K_r where

$$K_r = K_I/K_{Ic} = L/L_K \qquad (2.105)$$

where L_K is the load or pressure to failure according to LEFM.

The assessment line for safe behaviour is shown in Fig. 2.71. This is in essence the same line as in Fig. 2.68, and all the data from Fig. 2.68, are shown in Fig. 2.71. In applying the method a factor of safety is applied to both K_r and S_r, the values of the factors being chosen to suit the problem, and for safety the resulting point should fall below to the left of the assessment line. It is recommended that the procedure for assessing a structure be conducted before and after allowance for fatigue crack growth.

The case of a buried or part-through crack is taken further if, on the above analysis, the result appears unsafe. Whereas, for a through-crack, such an unsafe result would indicate the likelihood of failure (with due regard to the safety factors, of course), for the part-through defect it is

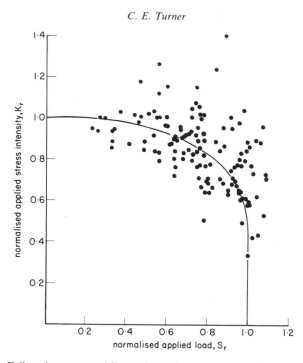

FIG. 2.71.   Failure data expressed for an alternative presentation of the two-criteria method in terms of normalised applied load $S_c = L/L_u$ and normalised applied stress intensity, $K_r = K_I/K_{Ic}$; eqns. (2.104) and (2.105) [152].

argued that the lack of satisfying Fig. 2.71 implies the crack will develop, a buried crack perhaps emerging on one side to become a semi-elliptical part-through crack and a semi-elliptical part-through crack growing to a through-wall crack.

Such cracks are re-characterised for the purpose of showing whether the act of development is likely itself to be catastrophic or to cease with a leak situation. Referring to Fig. 2.72, a buried crack, length l, is taken to grow to a semi-elliptical crack with dimensions a', l', where $l' = 4a'$ or $2l$, whichever is greater. An original semi-elliptical crack in tension of face length l is taken to grow to a through-crack length $2a = 2l$. Yet further factors are introduced if the defect is in bending. Flaws not perpendicular to the maximum principal stress are treated by projection (as in ASME XI IWB 3300). The redefined flaw is again assessed. It is suggested that the factors incorporated in the redefinition will allow adequate safety for the dynamic event of 'snap'-through. If the redefined flaws appear 'safe' then the original flaw, if it fails, will stay as a surface or leak case. If the redefined flaw appears unsafe then the failure of the original flaw may cause complete failure.

The test data in Fig. 2.71 were calculated on most realistic assessments of properties, stresses, etc. It is emphasised that in applying the method safe bounds must be used at all stages, e.g. lower bound for collapse, lowest values for K_{Ic}, highest values for stresses and K, etc. The authors freely admit that evidence for the general trend of the assessment curve is stronger

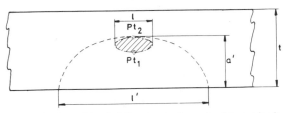

FIG. 2.72. Recharacterisation of buried flaw in the formalised two-criteria method [152, 153]. Pt_1 is critical point on original buried crack which breaks through to lower surface. Pt_2 becomes critical point on new enveloping crack of size a' by l'.

than for its absolute value as a safe lower bound. It is obvious that the procedure is largely pressure vessel oriented but some non-pressure vessel data are included in the test results of Fig. 2.68. As remarked in connection with Fig. 2.68, it is not clear to the writer how much of the scatter could be removed by the use of improved techniques of analysis. However, uncertainty over toughness and collapse load is undoubtedly an important factor. Estimates of collapse loads (eqn. (2.104)) may be difficult for some problems and the use for all geometrical configurations of the 'ln sec' plasticity factor (eqn. (2.7)), is an approximation discussed in Section 2.3.1. For example, an estimate for a crack emanating from a stress concentration (a value of J for such cases was discussed in Section 2.2.2.1.4) might differ appreciably from the ln sec value. Residual and thermal stresses are included [152] in the estimate of K_1 (eqn. (2.105)). No specific allowance appears to be made for the notional post-yield value that may be found if both, say, residual and applied stresses are a significant proportion of yield. For a given K_{Ic}, the LEFM analysis may then appear unsafe, since plasticity correction factors are specifically forbidden. However, the 'K_{Ic}' value in question will have been determined at the relevant temperature where the test may well not have been valid, and the equivalent energy method (Section 2.3.5) is recommended for determining a lower bound K_{Ic}. In so far as this value represents a post-yield situation, then a rigorous LEFM approach is not in question, and the method appears more nearly analogous to the COD design curve method where a stress $(\sigma + \sigma_r)$ would have been used to allow for residual stress (probably with $\sigma_r = \sigma_Y$) with a toughness expressed as δ_c. If δ_c translated into the same 'K_{Ic}' value as found by the equivalent energy procedure then the two methods would seem identical in the fracture regime. (The COD method does not of course discuss the limit-load regime other than by extension of COD procedure.) If there is a significant underestimate in the equivalent energy lower bound 'K_{Ic}', then the combined applied and residual stresses might appear unsafe, whereas by the COD design method they might not. The writer is not aware of any direct comparisons on this point. In his view the equivalent energy procedure is somewhat akin to the J concepts, at least in principle. If this is so then the two-criteria method is in effect an approximate combination of J, and collapse techniques (with many qualifications on points of detail) specially directed at load-controlled situations. In so far as there is great similarity in outline between J and COD, no major differences should exist between COD and the two-criteria method, though it seems likely that the respective treatments of slow stable growth (by choice of δ_i or δ_c or δ_m for critical COD or the re-assessment of part-through cracks described above but toughness based on K_{Ic}) would lead to different predictions.

It should also be remarked that Ref. [152] outlines procedures for crack growth by fatigue using LEFM methods, but discussion of fatigue is beyond the scope of the present chapter. In summary, the technical content of characterising the severity of a defect is in accord with the COD design procedure to within such differences as the appropriate value for m in $G = m\sigma_Y\delta$, and differs from the J concept mainly in using the ln sec plasticity correction factor for all cases, perhaps in conjunction with the LEFM shape factors. Such differences as exist lie rather in the *philosophy of a design limit*—in terms of load (pressure), displacement (strain) or some combination of both, and the *empirical judgements* by which degrees of conservatism are introduced (lower bounds, factors of safety) or degrees of 'realism' allowed (tolerance of some slow crack growth, recognition of degradation in weldments). The resulting methodologies may well not be 'good' or 'bad' but 'more (or less) appropriate' to a particular area of structural integrity to which the philosophy and judgement have been tuned.

2.3.8 Stress Concentration Theory (SCT)

This originated as a simple equilibrium model applicable to wide plates or large diameter vessels with small cracks in the post-yield regime [154]. It is assumed that the load shed by the cracked region is carried as a stress of mean value σ_c over a region of width s adjacent to the crack tip. The stress on the net remaining ligament is taken to be the same as the remote stress. Thus

$$\sigma BW = 2\sigma_c B_1 s + \sigma B(W - 2a - 2s) \tag{2.106}$$

where B_1 is the reduced thickness of the region s due to slight necking. Therefore

$$\sigma Bs = \sigma_c B_1 s - \sigma Ba \tag{2.107}$$

whence

$$\sigma \frac{B}{B_1}\left(\frac{a}{s} + 1\right) = \sigma_c \tag{2.108}$$

For most practical purposes B/B_1 is taken as unity. It is suggested that failure occurs when the stress σ_c on the region s reaches a critical value, and that for a given temperature, strain rate, etc., s is a material constant, so that a plot of stress at fracture against σa gives a straight-line graph of slope $-1/s$ and intercept σ_c. Such a graph is shown in Fig. 2.73. It transpires that $\sigma_c \simeq \sigma_u$, the tensile strength. It is pointed out in Ref. [154] that results can

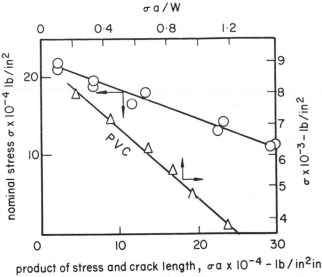

FIG. 2.73.   Linear plot of remote stress at failure, σ, against the product σa in the stress concentration method of analysis (Ref. [154], using data for SAE 4340 from Manning, G. K. (1962) ASTM–STP 302 and PVC data from Isherwood, D. P. & Williams, J. G. (1970). *Engng Fr Mech.*, **2**, p. 19).

easily be distorted by plates that are too small. If the length is inadequate
the stress over the cracked region that is transferred to region s, will be less
than the nominal gross stress σ acting over the crack length, the more so the
longer the crack relative to plate width. This length effect tends to make the
fracture data lie above the linear plot and the lack of width effect reduces
data below the linear. A width effect is not unexpected since net stress has
been approximated to gross stress, but it is perhaps surprising that for a
particular set of test data [154] a plate 2250 mm long with a 75 mm crack
shows pronounced length effects. This seems to preclude the use of the
analysis for other than very long plates of small a/W. For cylindrical vessels,
a correction is offered in which bulging effects at the crack tip induced by the
interruption of hoop stress by the crack are estimated. Equation (2.108) is
modified to become

$$\sigma\left(\frac{a}{s} + 1\right) + k\sigma a^2 = \sigma_c \qquad (2.109)$$

where the bulging correction is expressed in terms of an unknown factor, k,
and the gross stress σ.

This is rearranged to give

$$\frac{(\sigma_c/\sigma) - 1}{a} = ka + \frac{1}{s} \qquad (2.110)$$

and plots [154] show a straight line of slope k and intercept $1/s$.

It has already been pointed out that $\sigma_c \simeq \sigma_u$ and s is now estimated for
material data such as Charpy V-notch energy, V. It is found that

$$s \propto (V/\sigma_Y)^{1/3} \qquad (2.111\text{a})$$

where the proportionality constant depends on ductility. For a variety of
steels it is suggested

$$s = \text{const}(1 - 0.7R)^{-1}(V/\sigma_Y)^{1/3} \qquad (2.111\text{b})$$

where R is reduction of area in a tensile test. A typical value of $1/s$ for steels
is in the range 5–25 mm^{-1}.

The whole edifice at this stage seems rather empirical and directed mainly
at wide plate or vessel behaviour with small notches in the ductile or semi-
ductile regime. It is clear that, with some local necking, the maximum stress
that can be reached is the tensile strength and this value must extend over
the very considerable size of the region s until either a local strain is
reached for slow tearing or perhaps an overall instability is reached. This
rationalisation implies that the data so far analysed to support the concept

relate to quite ductile failures and that the geometrical effects of inadequate plate width and length, though not of local necking would probably be accounted for in a yielding fracture mechanics analysis of those situations. The writer has not attempted any such re-analysis of the test results used in Ref. [154] and it is likely that some of the data required for a COD or *J* analysis would not have been collected. The inclusion of small test piece data seems impossible if the 'process zone' is of the order of 25–125 mm, failure by plastic collapse rather than brittle fracture being anticipated at least for slow loading rates.

A relationship to other fracture mechanics approaches is pointed out and elaborated in Ref. [155] in what is essentially a modified form of the theory. By considering the form of the elastic and elastic–plastic crack tip stress fields in a Neuber-type manner, an equivalent stress $\bar{\sigma}$ over a distance ξ ahead of the crack is deduced

$$\bar{\sigma} = \sigma \left(\frac{2a}{\xi} + 1 \right)^{1/2} \tag{2.112a}$$

At fracture with $a = 0$, $\bar{\sigma} = \sigma = \sigma_u$ so that eqn. (2.112a) is re-written

$$\sigma_u = \sigma \left(\frac{2a}{\xi_c} + 1 \right)^{1/2} \tag{2.112b}$$

where ξ_c is the critical value of ξ at fracture.

For the LEFM case where the crack tip stress $\sigma_{yy} = K/\sqrt{2\pi r}$

$$\bar{\sigma} = \frac{1}{\xi} \int_0^\xi \sigma_{yy} \, dr = \sigma (2a/\xi)^{1/2} \tag{2.113}$$

If at fracture $\bar{\sigma} = \sigma_u$, then, using $K_c^2 = \sigma^2 \pi a$ and combining with eqn. (2.113),

$$K_c^2 = \sigma_u^2 \pi \xi_c / 2 \tag{2.114}$$

Equation (2.112a) reduces to eqn. (2.113) for $2a \gg \xi_c$. The three relationships (2.108), (2.112a) and (2.113) are plotted in Fig. 2.74. Clearly, the equivalent stress on a particle concept fits well between LEFM at low values of σ/σ_u (i.e. $\sigma_u/\sigma > 5$, say) and the simple stress concentration approach described above for $\sigma/\sigma_u \simeq 1 \cdot 0$. Taking σ arbitrarily as $\sigma_u/2$ and using eqn. (2.108) and (2.112b) with $B_1 = B$, $\sigma_c = \sigma_u$ then $a = s$, whence, from eqn. (2.112b), $\xi_c = 2s/3$ and approximately

$$K_c^2 = \sigma_u^2 s \tag{2.115}$$

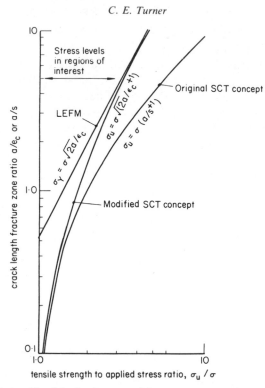

FIG. 2.74. Relationship of the development of the stress concentration method to a Neuber-type theory and to LEFM, eqns. (2.108), (2.112a) and (2.113) [155].

Overlooking the somewhat random excursions into plasticity or elasticity to simplify the analysis, an approximately physical basis for s is thus established. The modified stress concentration theory is thus seen [155] as a plasticity correction factor of the HSW-type (Section 2.3.2) related to it by

$$\frac{\pi s}{8a} = \ln\left[\sec\left(\frac{\pi}{2}\frac{\sigma}{\sigma_u}\right)\right] \qquad (2.116)$$

The variation of s with temperature is shown in Fig. 2.75 for A533B. Correlation of s with K_c follows if eqn. (2.115) is accepted. Figure 2.76 shows that compact tension data, analysed to obtain effective K_c (SCT), relate well to K_c (LEFM). In essence, s has been treated as the tensile ligament between the crack tip and the neutral axis and then eqn. (2.110) used to give K_c (SCT).

It seems to the writer that on this more recent basis, the theory is as well or

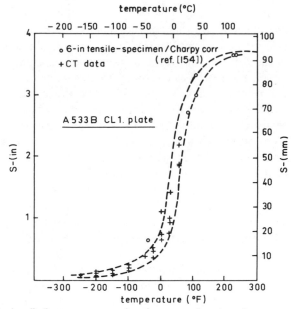

FIG. 2.75. Region S of stress concentration theory as a function of temperature for A533B [155].

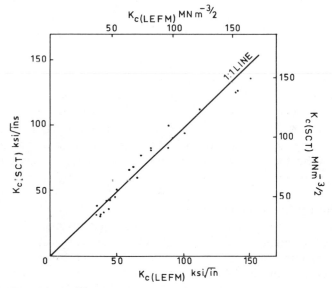

FIG. 2.76. Comparison of apparent toughness, K_c (SCT) derived from stress concentration theory and conventional K_c (LEFM) data from compact tension tests on A533B [155].

ill justified as practically any other fracture mechanics theory. The essential links are the treatment of K_c^2 or G or J as one entity or the breaking of such a term down to the form $\sigma_u^2 s$ in which σ_u is implied as an already known material property, the tensile strength. In the finite element calculations (Section 2.2.2.1.3) for full plane strain and little work-hardening the stress ahead of the crack tended to a value $m\sigma_Y$, where m was a constraint factor that depended on the geometry (i.e. the in-plane constraint) and work-hardening. In real situations with some lack of plane strain and some local necking deformation, σ_u may be as good a value of this stress as $m\sigma_Y$. In so far as there is little reduction in thickness in K_{Ic} testing, certainly far less than in the tensile test in which σ_u is established, it may well be that use of σ_u in eqn. (2.110) is artificial and the value of the process zone s correspondingly so, but provided the analysis is self-consistent the details are immaterial. For failure in a more ductile mode, as in Ref. [154], the use of σ_u may be advantageous. J theory would also have the plasticity correction factor a function of geometry, whereas both HSW [122] in its own right and as used for the two-criteria method [150] and also the SCT theories [155] are content to use a constant form (related by eqn. (2.116). Both these latter have the merit of tying the gross plasticity case to σ_u, whereas the J theory is a forward extrapolation from a rigorous elastic solution with the formalities of plastic analysis no doubt sometimes allowing the solution to drift away from reality by neglect of degree of constraint, large deformation, perhaps an unrealistic work-hardening law, and so on.

The writer is not aware of any treatment of residual stress, thermal stress, defects at stress concentrations, etc., by the stress concentration theory, nor of the question of crack initiation and slow stable growth. The analysis of K_{Ic} from compact tension pieces surely refers to initiation or negligible growth, whereas much of the ductile vessel data of Ref. [154] are likely to include a significant amount of slow stable growth before the final instability, but the effect may well be more marked in terms of local ductility than on the actual failure load.

2.3.9 Soete's Gross Strain Method

Soete has proposed a method primarily based on experiment, for assessing the required toughness of parent plates [156] or weld metal [157] in the presence of a small defect. The key suggestion is that the material should be tough enough to attain yield in the adjacent defect-free region. His criterion is, 'Each defect which initiates no fracture before plastic deformation occurs in the defect-free zone is acceptable', and he defines four intensities of loading conceptually very similar to those set out in the introduction (Section 2.1) and illustrated by schematic diagrams of the extent of deformation on which, indeed, Fig. 2.1 is based. Denoting the local crack tip stress by $_1\sigma$ and the stress at the boundary in the plane of the defect by σ_b (which must be just less than the mean net section stress σ_n), then four cases are discussed for a centre-cracked plate with applied stress, σ:

> *Intensity of loading* (*Soete's definitions*)
> Case (a) $\sigma_Y > _1\sigma > \sigma_b > \sigma$; this is LEFM
> Case (b) $_1\sigma > \sigma_Y > \sigma_b > \sigma$; this is an elastic–plastic situation
> Case (c) $_1\sigma > \sigma_b > \sigma_Y > \sigma$; this is gross (or full) yield
> Case (d) $_1\sigma > \sigma_b > \sigma > \sigma_Y$; this is general yield.

If Soete's own definitions are strictly adhered to then there seems no direct congruence with yielding mechanics as developed in this chapter and it seems likely that values of J or COD at the boundaries of his regimes would be geometry-dependent. If these distinctions are taken to be semantic then the definitions of Section 2.1 clearly allow Soete's concepts to be treated in terms of J or COD. Soete's own definitions are of course unambiguous, whereas the more general definitions allow some discussion of just how extensive yield can be before case (a) merges into (b) and (b) into (c).

Soete's proposals have emerged from an extensive series of tests and it is extremely interesting that he should have distinguished by that method the same regimes that have emerged from the computational studies of J described in this chapter. In particular he distinguishes between 'full' (or gross) yield where plasticity reaches the lateral boundary (Fig. 2.1c), which is the regime of much deep-notch low work-hardening test data, and the 'general yield' condition where, for shallow notches and with work-hardening, yield spreads along the length of the component. He terms 'subcritical' those cracks which, for a given material, allow such general plasticity, whereas cracks that cause failure before general yield are termed 'hypercritical'. In so far as there is no geometric factor introduced corresponding to the LEFM shape factor, then the method would have to be treated with caution for configurations other than those tested. The test results are extensive, covering some 1000 plates ranging from 100 mm to 1000 mm in width and 8 mm to 100 mm in thickness. Above a certain thickness (20 mm for the C–Mn steel tested) the critical crack appears to be a material constant for a given temperature, but the test data appear to be mainly for centre-cracked tension and part-through surface cracks. A typical critical crack length for a 14-mm-thick C–Mn steel

($\sigma_Y \simeq 250 \, \text{MN/m}^2$) tested at $-30\,°\text{C}$ is 14 mm. This critical crack length increases with temperature, much as does the conventional brittle–ductile appearance or the ductility transition curve (Fig. 2.57), i.e. there is a transition range of perhaps some $20\,°\text{C}$ extent, with, thereafter, little change in critical crack length either in brittle (below transition) or in ductile (above transition) regimes for at least another $20\,°\text{C}$ or so. More extensive temperature data are not shown. A specific definition of general yield is not given but reference is made to a strain of 2 % being easily obtained on gauge length 'sufficiently large . . . to reduce the contribution of local deformations'. For transverse welds (i.e. loaded perpendicular to the welding direction) overmatching strength is recommended, whilst for longitudinal welds (i.e. loaded parallel to the welding direction) it is 'essential to have weld metal with good ductility'. These recommendations are clearly consistent with the requirement that the parent plate enters into general yield. A severe criterion 'that fracture occurs in the base metal' is adopted for welds that are 'without defects'. This condition can also be met with a small weld defect if the tensile strength of the weld metal is greater than that of the plate and there is adequate notch ductility. For butt welds with defects, two further, less severe criteria are discussed, that either the base metal reaches general yield (thereby defining a critical crack in the weld that leaves the component as ductile as the weld-free case) or the weld metal reaches full (gross) yield, thus defining a larger crack that permits some limited ductility before fracture. The level of weld ductility required to meet these various recommendations is not defined other than by large-scale testing, and no toughness data are given for plate or weld metal.

For two-dimensional plane strain Fig. 2.48 specified the toughness required in non-dimensional terms for various configurations, with relatively small dependence on a/W ratio, as noted by Soete. There no allowance is made for weldments, since neither residual stress nor overmatching yield stress effect are included, but the general pattern of Soete's arguments and experiments appears entirely compatible with the J design curves entered at some specified strain level. The COD design method gives specific suggestions for what these levels might be and rule of thumb for incorporating residual stresses, whilst using experimental rather than computed estimates for the curve itself. Thus Soete's results and proposals seem quite compatible with COD and J procedures, provided a specific level is identified for the value of $\varepsilon/\varepsilon_Y$ to be reached in general yield, since the value of toughness required to provide a certain critical crack size increases linearly as strain to fracture continues to rise beyond $\varepsilon/\varepsilon_Y = 1$. Such a procedure for estimating toughness as well as requiring 'overmatching strength' or 'good ductility' seems essential for selecting base or weld metal without conducting near full-size tests, and lack of it appears to be an important weakness of Soete's method as at present developed.

2.4   GENERAL PROBLEMS

2.4.1   Slow Stable Crack Growth and Instability

In the preceding text, notably in the discussion on the fracture criteria $\delta = \delta_c$ (Section 2.2.1.1) and $J = J_c$ (Section 2.2.2.1.2), mention was made of the phenomenon of slow stable crack growth. Slow growth is said to occur if an (infinitesimally) small increase in load causes a correspondingly small increase in crack length. It has long been studied in plane-stress situations. Krafft *et al.* [13] expressed the growth of resistance, R, with crack extension as a rising curve (Fig. 2.77) to which the curve of energy rate available (or crack driving force), G, must become tangential for unstable growth. It seems generally accepted that the picture is valid. It implies that the final instability, as measured by an apparent toughness, G_c, is not independent of size, even for a given thickness [158], since the origin from which G is drawn (Fig. 2.77) depends on the initial crack length, and the precise shape of the curve depends on the LEFM shape factor for the

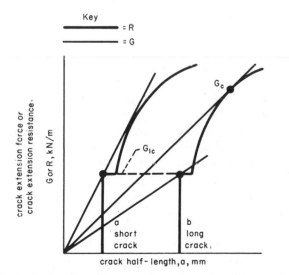

FIG. 2.77.   Driving force, G, and crack growth resistance curves, R, as used in LEFM. (From Heyer in Ref. [159].)

configuration in question. An earlier picture of an R curve falling with crack growth, also outlined in Ref. [13], is not now commonly used but will be discussed later. The resistance curve concept has also been extended in a descriptive manner to cases of more widespread plasticity by Rice [54] but expressed in terms of growth of plastic zone. Proposals to determine R curves expressed in terms of K are described in Ref. [159]. In the following discussion of the application of the R curve concept to plasticity problems it is assumed that the slow growth of the crack is not influenced by the surrounding environment. Clearly, there are cases where the environment would cause additional effects.

It is generally accepted (see, for example, Refs. [15] and [16] and the recently developed methods of J testing (Chapter 3)), that slow growth occurs in at least partly plane strain situations, although it is still arguable that growth of shear lip is the main cause of the increase in toughness. What is less clear is whether there is a growth of resistance in full plane strain, where the shear lip effect is negligible, and, if there is, how to characterise it and the final unstable behaviour that may subsequently occur. Not unnaturally, experimentalists have turned to accepted techniques such as COD and J to see whether they can be extended to the stable crack growth regime.

What is virtually a COD method is described in Ref. [160] under the name 'crack opening stretch' (COS). Tests on WOL or compact tension pieces are analysed by a simple hinge rotation picture.

$$\text{COS} = 0 \cdot 45(W - a)\theta \qquad (2.117)$$

This equation is identical in concept to that on which BS DD 19 [8] bases the analysis of COD in three-point-bend tests. Equation (2.117) has no elastic term and is thus more approximate, but by postulating a hinge rotation mechanism, plastic collapse, and thus a condition of extensive plasticity, is implied, at least in the test piece. The main experimental technique is the so-called double compliance method, in which the ratio of the readings of two displacement gauges, one near the crack mouth and one nearer the crack tip, are taken in a calibration test as a function of a/W. In crack growth testing, the two gauge readings are observed and crack length, a, inferred. The detailed technique is discussed in Ref. [160]. The main point of interest here is whether the R curve so obtained is independent of specimen geometry and initial crack length. There is some evidence that the curves, expressed in terms of K versus Δa, have a measure of independence for the near LEFM regime (see, for example, Fig. 2.78, where tests on centre-cracked tension (CCT) and crack-line wedge-loaded (CLWL) pieces—generally similar to compact tension—are compared). The degree of yielding is reported to be adequately covered by the LEFM plastic zone correction factor. No evidence is shown for *R* curves developed with extensive plasticity.

A very simple test based on energy measurement was adopted by Judy & Goode [161] who tested three steels in each of three thicknesses and four

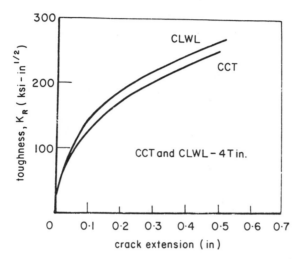

FIG. 2.78.   Elastic crack growth resistance curves for two different geometries by the double compliance method [160]. Stainless steel sheet 0·050-in thick, $\sigma_Y = 206 \times 10^3 \text{ lb/in}^2$; CLWL = crack line wedge-loaded, CCT = centre-cracked tension.

ligament lengths. The assessment of toughness was made in terms of energy to fracture/ligament areas in the DT (dynamic three-point-bend) test, so the test is essentially qualitative. The trend is for much heavier shear lip to develop as the fractured ligament becomes longer and all the data follow a linear log–log plot of energy versus crack length (i.e. ligament length). For the materials reported on, the specific energy per unit area did not alter as thickness decreased, but for small ligament lengths ($W - a \simeq B$) the data points tend not to fall on the main family curve. A typical result is shown in Fig. 2.79. It is clear from photographs of the fractured pieces that shear lip dominates, there being practically no flat fracture region in many tests beyond a length some fraction of *B* in which the original flat initiating crack dominates. Schematically, it is pictured that the development of *R* curves is essentially due to loss of constraint, and the test results, by intention, are representative of a near plane-stress behaviour in the plastic regime.

The onset of failure has also been discussed in Ref. [162]. Slow growth was not considered and the problem was indeed stated to be where there was no stable crack extension prior to instability. This initial instability was discussed in terms of the derivative of the area under a non-linear load extension diagram and an 'extended Griffiths criterion' postulated for application in both elasticity and plasticity. The J and equivalent energy

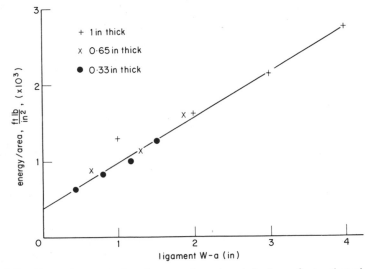

FIG. 2.79. Dynamic tear test data for near plane stress behaviour of a tough steel. (Data from Ref. [161].)

concepts were rationalised much as in Section 2.3.5 here, and the final conclusions suggested that J and the 'extended Griffiths equation' are indeed identical, at least for certain pieces, and that K_{1c} is the fracture criterion even with extensive plasticity. For the present discussion the important point seems to be the statement: 'The authors suggest that the errors introduced by neglecting this stable crack growth are, in practical terms, insignificant.' This seems contrary to much current thought unless it is taken to imply that if unstable fracture is to be avoided slow growth cannot be tolerated at all. This would be a very restrictive conclusion, although slow growth must surely be regarded as a safety feature against overload or the like, rather than a design margin that can be used repeatedly, and in that sense cannot be included in routine design methods.

COD concepts have been applied to slow crack growth by several workers. Three concepts can be distinguished: COD, δ_F at the tip of the original crack (probably itself a fatigue crack); COD defined from a clip gauge reading by, for example, the BS DD 19 formula in Ref. [8]; and COD, δ_a, at the real tip of the advancing crack. Green & Knott [163] studied the first of these. Clearly COD at the original crack tip must increase as a crack grows; if this takes place under rising load then some increase in toughness is inferred but as the tip grows well beyond the original crack length no direct measurement of the new crack toughness can be implied. Green & Knott [164] and Tanaka & Harrison [165] have also used the second method, in which the actual crack length is used in the BS DD 19 COD test formula. This seems to denote the COD that would be found by a calibration piece of the same final crack length bent to the same displacement, so again does not strictly relate to the growing crack situation. The crack infiltration technique [11] has been used by Garwood [166] to measure COD at the advancing crack tip for a mild steel on 10-mm-square three-point bend pieces for which the initiation COD, δ_i (itself measured by infiltration so not strictly comparable to a standardised test value), was 0·2 mm. He found that COD, δ_F, at the mid-section of the original fatigue crack tip increased linearly with Δa, and the COD, δ_a, at the advancing tip was smaller (here by a factor of four) but remained constant over a near ten-fold increase in growth (Fig. 2.80a). Using $(\delta_F - \delta_a)/\Delta a$ to define a flank angle, the picture is thus of an apparent angle that decreases slightly (from 51° to 32° for the whole angle) with a crack growth of a few millimetres. At small amounts of growth $(\delta_f - \delta_a)/\Delta a$ is not a good measure of flank angle, as is clear from Fig. 2.80b(ii), but for larger growth (Fig. 80b(iii)) it is an adequate measure. The true angle, based on $(\delta_m - \delta_a)/\Delta a$ where δ_m is the opening of the mouth of the growing section of the crack, may well be independent of Δa. Presumably δ_a is not itself measured 'at' the real tip but, rather like Fig. 2.5 repeated on a smaller scale, at some finite distance, r_a, behind the tip. This detail is lost in the difficulties of the technique and the microstructural irregularities but the uncertainties imply r_a is not more than δ_a and may be vanishingly small. Results on other sizes and geometries have not yet been reported. With such a small amount of data it is open to question whether the change from δ_i to δ_a is sudden (within 0·2 mm: Fig. 2.80a). Presumably the two quantities are strictly discontinuous since δ_i merges into δ_f whilst δ_a is non-existent for a growth $\Delta a < 0(r_a)$.

There have been several recent attempts to describe resistance curves in terms of J. The definition of J relevant to this usage is not clear and a generic term J_R is here employed. Each separate study may be regarded as an attempt to define J_R in a relevant way.

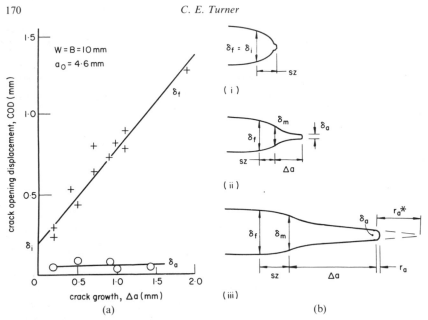

FIG. 2.80.   COD by infiltration measurement during stable crack growth in steel (a) δ_a at the advancing tip and δ_f at the original fatigue tip. En 32 structural steel, $\sigma_Y = 285 \, MNm^{-2}$; three-point bending, $B = W = 10 \, mm$. (b) Development of slow crack growth; schematic; sz = stretch zone. (i) Initiation, (ii) first development of slow crack growth, (iii) well developed slow crack growth; $\bigcirc$ at advancing tip; $\times$ at original tip [166, 178].

Garwood *et al.* [167] suggested a modification to the technique for measuring *J* in order to overcome at least the most serious of objections to the use of *J* after initiation, when material is unloading, and this modified *J* has also been used by Garwood & Turner [168] to express *R* curves for growth in terms of *J* concepts. The test technique attempts to express *J* in terms of work done (eqn. (2.30)) as commonly used for three-point-bend pieces, but implies an unknown loading curve rising from the origin to the point to which slow tearing has spread (Fig. 2.81). This curve is thus a fictitious loading curve matching load, displacement and crack length to those of the real test piece at the point under study to which the real crack has grown whilst under load. This fictitious loading curve defines the value of work, *U*, to be used in the formula for *J*. If the steps of crack growth are small and the technique is used incrementally, then, as shown in Ref. [167], the fictitious curve need not be known precisely for a good estimate of the modified *J* to be found. The equation for the *n*th step is [166]

$$_{(n)}J_r = {_{(n-1)}}J_r \frac{[W - _{(n-1)}a - (\eta\Delta a/2)]}{[W - _{(n)}a + (\eta\Delta a/2)]} + \frac{\eta U_4}{B[W - _{(n)}a + (\Delta a/2)]} \qquad (2.118a)$$

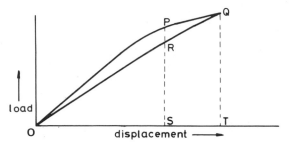

FIG. 2.81. Application of J to slow stable crack growth by construction of a fictitious loading curve; OPQ = actual loading curve, crack growth starting at P, crack length a_0; ORQ = postulated loading curve; new crack length a_1. U_4 is the area PQTSRP [167].

The terms are defined in Fig. 2.81. For deep-notch three-point bend, the case of immediate interest and the only one referred to in Ref. [167], $\eta = 2$, whence

$$_{(n)}J_r = {}_{(n-1)}J_r \frac{[W - {}_{(n)}a]}{[W - {}_{(n-1)}a]} + \frac{2U_4}{B[W - {}_{(n-1)}a]} \qquad (2.118b)$$

In general, writing U_4 as the increment of work, ΔU, and noting η may not equal 2 but neglecting changes in η with crack length

$$\Delta J_r \equiv {}_{(n)}J_r - {}_{(n-1)}J_r = \frac{\eta \Delta U}{B(W - {}_{(n-1)}a)} + {}_{(n-1)}J_r \frac{(1-\eta)\Delta a}{(W - {}_{(n-1)}a)} \qquad (2.118c)$$

and

$$J_r = J_i + \sum \Delta J_r \qquad (2.118d)$$

This technique has been applied to a range of three-point-bend geometries, some with deep side grooves so that flat fracture resulted with no shear lip. Over a modest range of geometric variables the resulting R curve [168] was independent of initial crack length and size (Fig. 2.82), thus holding out some hope of a geometry-independent behaviour. Note η is geometry-dependent (Section 2.2.2.1.3 and Appendix 4) so that if the resulting $J_r - \Delta a$ curve is indeed independent of geometry, different amounts of plastic work, ΔU, have been expended in the tests on different configurations to cause a given amount of crack growth, Δa, in flat fracture. This usage of J is here called J_r. If, by analogy with $BJ = -\partial U/\partial a$ (where a is strictly the *initial* crack length a_0), a term was envisaged as $BJ_R = -\partial U/\partial a$ (where a is the current crack length) then J_r in eqn. (2.118) is not that term. J_r is a particular attempt to define a J-like term suitable for use with resistance curves. It involves a_0 and W via the variation of η and $(W - a)$ from one configuration to another in a manner that is hopefully

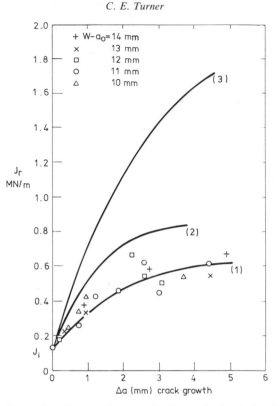

FIG. 2.82. Crack growth resistance, J_r, curves in three-point-bend pieces of En 32 steel, $\sigma_Y = 285\,\mathrm{MNm^{-2}}$. Curve 1: flat fracture; side groove ratio $(B - B_n)/B = 0\cdot56$, $B = 12\,\mathrm{mm}$, $W = 24\,\mathrm{mm}$, showing independence of initial crack length a_0. Curve 2: full thickness, $W = B = 10\,\mathrm{mm}$ and $W = B = 24\,\mathrm{mm}$. Curve 3: full thickness, $B = 10\,\mathrm{mm}$, $W = 24\,\mathrm{mm}$. (After Refs. [166, 168 and 178].) Note that Curve 2 relates to Fig. 2.80a.)

independent of in-plane geometry in flat fracture (Fig. 2.82). For steady-state growth of a tear or crack with its associated contained plastic zone, $dJ_r/da \to$ constant, not zero, since a uniform rate of working is being described.

It is thus being stated that the increment of work, ΔU, can be related to a J-like term, J_r, by the well known η relation (eqn. (2.30)), so that for circumstances in which $J_r - \Delta a$ is known the increment of work to propagate the crack for an amount Δa can be assessed as $\Delta U/B = (W - a)\Delta J_r/\eta - J_r(1 - \eta)\Delta a/\eta$. Clearly $J_r - \Delta a$ is material-dependent. It is also seen (Fig. 2.82 (1) and (3)) to be thickness-dependent if shear lip contribution is not negligible.

Despite the use of the relevant value of η in analysing a crack growth curve to obtain $J_r - \Delta a$, there is neither *a priori* proof nor experimental evidence, as yet, that the same curve will result from tests on different configurations giving different degrees of in-plane constraint. Indeed, the uncertainty over the use of J as a criterion for crack initiation is here repeated. It is being asked whether, for a given material thickness, there is constant 'tearing toughness' to which all plastic tearing can be related (for flat fracture in the first instance but hopefully with added shear lip components which may well be a function of in-plane configuration as well as of thickness), and if so whether an energy term is characteristic of this 'tearing toughness' for which J_r is a workable approximation.

A weakness of the argument is that although a J-dominated stress field during growth is not envisaged, J_r seeks to measure a geometry-independent or reference component of the plastic work rate that is relevant to the cracking process, by use of the η factor that is related to J theory (Appendix 4). The justification [167] is that J_r is by definition a relevant measure of energy absorption (though not of energy release rate) even for the growing crack case. This justification is rather different from that where the characterising role was emphasised for the monotonic case (Section 2.2.2.1.3). In deriving J_r, there is, however, the implicit assumption that the numerical value of η is governed by geometry and gross plastic flow, and is not closely dependent on the details of the crack tip field that alter subsequent to initiation.

The effect of shear lip can be added to the flat fracture case, at least approximately, to show the full R-curve effect, on the assumption that there is an energy per unit volume, E_r, for shear lip formation that can be found from plane stress tests and applied to the shear lip part only of a mixed (part flat, part shear lip) fracture. The width of shear lip is itself a complicated function of test piece width and thickness that is not fully understood. Over a rather small range of sizes, Garwood [166] found that the quantity $(B/2)(2s/B)^2$ remained roughly constant after a few millimetres of growth, where B is thickness and $2s$ is the total shear lip width. Thus the total toughness value tends to

$$R = R_1 + E_r \frac{B}{2}\left(\frac{2s}{B}\right)^2 \qquad (2.119)$$

where R_1 is the plane strain value found from the side-grooved J–R tests (in units as for G), increasing according to Fig. 2.82 with Δa. The whole curve cannot be predicted unless the growth of s with Δa is known or observed from experiment, and s appears to be a function of B and W separately, in a manner not yet understood.

The intention, of which the realisation still needs considerable further demonstration, is to be able to predict the R curve for a given application from the flat and oblique components, or at least from an overall experimental determination for a given thickness, in the hope that this R curve will be substantially independent of in-plane geometry.

Apart from the shear lip effect, the evidence of constant COD at the tip of the growing crack suggests that tip toughness is not itself increasing and that the apparent increase in toughness in the flat fracture regime is due mainly to the remote plastic work that is necessary to deform the structure as the crack moves forward under rising load.

Assuming that plasticity away from the plane of the crack is a major feature of the R curve, it is suggested that J_r as here defined is a more useful measure of tearing toughness than a conventional J test. The two can be related. From eqn. (2.30)

$$\frac{dJ}{da} = \frac{J}{\eta}\frac{d\eta}{da} + \frac{J}{U}\frac{dU}{da} + \frac{J}{W-a} \tag{2.120a}$$

If $d\eta/da \ll \eta/J(dJ/da)$ then the first term on the right can be neglected and it is known (Section 2.2.2.1.3), that at least for deep-notch cases η is near invariant with crack length. Thus, neglecting the η term for simplicity but noting that it can be incorporated if desired since estimates of η are available (Appendix 4)

$$\frac{dJ}{da} \simeq \frac{\eta}{B(W-a)}\frac{dU}{da} + \frac{J}{W-a} \tag{2.120b}$$

Furthermore

$$\frac{dJ}{da} \simeq \frac{\eta}{B(W-a)}\frac{dU}{da} \tag{2.120c}$$

if

$$J/(W-a) \ll dJ/da \tag{2.121a}$$

or, as expressed later, if

$$\omega \equiv \frac{(W-a)}{J}\frac{dJ}{da} \gg 1 \tag{2.121b}$$

Using (2.118c) and assuming the inequality (2.121) applies to J_r as well as to J, then

$$dJ_r/da \simeq dJ/da \tag{2.122}$$

(provided (2.121) is satisfied).

Thus the conventional estimate of $J - \Delta a$ from the normal procedure as used to define the initiation value, J_i, gives a satisfactory estimate of dJ_r, and thus of dU only if inequality (2.121) is satisfied. The resistance curve defined by the conventional J test procedure is here referred to as J_{0r}. It is tangential to J_r (Fig. 2.82) at $\Delta a \rightarrow 0$ and rises at first linearly with Δa. Several examples of $J_{0r} - \Delta a$ curves are given by Logsdon [169].

The inequality [121], when expressed in terms of J_r, can be broken by decrease of dJ/da with crack growth, or by a ligament $(W - a)$ too small in relation to J. Note, however, that the latter restricts the use of J as a measure of J_r, whilst the meaning of J_r as a measure of work done remains intact, although there must be limits of size beyond which eqns. (2.118) are invalid. In the present context it could be reasoned that if J tests for measurement of initiation toughness are valid for sizes down to $(W - a) > 25 J/\sigma_Y$ where J/σ_Y may be interpreted as δ_i, then for the tearing crack the limiting size restriction might be $(W - a) > 25\delta_a$. In the present instance Garwood's J_r data are valid at initiation according to $(W - a) > 25\delta_i$ and during growth, according to $[W - (a_0 + \Delta a)] > 25\delta_a$, but clearly more study of the necessary restrictions is required.

As noted above, the J_r concept does not postulate the maintenance of the characteristic J monotonic stress field. The observance of the inequality (2.121b) and the use of J_{0r} would seem to maintain a J field by keeping the amount of crack growth small in relation to the critical J-dominated region.

A completely different viewpoint is taken by Mai *et al.* [170]. They use side-grooved pieces of DCB type in which the arms are encompassed in a 'rigid' framework so that tearing occurs with plasticity restricted to the side-grooved region that is not encompassed by the frame. An R curve is indeed found, i.e. there is an apparent increase in toughness despite negligible 'remote' plasticity. In the present writer's opinion this is not a measure of the dissipations that would be present in the plastic tearing failure of a component, since the major term, i.e. 'remote' plasticity in the normal crack slip-line field region, is prevented. As noted in Ref. [170], the tear changes from flat to oblique, or rather to a necking-type action in the very thin pieces tested. This seems to be the reason for the R curve effect found and it is doubtful whether it is representative of the material in a more general configuration. In short, the real component would have to allow necking of a thin panel (perhaps, for example, in a bursting diaphragm unit) whilst itself remaining elastic by virtue of its own massive thickness or external constraint, as in the test method. It seems the technique takes the use of side grooves beyond the arguably useful condition of enforcing flat fracture to simulating an extremely deep side groove sufficient to enforce necking through the thickness, thus causing elastic behaviour away from the crack plane and thereby suppressing the main cause of the R curve effect.

De Köning et al. [171] have modelled slow stable growth by computational methods. They note that G is an inadequate term for the elastic energy rate available in the elastic–plastic case. R is estimated by three methods. In one method it is stated that (in terms of Fig. 2.28 with A now denoting the point of first crack growth): 'In deriving the crack driving force and hence the R curve, it is assumed that the work dissipated, U_R, is represented by the area OALO: whereas for ductile material behaviour the actual work dissipated in the plastic zone U_{pl}, is represented by the area OALCO. It is obvious that $U_{pl} \geq U_R$.' dU_R/da' is then found from an experimental load–deflection record where a' is the effective crack length (adjusted to suit the compliance) and is close to $(a + r_{pl})$. A near-linear plot of energy rate versus Δa for two ratios of a/W for centre-cracked plates of aluminium alloy is found, with near agreement between the above and the other two methods of finding R that seem to be essentially elastic. Repeated unloading decreased U_{pl} and U_R by nearly 50 % in terms of real crack length but by only small amounts in terms of the adjusted crack length a'. A finite-element model followed these results by adjusting the release of the crack tip node to follow the experimental load–deflection record, this then giving good representation of the various energy rates. The statement that, 'For engineering applications the crack driving force G can now be obtained from $G = (F^2/2B) [dC/d(2a')]$', admits an essentially elastic approach that does not reflect the additional term in eqn. (2.28) here. The final conclusion that, 'The work U_R associated with the R curve is much less than the plastic work dissipated', would in the writer's view be better stated in terms of energy rates, and the conclusion, 'The elastic work is much smaller than $F^2C/2$', seems contrary to eqn. (2.28) and indeed to de Köning and co-workers' own arguments that G is inadequate in the plastic case. The physical basis of U_R is not clear to the present writer and may be an attempt to divide the total plastic work into two components, one associated with the crack growth and the other with general (remote) plasticity. This latter would alter with load–unload cycles, leading to the decrease in U_{pl} and U_R noted previously, but, following the discussions in Section 2.2.2.1.3 (iii), it is not clear whether this distinction can be drawn in a meaningful way.

Wnuk [172, 173] has modelled slow stable crack growth and described the model more fully [174]. His analysis is for a rate-dependent material with the rate-independent elastic–plastic problem emerging as a special case. A criterion of fracture called the 'final stretch' is postulated, in which a critical amount of stretching occurs within a fracture process zone, Δ, 'just prior to fracture'. The criterion is history-dependent and is stated to correspond to a criterion of critical strain attained over a 'Neuber particle'. It reduces to COD for steady-state propagation. An alternative statement of the criterion is in terms of 'the essential energy of fracture', $\hat{w}_c = \sigma_Y \hat{u}_c$. The model is analysed in terms of the BCS–Dugdale line plasticity model for plane stress. The problem is not restricted to small-scale yielding but numerical computation is required for other than contained yield and only this latter is referred to here. In order to maintain the criterion that $\hat{u}_c$ (or $\hat{w}_c$) is constant during slow crack growth, the growth of plastic zone size, $\mathrm{d}R/\mathrm{d}a$, is related to the final condition, R_α and current condition, R (and initial condition when $R = R_i$) by

$$\hat{u}_c = \frac{\Delta \delta}{4R} [\ln (4R_\alpha / \Delta) + 1] \tag{2.123}$$

$$\frac{\mathrm{d}R}{\mathrm{d}a} = \tfrac{1}{2} \ln (R_\alpha / R) \tag{2.124a}$$

The measure of R is perforce the Dugdale model term $(a_1 - a) = [\sec(\pi\sigma/2\sigma_Y) - 1]a$, as in Section 2.2.1.1, whence $R \simeq (\pi E/8\sigma_Y)\delta$, and J is related by $J = \sigma_Y \delta$.

The material characteristics $\hat{u}_c$ or $\hat{w}_c$ are written for the Dugdale model in terms of plastic zone size $\hat{R}_c (= \hat{u}_c \pi E/4\sigma_Y)$, crack-opening $\hat{\delta}_c (= 2\hat{u}_c)$ or $\hat{J}_c (= \sigma_Y \hat{\delta}_c)$, just as for monotonic loading.

Expressions corresponding to eqn. (2.124a), for contained yield are[18]

$$\frac{\mathrm{d}J_{RW}}{\mathrm{d}a} = \frac{\sigma_Y \delta}{2R} \ln (J_{\alpha W} / J_{RW}) \tag{2.124b}$$

$$\frac{\mathrm{d}\delta_R}{\mathrm{d}a} = \frac{\delta}{2R} \ln (\delta_\alpha / \delta) \tag{2.124c}$$

The fracture criterion for the final separation is expressed as $\hat{u}_c$ or $\hat{w}_c$ or the equivalent extent of plastic zone $\hat{R}_c = \Delta/2[\ln (4R_\alpha/\Delta + 1]$ with $\hat{\delta}_c - 8\sigma_Y \hat{R}_c \pi E$. Unless $R_\alpha \to \infty$, with Δ restricted to the lattice spacing or perhaps much more, $[\ln (4R\infty/\Delta) + 1]$ must be of the order 2 to 20, so that $\hat{R}_c = (0)\Delta$ (or up to 10Δ). Thus $\hat{\delta}_c = (0)(\sigma_Y/E)$ and $\hat{w}_c = (0)(\sigma_Y/E)\sigma_Y\Delta$. Thus, if $\Delta = n_i$ and $G_i = \sigma_Y \delta_i$, $\hat{w}_c = (0)n(\sigma_Y/E)G_i$ where $n \ll 1$ for Δ equal to a few lattice spacings and $n > 1$ for a large process zone ($n = (1/2\pi)(E/\sigma_Y)$ if the process zone $\Delta = r_{pl}$ as in the LEFM expressions in eqn. (1.10). This seems consistent with Wnuk's remarks for brittle material

---

[18] J_{RW} denotes a measure of J_R by Wnuk's usage.

where, he suggests, w_c approaches the surface energy for a classic Griffith solid. For contained steady-state propagation Wnuk suggests that w_c can approach the Irwin–Orowan value for quasi-brittle failure, presumably G_{Ic} or even G_c, if the plastic zone, large in relation to the crack tip size, is still small in relation to plate size so that the whole plastic dissipation is treated as an equivalent surface energy. Wnuk [174] quotes, but does not use, an estimate of the essential work of fracture, made in terms of tensile strength and COD (in plane stress) as $\hat{w}_c = 7 \cdot 7 \, 10^{-1} \sigma_u \delta$, which appears to be comparable to $G_c = \sigma_Y \delta$, thus implying $\hat{w}_c$ is to be identified with G_c for plane stress.

If a 'local fracture work rate' is written as $\bar{r} \, dJ_R/da$, where $\bar{r}$ is a process zone or step size (i.e. Δ in Wnuk's notation) then, if $\bar{r}$ is constant, and since any formulation of J_R (be it J_r or other) envisages dJ_R/da decreasing with crack growth, such a term is not compatible with an 'essential work of fracture' taken to be constant at initiation and during slow growth. Wnuk's basic argument is formulated in terms of $\hat{u}_c$, $\hat{w}_c$ and plastic zone size, R. It is re-expressed in terms of J_{RW} and δ_R (eqns. (2.124b, c), which, it must be recalled are for contained yield).

From Figs. 2.80 and 2.82 it is seen that δ_f increases; δ_a reduces from δ_i and then remains substantially constant while J_r increases, so that eqns. (2.124a) and (2.124b) cannot be satisfied, even in overall trend, if J_{RW} in eqn. (2.124b) is identified with J_r and δ_R with δ_a. If, despite this discrepancy, J_{RW} in eqn. (2.124b) is taken to be J_r as defined here, the term $\delta/2R$ is of the order of unity rather than $4\sigma_Y/\pi E$.

Thus, conceptually it seems that $\hat{w}_c$ is that fracture energy that is not included in other estimates (such as plasticity) of the dissipative terms. Its value will depend on the terms included or omitted from the model used, although the writer cannot reconcile the expressions quoted above in terms of J_{RW} with the formulation in terms of J_r as outlined here.

A key to these differences seems to lie in the formulation of the model as discussed in Section 2.2.2.1.3 (iii), where it was suggested that if the model was phrased so as to allow a non-continuum energy-rate term then surface energy or 'essential work of fracture' can be found in a meaningful way. LEFM solutions and HRR solutions (J for monotonic loading) are so phrased. Wnuk's model, with its finite COD, δ_c and step length Δ, is also so phrased, even for the advancing crack, so that notionally a constant work of fracture is possible. However, it appears to the writer that a relationship between R, J and δ based on the BCS–Dugdale model is not valid after initiation if J is to be interpreted as J_R, unless δ_R is related to the original crack tip rather than to the advancing tip. However, as seen in the following, once a steady state is established a constant relationship between δ_a at the advancing tip and dJ/da may emerge where J_r is the measure used for J, and the reduction of Wnuk's criterion to COD was only envisaged for the steady state. For thin sheet, predominantly in plane stress with contained plasticity that thus comes near to the steady-state model, the double COD technique [160] has been employed by Wilhelm *et al.* [175] with J formulated in the elastic sense of G but using an equivalent crack, length including crack growth and plastic zone correction. This may well extend LEFM-based methods to a limited amount of plasticity but does not seem adequate for extensive plasticity.

Rice has also suggested the form for a growing crack [176]. A steady state is envisaged in plane strain in which a Prandtl-type field is moved along with the tip implying contained, or at least highly constrained, yield. The opening of the advancing crack is found to be

$$\delta = \frac{\alpha r}{\sigma_Y} \frac{dJ}{da} + \frac{\beta_r \sigma_Y}{E}\left[\ln\left(\frac{R}{r}\right) + 1\right] \qquad (2.125a)$$

where α is an undetermined, dimensionless term not necessarily constant, $\beta = 4(2 - v)/3^{1/2} \simeq 3\cdot 93$ for $v = 0\cdot 3$, r is the polar distance from the tip, and R is an undetermined parameter with dimensions of length, scaling with JE/σ_Y^2 for dimensional reasons. Some computed data led Rice to the suggestion[19] that $R \simeq 0\cdot 16\, JE/\sigma_Y^2$, comparable to the maximum extent of plastic zone for that case, and $\alpha \simeq 1/m$ where m is the constant in the monotonic relation $J = m\sigma_Y\delta$.

Rice points out that as $r \to 0$, $d\delta/dr \to \infty$, so the crack has a vertical tangent at its apex. He suggests that this is different from the concept of 'crack tip opening angle' found elsewhere [74, 163]. As discussed in relation to Fig. 2.80, Garwood & Turner's results [177] allow both features to be accommodated, i.e. a flank angle substantially constant with r according to the infiltration casts, and a final small 'nose' characterised by $\delta_a \simeq 0\cdot 05$ mm occurring at some undetermined distance, r_a, from the actual tip at which the detail is lost in the uncertainty of the experimental measurements and microstructural irregularities.

If eqn. (2.125a) is evaluated at $r = \Delta$ the expression has a close similarity to those of Wnuk, despite the very different a umptions. An estimate of the unknowns in eqn. (2.125a) from the data of Garwood appears to require further assumptions, since both α and R are linear with r. If the estimates for

---

[19] Unfortunately, Ref. [176] does not give the derivation of eqn. (2.125a) in detail but refers to work not seen by the writer.

α and R quoted previously are used, it is noted [177] that the term $(\beta r \sigma_Y / E)[\ln (R/r) + 1]$ appears negligible in relation to $(\alpha_r / \sigma_Y)(dJ/da)$ for values of r of the order of δ_a to δ_f. Indeed, for this term to be comparable with the first, with $\alpha \simeq 1/m(m \simeq 2)$ requires r to be around $(\sigma_Y / E)\delta_a$ or even less, implying that the 'final nose', r_a, beyond the point at which δ_a is measured is only of the order of $r_a \simeq (\sigma_Y / E)\delta_a$ or less. If r_a is indeed this small then for continuum purposes, where, at distances less than r_a, the details are lost in the microstructure, the term in $\ln R/r$ can presumably be omitted and eqn. (2.125a) reduces (for this tough material in which the dJ/da term dominates) to

$$\delta \simeq \frac{\alpha r}{\sigma_Y} \frac{dJ}{da} \text{(for } r \geq r_a^*) \tag{2.125b}$$

where, to fit the present data, r_a^* must define an apparent origin of r (Fig. 2.80b), by simple extrapolation of the flanks of the advancing crack. This notional crack tip appears to be situated at a fictitious point $r_a^* = \Delta a \delta_a / (\delta_m - \delta_a)$ beyond the point at which δ_a is measured. Taking a well developed crack, $\delta_m \simeq \delta_f - \delta_i$ (Fig. 2.80b(iii)), and using the data of Figs. 2.80 and 2.82, $r_a^* \simeq 1.9 \, \delta_a$. Clearly, if at $r = r_a^*$, $\delta = \delta_a$ for all values of Δa then

$$\delta_a = \alpha r_{/a}^* \frac{1}{\sigma_Y} \frac{dJ}{da} \tag{2.125c}$$

It is not clear that J as used by Rice for the advancing crack can be identified with J_r used here, but using the data of Fig. 2.82, $1/\alpha \simeq 1.9$ for flat fracture and 5.4 for full thickness at $\Delta a = 0.2$ mm, reducing to $1/\alpha \simeq 0.64$ for flat fracture and 1.0 for full thickness at $\Delta a = 2.0$ mm, and to about 0.16 for both cases as the steady state is approached (as far as it is, in these tests). It is not implied that the steady-state conditions are identical for both cases but the data do not justify quoting more precise values and the amount of shear lip is not large here (about 20%). Note that for flat fracture 'just after' initiation $1/\alpha(\simeq 1.9)$ is fairly closely equal to $m (\simeq 2.1)$.

As already noted, if a process zone or step length, $\bar{r} = $ constant, is envisaged, then the 'local fracture work rate' $\bar{r} \, dJ_r/da$ must diminish as Δa increases, to some steady-state value $\bar{r} \, dJ_r/da|_{ss}$ that is not zero, and $1/\alpha$ decreases in the same way, with $\sigma_Y \delta a$ remaining constant. For very small amounts of growth where the inequality (2.121a) is satisfied and the monotonic J stress field still dominates the crack tip, it seems reasonable to equate the rates J_i and $\bar{r} \, dJ_r/da|_i$. This gives

$$\delta_a \sigma_Y \bar{r} / \alpha_i r_a^* = \bar{r} \, dJ_r/da|_i = J_i = m \sigma_Y \delta_i \tag{2.126a}$$

whence (since $m\alpha_i \simeq 1$ and $\delta_i/\delta_a \simeq 4$ here),

$$\bar{r} \simeq \delta_i r_a^*/\delta_a \simeq 4r_a^* \simeq 2\delta_i \simeq 8\delta_a \qquad (2.126b)$$

For the present data $\bar{r} \simeq 0.4\,\text{mm}$. No physical interpretation is offered at this stage beyond noting that $\bar{r}$ is comparable to the stretch zone or δ_i rather than to grain size or inclusion spacing, some $20\,\mu\text{m}$ and $50\,\mu\text{m}$ respectively for Fig. 2.82. If $\bar{r}$ were to vary from one case to another (as seems likely if it is related to microstructure), then even for a given J_i crack growth resistance $dJ_r/da|_i$ could be quite different. There is some evidence for this in Ref. [178], where Willoughby *et al.* report very similar values of J_i but quite different J_r curves for tests on anisotropic steel in different directions, but there is no direct evidence on $\bar{r}$. In so far as J_i values are similar, so presumably are the stretch zones to which $\bar{r}$ may be related.

Broberg [179] has discussed slow stable growth. He writes:

(i)   In terms of the J integral (for the case of small-scale yielding) $J = J_i$ at the start of slow crack growth; $\alpha_B J_{ss} = J_i$ at the end of slow crack growth, where the subscript, ss, denotes steady state, and α_B is the fraction of the energy flow to the plastic region that reaches the end region (i.e. the fracture process zone) for steady state growth. α_B is called the screening function.

(ii) J equals the energy flow to the plastic region during the whole of slow crack growth.

(iii) The J integral for a certain position of the crack tip ought to be approximately the same as the J integral calculated for the hypothetical case of a non-moving crack with the same position of the crack tip and the same load, calculated on a path completely outside the plastic region.

Statement (iii) is precisely the viewpoint used to derive eqn. (2.118) providing the hypothetical loading curve ORQ is used (Fig. 2.81). This leads, as seen, to the statement that dJ_r (not J_r) is related to dU. The energy flow rate to the process zone is $\bar{r}\,dJ_r/da$ and decreases (for constant $\bar{r}$) as Δa increases, whereas the descriptive term J_R increases. Thus, although statement (ii) may be admissible for small-scale yielding (for which it was formulated), it does not seem relevant to appreciable growth in extensive plasticity where J is no longer a measure of the energy rate at the tip singularity. Moreover, Broberg's J increases as a increases in a way that is conceptually similar to J_R, which, as noted, does not seem to be a measure of energy flow rate except for small amounts of growth.

Statement (i) contains two concepts. The first, here called 'the screening ratio, is contained in the statement $\alpha_{sr}J_{ss} = J_i$ and the second, the (steady-state) 'screening function', here denoted α_{sf}, is the fraction of the overall work to reach the tip. These two terms do not seem to be the same. The latter is clearly dependent on absolute size since it is a volume–surface ratio. In the present terminology based on *plastic* work it is

$$\alpha_{sf} = (\bar{r}\,dJ_r/da)/(d\,U_{pl}/B\,da) \qquad (2.127a)$$

$$= \bar{r}\eta_{pl}/(W - a) \qquad (2.127b)$$

This term varies with the size (if $\bar{r}$ is constant) according to $1/(W - a)$, and with geometry according to η_{pl} but is constant for all degrees of growth, not just steady state. If elastic plus plastic terms are used, the overall η term is not constant unless $\eta_{el} \simeq \eta_{pl}$ (Appendix 4). This total usage does not seem desirable. Clearly, for elastic behaviour if growth occurs during loading energy is absorbed both by the surface term and as an increase in strain energy, whereas, if growth occurs during unloading, it is the decrease in strain energy that propagates the crack. Thus in elastic–plastic situations there are the two separate effects, the reversible elastic and the always dissipative plastic, that are better treated separately. Equations (2.127) suggest that for a given size and geometry some constant function of the latter is always dissipated at the surface. (Again, following the discussion in Section 2.2.2.1.3 (iii), this argument may depend on the way in which the problem is formulated.) From the data of Fig. 2.82, $\alpha_{sf} \simeq 0.08$ for a 10-mm ligament (with $\bar{r} = 0.4$ mm as estimated above), falling to $\alpha_{sf} = 0.06$ for a 15-mm ligament. Taking the former definition that implies the term α_{sr} (but inverting it since Broberg's measure of dissipation, J, increases as does J_R whereas the measure $\bar{r}\,dJ_r/da$ used here decreases)

$$\alpha_{sr} = (\bar{r}\,dJ_r/da|_{ss})/J_i = (\bar{r}\,dJ_r/da|_{ss})/(\bar{r}\,dJ_r/da|_i) \qquad (2.127c)$$

giving $\alpha_{sr} = 0.085$ for flat fracture independent of size, if the J_r curve concept is valid. This value is in fair agreement with Broberg's estimate of $\alpha_B = 0.1$ for plane strain. The two are related by

$$\alpha_{sf}/\alpha_{sr} = J_i/(d\,U_{pl}/B\,da|_{ss}) \qquad (2.127d)$$

α_{sr} (eqn. (2.127c)) is the ratio of energy passing to the tip under the steady-state (non-J dominated) singularity to the tip energy 'at' or 'just after' initiation (which is J dominated). (Note the term α in eqns. (2.125) and (2.126) following Rice [176] is not the same α as used by Broberg [179]. In the $1/\alpha$ notation of eqn. (2.126a) $\alpha_{sr} = 1/\alpha_{ss}/1/\alpha_i$.) These differences arise because in plasticity $\partial/\partial a_0 \neq \partial/\partial a$, where a_0 is initial and a current crack length, whereas in elasticity they are identical and presumably nearly so for small-scale yielding. Clearly the numerical values discussed in the foregoing may be pressing the scanty data beyond a reasonable limit.

In summary, some understanding, both theoretical and experimental, of stable and crack growth is emerging. The models are for small geometry change and are most easily interpretable for steady-state contained yield. The data include effects of large geometry change, non-steady state and extensive yield. These differences are severe, yet some link between the results seems feasible. A yet more serious drawback to the whole problem is the addition of shear lip effects that may well be a function not only of thickness but also of in-plane constraint and width–thickness ratio. The writer is not aware of any suitable analysis of the shear lip problem.

Advances have also been made recently in examining the final instability. In Ref. [180] Paris *et al.* accept J_{or} versus Δa as a measure of resistance to propagation and equate the axial contraction of the elastic region of a component, as the load relaxes consequent upon crack spreading, with the extension of the notched plastic region. The model is presented for deep notches (or non-hardening), in that the plastic extension is equated to COD, δ (i.e. there is no plastic extension remote from the notch slip-line field). In so far as the argument is based on δ there is no need to invoke J, but the argument is in fact translated to J by writing $J = \sigma_Y \delta$ so that the J_{or} versus Δa curve can be used as the measure of crack growth toughness. For centre-crack tension, neglecting an elastic compliance term for simplicity, onset of unstable behaviour is found to occur when[20]

$$T \equiv \frac{\partial J_{\text{or}}}{\partial a} \frac{E}{\sigma_Y^2} \leq \frac{2D}{W} \qquad (2.128a)$$

For deep-notch three-point bend it is found that

$$T \equiv \frac{\partial J_{\text{or}}}{\partial a} \frac{E}{\sigma_Y^2} \leq \frac{4(W-a)^2}{W^3} S - \frac{\theta E}{\sigma_Y} \qquad (2.128b)$$

The term $(E/\sigma_Y)(dJ/da)$ is termed the 'tearing modulus' in Ref. [180] and in so far as it is based on J_{or} for full thickness, it appears as a material constant, at least for that thickness. The inequality

$$\omega \equiv \frac{W-a}{J} \frac{dJ}{da} \gg 1 \qquad (2.129)$$

---

[20] The present writer associated the factor of 2 in eqn. (2.128a) with the centre-cracked geometry where the crack length of $2a$ is being treated, and would suggest writing $d/d(2a) \leq D/W$. In eqn. (2.128b) S is the half span.

has also been introduced in related work[21] as a requirement for valid J resistance curve data. It will be recognised that eqn. (2.121) expresses the same relationship. The basis of the derivation of eqn. (2.129) is not known to the present writer but presumably differs from eqn. (2.121) since that expresses a requirement such that $J_{or} \simeq J_r$, whereas Ref. [180] is based only on J_{or}. It thus appears as a restriction in the use of J_{or}, presumably to maintain a J-dominated stress field, which, it has already been argued, is not necessary for use of J_r as a measure of remote work. A restriction that $d\eta/da$ be small was also noted after eqns. (2.118b) and (2.120) as an additional requirement for the dJ/da term to dominate the description of slow growth.

Results similar to eqn. (2.128) have also been obtained by an energy balance approach [181]. Recognising that J_r as here defined (eqn. (2.118)) is not a measure of work rate as implied by the definition $BJ = -\partial U/\partial a$ (eqn. (2.29))[22] but rather that the derivative dJ_r/da is the measure of total dissipation rate, $\Delta U_T/\Delta a$, normalised by means of $\eta/B(W-a)$ to a notionally geometry-independent term (at least in flat fracture and hopefully for full thickness if the shear lip term is constant or quantifiable) then, provided inequality (2.121) is satisfied, a balance between energy rates can be written in the simple form

$$BI = dU_{pl}/da = f(dJ_r/da) \qquad (2.130)$$

If the inequality is not satisfied a further term in J arises from eqn. (2.118). This is omitted here for simplicity. The left side states the rate at which elastic energy is released from the elastic–plastic situation (eqn. (2.28)) at *constant displacement*, and the right side the rate of work absorption (including surface work) as a crack is grown at *constant load* (in fact slightly rising). The function on the right side must not only contain the geometric normalising term $(W-a)/\eta$ for plastic work but must also subtract off the increase of elastic strain energy that is included in growth at constant load while J_r is measured. Thus, splitting $\eta\Delta U_T$ into $\eta_{pl}\Delta U_{pl} + \eta_{el}\Delta U_{el}$, and noting that the elastic energy rate component $\Delta U_{el}/B\Delta a$ that has to be

---

[21] This usage of ω to describe the existence of a J-dominated field is understood to be derived in a paper that the writer has not yet seen (Hutchinson, J. W. & Paris, P. C. (1977). Stability analysis of J-controlled crack growth. *Ibid.* Ref. [98]).

[22] It is clear from the present discussion that, for plasticity, eqn. (2.29) implies only $\partial/\partial a_o$ (i.e. increase in initial crack length) and not $\partial/\partial a$ (where a is a variable current crack length) whereas in elasticity these two terms are the same.

subtracted[23] is just G, then for crack growth to be possible at constant overall displacement

$$I \geq \frac{W - a \, \mathrm{d}J_r}{\eta_{pl}} \frac{}{\mathrm{d}a} - \frac{\eta_{el}G}{\eta_{pl}} \qquad (2.131)$$

This analysis assumes that the remaining plastic plus surface components of dissipation are the same for an increment of crack growth whether induced at constant load or at constant displacement. This growth is not necessarily unstable. Stability is governed by the second derivative

$$\frac{d^2 U_{el}}{da^2} \geq \frac{d^2 U_{pl}}{da^2} \qquad (2.132a)$$

$$\frac{\mathrm{d}I}{\mathrm{d}a} \geq \frac{f(\mathrm{d}^2 J_r)}{\mathrm{d}a^2} \qquad (2.132b)$$

The argument is not here pursued in detail beyond noting that if the J_r curve is convex—as Fig. 2.82—$\mathrm{d}^2 J_r/\mathrm{d}a^2$ is zero or inherently negative. (A concave region might exist for full-thickness R curves as the initial resistance in flat fracture is augmented by a rapidly increasing proportion of shear lip.) Thus, slow growth at constant displacement according to eqn. (2.131) will be inherently unstable when the fracture initiation condition (Section 2.2.2.1.2) is satisfied and also $\mathrm{d}I/\mathrm{d}a$ is positive. A first guide to this is offered by examining $\mathrm{d}G/\mathrm{d}a|_q$ to which an assessment of $\mathrm{d}F_L/\mathrm{d}a$ can be added from eqn. (2.28) including the variation of constraint factor with crack length, if not negligible. The requirement that $\mathrm{d}G/\mathrm{d}a|_q$ be positive is that $\eta_{el} < 1$. This follows from eqn. (2.30b). For a material in which toughness does not increase with crack speed (or if it does, some average value of toughness is known), the work to fracture the ligament $U_f = G_c B(W - a)$. If the stored strain energy, $U_{el} = Fq/2$, is evaluated at the condition $G = G_c$ then $U_{el} = G_c B(W - a)/\eta_{el}$ so that if $\eta_{el} \leq 1$ the satisfaction of $G = G_c$ automatically satisfies the requirement that the total stored energy $U_{el} > U_f$; thus, in principle, initiation will be followed by propagation. In a rigorous fixed displacement situation with no other dissipations, propagation then appears to be unstable. (In reality there is an interplay

---

[23] In measuring elastic energy changes under constant (rising) load the total change in potential energy $\mathrm{d}P/\mathrm{d}a = -BG$ but the internal energy component $\mathrm{d}U_{el}/\mathrm{d}a = +BG$ since the loss of external potential energy is $-2BG$. Clearly J_r could be redefined to relate to U_{pl} only with this elastic component already subtracted as for the original proposed usage of eqn. (2.30).

between kinetic energy of separation, compliance effects and inertia of the loading system which controls whether or not propagation will be arrested.) Since $-\mathrm{d}F_{\mathrm{L}}/\mathrm{d}a$ is inherently positive, thus adding to G, crack growth at constant overall displacement will be unstable for all such cases. Indeed other cases for which η_{el} is not too much in excess of unity may still allow I to decrease at a lesser rate than $\mathrm{d}J_{\mathrm{r}}/\mathrm{d}a$ and thus also be unstable.

If eqn. (2.131) is evaluated for tension using I from eqn. (2.28), $q_{\mathrm{el}} = (\sigma D/E)(C_{\mathrm{n}}/C_{\mathrm{o}})$ where D is gauge length, C_{n} compliance of the notched piece and C_{o} the compliance of the unnotched piece, D/EBW. The gross stress σ must be expressed in terms of σ_{Y} via the relationships $\sigma = L\sigma_{\mathrm{Y}}(W - a)/W$ where L is the constraint factor, and $\mathrm{d}F_{\mathrm{L}}/\mathrm{d}a$ is evaluated from $F_{\mathrm{L}} = L\sigma_{\mathrm{Y}}B(W - a)$. The variation of L with a is discussed in Appendix 4. For simplicity $\mathrm{d}L/\mathrm{d}a$ is here neglected whence eqn. (2.131) becomes

$$\left(\frac{\eta_{\mathrm{pl}} + \eta_{\mathrm{el}}}{W - a}\right)G + \frac{\sigma_{\mathrm{Y}}^2 DLC_{\mathrm{n}}}{EWC_{\mathrm{o}}}(\eta_{\mathrm{pl}} - \eta_{\mathrm{el}}) \geq \frac{\mathrm{d}J_{\mathrm{r}}}{\mathrm{d}a} \qquad (2.133)$$

The two terms in G include the component G in the expression for I and the transference of $(\eta_{\mathrm{el}}/\eta_{\mathrm{pl}})G$ from the right side of eqn. (2.131). In Ref. [180] these terms do not arise and the term $C_{\mathrm{n}}/C_{\mathrm{o}}$ is also omitted for simplicity. The remaining terms, essentially $f(\sigma_{\mathrm{Y}}^2 D/EW) \geq \mathrm{d}J_{\mathrm{r}}/\mathrm{d}a$ correspond closely with eqn. (2.128a)[24] but the factor $(\eta_{\mathrm{pl}} - \eta_{\mathrm{el}})$ appears to have no direct counterpart in eqn. (2.128). To the extent that the terms in G and the factor $(\eta_{\mathrm{pl}} - \eta_{\mathrm{el}})$ tend to cancel, the final result might often be similar, but whilst the arguments leading to eqns. (2.128) and (2.131) are broadly compatible they do not appear to produce absolutely identical terms. A result similar to eqn. (2.133) with a similarity to eqn. (2.128b) can also be shown for three-point bending using $q_{\mathrm{el}} = (\sigma S^2/6EW)(C_{\mathrm{n}}/C_{\mathrm{o}})$ and $F_{\mathrm{L}} = L\sigma_{\mathrm{Y}}B(W - a)^2/S$, although for deep notches $\eta_{\mathrm{el}} \simeq \eta_{\mathrm{pl}} \simeq 2$ and the second term is small, leaving the term in G to dominate.

---

[24] The factor 2 arising in eqn. (2.128) from treating a centre crack of length $2a$ would arise here in forming $\mathrm{d}F/\mathrm{d}a$ from the ligament area $(W - 2a)$. Equation (2.128) [180] also assumes inequality (2.129) is satisfied. Strict usage of inequality (2.121) applied to J_{r} requires further study. If it is met the term in G (eqn. 2.131) appears negligible in relation to $(W - a)\,\mathrm{d}J/\mathrm{d}a$ provided η_{el} is not large) since $G \leq J \ll (W - a)\mathrm{d}J/\mathrm{d}a$. However, in eqn. (2.133) the corresponding term $\eta_{\mathrm{el}}G$ may well be comparable to $\eta_{\mathrm{pl}}G$ that arises from the first term in I(eqn. 2.28)). Hence, the combined term $(\eta_{\mathrm{el}} + \eta_{\mathrm{pl}})G$ might dominate eqn. (2.133) if $\eta_{\mathrm{el}} \simeq \eta_{\mathrm{pl}}$, thus causing the second term to be small, as, for example, in deep-notch bending with $S/W = 4$.

If, following Appendix 4, C_n/C_o is written

$$\frac{C_n}{C_o} = \frac{2Y^2 a(W - a)}{\eta D W} \qquad (2.134)$$

(replace D by $S/9$ for bending), where Y is the LEFM shape factor, then, still neglecting dL/da, eqn. (2.131) reduces to

(tension)

$$Y^2 L^2 \frac{a}{W} \cdot \frac{(W - a)}{W} \left[\eta_{pl} \left(1 + \frac{2}{\eta_{el}} \right) + \eta_{el} - 2 \right] \geq \frac{E}{\sigma_Y^2} \frac{dJ_r}{da} \qquad (2.135a)$$

(bending)

$$Y^2 L^2 \frac{a}{W} \left(\frac{W - a}{W} \right)^3 \left[\eta_{pl} \left(1 + \frac{9}{\eta_{el}} \right) + \eta_{el} - 9 \right] \geq \frac{E}{\sigma_Y^2} \frac{dJ_r}{da} \qquad (2.135b)$$

In all the foregoing the effect of external compliance can be included by taking an effective gauge length D_e or span S_e adjusted to suit any known machine or structural compliance. Typical values for testing machines might give an effective gauge length in tension of perhaps 50 or 200 W, according to size, thickness and modulus. Values of η_{el} can be assessed at least approximately, as outlined in Appendix 4. A few values of the left side of eqn. (2.135) have been assessed [181], ranging from about 25 for deep-notch bending to 4 for deep-notch, short-gauge-length tension, rising to 11 for moderate notch-depth, long-gauge-length tension. The deep-notch bending estimates show negligible effects of crack length whereas for tension the left side increases with crack length for the particular case studied. The right side of eqn. (2.125) might be of the order of 80 for low-strength steel near steady-state growth, or nearer 800 soon after initiation in flat fracture, or yet more full thickness (but correspondingly lower for high-strength steel based on the same J_r values, although for such materials the J curve may well be much lower than for the structural steel of Fig. 2.82). Clearly, for these figures slow growth at constant displacement is not possible and a sustained load to 'drive' the crack is necessary. If compliance is introduced to give an effective gauge length of perhaps 50 or 200 W the left side increases to a number such as 100 and growth at constant displacement is possible as $(E/\sigma_Y^2)(dJ_r/da)$ reaches a similar value, perhaps after a few millimetres of growth for the material shown here. By the previous arguments such growth at constant displacement would be unstable. The right side of eqn. (2.135) is again the 'tearing modulus' but in so far as it can relate to flat fracture or to full thickness and to any condition between initiation and quasi-steady-state growth it is not a material 'constant', even for a given thickness.

Clearly, further study of this problem is required, both in principle and in detailed evaluation of the theories that are now coming forward. In particular, comparison with experiments including known compliances is desirable. However, if the problem is restricted to growth with no change in micro-mode, techniques for making assessments of slow growth and final instability seem to be becoming available.

The use in these discussions of $\bar{r}\,dJ_r/da$ as the energy dissipation rate, initially equal to J_i but then decreasing as Δa increases (subject to some uncertainty on whether the 'process zone' $\bar{r}$ is constant), suggests that the picture of an R curve reducing with Δa, previously rejected, may after all be relevant. The minimum final absorption rate reached after appreciable growth would correspond to the rate for steady-state growth with plasticity in an infinite plate. In such a model it must be recalled that the dissipation rate at initiation, J_i, is provided mainly by external work rather than by the release of internal (strain) energy. The stable nature of the growth is expressed by the increase in the total work, U, as the crack grows, although the rate of dissipation decreases—for a given geometry—after initiation, given that increments of work, ΔU, cause successively larger increments of crack growth. The effect of time-dependent processes is ignored in these discussions but may be relevant even at room temperature in the final stages of crack growth in a ductile material.

In summary, if $J > J_i$ (or $\delta \geq \delta_i$) or some further improved form of such a statement) crack initiation occurs. If, in addition, $I < [(W - a)/\eta_{pl}]\,dJ_r/da$ slow growth occurs only under rising load, but if $I \geq [(W - a)/\eta_{pl}]dJ_r/da$ then growth occurs at fixed displacement and if the J_r curve is convex the growth is unstable, certainly for configurations for which $\eta_{el} < 1$ and probably for some other cases not here defined where eqn. (2.132b) is met with dI/da less negative than $f(d^2J_r/da^2)$. To be effective, these inequalities must be met collectively since, for example, in satisfying eqn. (2.131) without meeting $J \geq J_i$ or the equivalent, crack initiation (from an existing defect) would not occur. The possibility of slow growth in one micro-mode leading to another exists if, as the crack accelerates, strain rate effects modify the material toughness, yield stress or modulus. In particular a more catastrophic growth (e.g. tearing in micro-void coalescence leading to rapid propagation in cleavage) is not precluded. As discussed in Section 2.4.2, the balance of the factors such as hardening and loss of constraint after yielding cannot yet be quantified. In terms of the resistance curve, such a change in mode would imply a sudden decrease in dJ_r/da, presumably at any instant during growth, i.e. at any value of Δa, so that stable slow growth would more readily become unstable. It appears that if this is to happen without an actual increase in load the given mode of slow growth must be on the margin of stability in order that a small quasi-static growth step can allow crack acceleration sufficient to induce a strain rate effect. Alternatively, a crack could grow into a zone of less tough material that might be sufficient to tip the balance of stability.

2.4.2 Size Effect

A true understanding of the role of size in fracture processes requires an insight into micro- as well as macro-aspects of fracture. Although a micro-mechanistic study of fracture is outside the scope of this book (see for example Ref. [182]), some attention will be paid in what follows to micro-structural features notably to explain the well-known size dependence of the cleavage–shear transition briefly touched upon in Section 1.2.

If a term such as G (interpreted very loosely as K^2/E'), J, $m\sigma_Y\delta$ or any other similar fracture mechanics' parameter is to be related to local stress at the crack tip then, for reasons of dimensional analysis, the grouping of terms to express the local stress, $_l\sigma$, in terms of G (or J or δ) is

$$\frac{_l\sigma}{\sigma_Y} = f\left(\frac{GE}{\sigma_Y^2 a}\right) \quad \left(\text{or} \frac{JE}{\sigma_Y^2 a} \text{ or} \frac{m\delta}{\varepsilon_Y a}\right) \qquad (2.136a)$$

where a is some representative length, not necessarily crack length itself. The form of the function f has been discussed in the foregoing for the various concepts of concern. The essential point here is that if it is demonstrated experimentally that, to within some agreed variation, $G = G_c$ at fracture (or, correspondingly, J_c or δ_c) then, for a given modulus, yield strength and degree of plane strain,

$$_l\sigma = f(1/\text{linear size}) \qquad (2.136b)$$

If the criterion of separation is indeed a local condition, such as stress, if the length dimension indeed scales with size then this critical stress must itself vary with test piece size. This seems unlikely. A more rational alternative is that there is some micro-structural unit of size or 'process zone', say $\bar{r}$, over which the criterion has to be met in some average sense. This concept is not new. Such ideas were at the heart of Neuber's analysis [183] and more recently Krafft [184] has used the term 'process zone'. Ritchie *et al.* [185] also write 'one must consider not only the value of σ_f but also the size scale over which the criteria is to be met', where σ_{f} is the value of a local stress to cause cleavage fracture. The philosophical implications are not pursued but the practical consequences are outlined below.

Let there be a process zone, extending $\bar{r}$ in front of a crack tip. In a material fracturing with a small plastic zone at the tip, even in the LEFM regime the average stress over the plastic zone, r_{pl}, is $m\sigma_{\mathrm{Y}}$. The precise value of m depends on the constraint and will not be argued here. If it is argued that fracture could occur with $\bar{r} \geq r_{\mathrm{pl}}$ then the mechanism for separation must be occurring, at least in part, in elastic material outside the plastic zone. There may perhaps be circumstances where, for very brittle materials, the micro-plasticity or occasional dislocation movement in a so-called elastic region could be an adequate mechanism for fracture. If it is accepted that most fracture mechanisms, even cleavage, require some prior plastic deformation then the implication is that at fracture $\bar{r} \leq r_{\mathrm{pl}}$ (for LEFM $r_{\mathrm{pl}} = K^2/2\pi m^2\sigma_{\mathrm{Y}}^2$) and the mean stress at fracture in the zone $\bar{r}$ is

$$\bar{\sigma}_{\mathrm{f}} = m\sigma_{\mathrm{Y}} \qquad (2.137)$$

If this stress were too small to cause fracture then the particular micromechanism involved could not occur. Obviously σ_{Y} may be a function of strain rate and temperature so a particular mechanism could occur in some circumstances but not in others. In the present context, for LEFM conditions K describes not so much the square root singularity, since that is elastic and hence not directly responsible for the fracture process, but rather K describes the size of the plastic zone which must exist for the particular fracture mechanism to operate.

Consider increasing the load on two test pieces, one large, one small, for otherwise identical conditions. A stress, σ, local to the tip will be established early in the test at the value $m\sigma_{\mathrm{Y}}$ where m is appropriate to the constraint (assumed the same for both sizes). If this is equal to σ_{f} for some mode of separation, then fracture will occur by that mode when $r_{\mathrm{pl}} \geq \bar{r}$. For an estimate by LEFM this will be when

$$K^2/2\pi m^2\sigma_{\mathrm{Y}}^2 = EG/2\pi m^2\sigma_{\mathrm{Y}}^2 = r_{\mathrm{pl}} \geq \bar{r}$$

i.e. when

$$G_{\mathrm{c}} = 2\pi m^2\sigma_{\mathrm{Y}}^2\bar{r}/E \qquad (2.138)$$

This value of G_{c} will be reached earlier in the test on the large piece (i.e. at a lower remote stress, $_{\mathrm{r}}\sigma$) than on the small since $G \sim {}_{\mathrm{r}}\sigma^2 a$. Similar trends emerge if other parameters, J or COD are used, although the particular relationships obviously differ. For the large pieces $_{\mathrm{r}}\sigma$ may be below the gross yield level and in the small piece, above it so that a 'ductility transition' (i.e. from below to above gross yield behaviour with no change in fracture mode) is seen to be size dependent, as is well known. If the value $m\sigma_Y$ is itself less than $\bar{\sigma}_f$ then there appears to be four conflicting actions as deformation continues: σ_Y is increased by work-hardening; m is reduced slightly with increased degree of yielding (for full plane strain—see for example Section 2.2.2.1.3); in practice some degree of plane strain may be lost because of large through thickness deformation, and $\bar{\sigma}_f$ might itself be a function of degree of deformation. If the above argument were then conducted on the edge of a regime where the micro-mode of fracture might change, it seems that the balance of the four uncertainties would determine whether or not the one piece failed in one micro-mode and the other in another mode in addition to passing through the ductility transition. In short, fracture mechanics tells that if other variables such as temperature and strain rate remain fixed the ductility transition has a primary dependence on size but the appearance transition only a secondary dependence, with work-hardening and perhaps loss of constraint or change in local strain rate deciding the balance.

It may therefore be concluded that the various versions of fracture mechanics, in so far as they are couched in terms of the same dimensional groups, predict similar broad trends of behaviour and cannot show *a priori* whether a change in micro-mode will be size dependent. Indeed, it points to an interaction of events that may have no unique answer. Comment on this has been offered in the full light of recent advances in fracture mechanics by Ref. [186], where the balance of evidence is taken to show that the shear cleavage transition is size dependent, and Ref. [187], where it is taken that the effects are of the 'weakest link' type so that a sample of small pieces will inevitably show a wider scatter than a sample of large pieces and that the differences found are indeed encompassed by this scatter rather than by a trend with size.

It may also be remarked that the one-parameter versions of fracture mechanics discussed in the foregoing sections cannot be relevant if the process zone, $\bar{r}$, is greater in size than the regime defined by the parameter in question. The core of the interaction between the micro and macro views of fracture thus seems to be, firstly, is there a unique condition at the tip of crack for a range of various geometries and stress fields and, secondly, does the unique condition extend a sufficient distance from the crack tip so that reaching a critical magnitude of the field parameter ensures that a certain local intensity is satisfied over the extent necessary for the fracture mechanism to operate. If the condition being considered is not severe enough then deformation can be continued until another mechanism operates. Thus if the yield stress is low such that eqn. (2.137) cannot be met, then even if $r_{pl} > \bar{r}$, fracture will not occur by that mechanism, However, as seen on p. 85, local strain is still increasing rapidly with continued deformation so that some strain controlled criterion, again for dimensional reasons averaged over a now different process zone, may become operative. When that condition cannot on the average be met, some yet larger scale of failure such as plastic collapse of the remaining ligament ensues. Only micro-structural evidence can say whether or not any particular material has the micro-strength and micro-structural mechanisms relevant to some particular scale of event controlling separation. In this broad way, all sharp notch fracture mechanics concepts tell the same story and it is only when the finer details of geometrical or stress field effects are looked at, for a given micro-mode of fracture, that one theory may appear to be more relevant than some other.

So far the discussion has been qualitative only and the value of the constraint factor m in eqn. (2.137) left unspecified. On the other hand, Chapter 1 has brought out the need to define a degree of plane strain sufficient to give apparent toughness values that will not reduce further with increasing thickness and has quantified the resulting minimum thickness requirement by the relationship $B \geq 2 \cdot 5\ (K_{Ic}/\sigma_Y)^2$.

Similarly, a minimum thickness requirement $B \geq (25\ \text{or}\ 50)\ (J_{Ic}/\sigma_Y)$ has been mentioned for the J tests to be described in Chapter 3. If J_{Ic} is identified with G_{Ic} then this criterion becomes $B \geq (25\ \text{or}\ 50)\ (K_{Ic}/\sigma_Y)^2$ (σ_Y/E) which is roughly 10 σ_Y/E smaller than the LEFM requirement. In other words the two requirements differ roughly by an order of magnitude. The difference in test details may account for some of this; namely, for J testing the first sign of crack growth is sought at the mid-section only whereas complete fracture instability or at least a measurable 'pop in' is sought in LEFM and some slow growth, small on a percentage scale but significant in terms of absolute value of δa, is permitted. Yet the extent of the discrepancy between the size requirements and the fact that the two tests apparently do not compare like to like foster doubts about the exact nature of the size effect on the fracture processes, underlining the need for understanding on the micro-scale cited above.

Size effects also appear to influence the extent of slow stable crack growth, as evidenced by the study on COD by Griffis mentioned before [65], in which attention was given to the use of small test pieces. The results for δ_i are shown in Fig. 2.83a for HY–80 ($2\frac{1}{2}\%$ Ni–Cr–Mo steel) as a function of thickness B for three-point bend tests ($a/W = 0 \cdot 5$; $S/W = 4$).

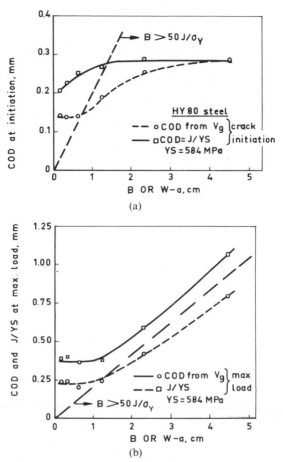

FIG. 2.83. Effect of specimen size on critical COD: (a) at initiation; (b) at maximum load as inferred from surface clip gauge reading; V_g or from J/σ_Y. (Data from Ref. [65].)

For the four thicker pieces, $B \simeq a$. For the two thinnest $B \sim a/2$ and $a/4$, respectively. COD was derived from surface clip gauge measurements not in strict accordance with Ref. [8] but using a rotational factor of 0·4. COD was also derived from $J = m\sigma_Y\delta$ with $m = 1$. In both methods δ was evaluated for no slow crack growth.

Figure 2.83b from the same reference shows δ_m at maximum load. It will be noted that δ_i is roughly constant for the thicker section but, perhaps surprisingly, rather lower for the thinner sections, the more so for the clip gauge derivation. If the two derivations were to be made coincident a value

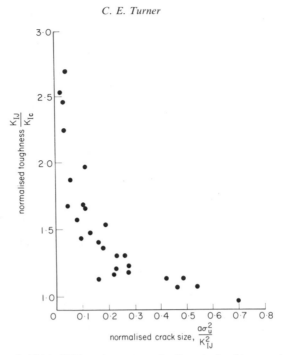

FIG. 2.84.   Effect of width in SEN tension tests on J_{Ic}. Quenched and tempered bainitic steel near the ductile–brittle transition temperature [188]. $K_{IJ} \equiv R = \sqrt{E'J}$ (eqn. (2.25), evaluated at fracture).

of m of about 1·6 would be required for the thinner pieces, dropping to unity for the thicker ones. Conversely, δ_m appears constant for the thin pieces and rises nearly four-fold for the thick pieces implying a factor of $m = 1·6$ for thin pieces dropping to 1·4 for thick, if the two results are to be in agreement. It will be seen that the amount of slow growth as indicated by δ_m/δ_i is greater for the thick pieces ($\delta_m/\delta_i \simeq 2·5$) than for the thin ($\delta_m/\delta_i \simeq 1·7$).

The effect of size has also been studied in Ref. [188] where J_{Ic} appears to be a function of width (Fig. 2.84). The trend is related at least qualitatively to a greater loss of plane strain in small pieces that fail after a gross yield than in large pieces that fail after only moderate yield with correspondingly less loss of plane strain, much as outlined earlier. It may well be that the evidence that plane-strain deformation is maintained over the central region of a small test piece [11] may be adequate for certain fracture modes but not sufficiently rigorous at the borderline of transition from one mode to another. This effect of size is also presented as a change in the ductile–brittle transition temperature. It is concluded [188] that attainment of J_{Ic} is a necessary macroscopic condition for fracture but not necessarily sufficient at the microscopic level. It is not, however, clear to the present writer how closely the J_{Ic} values in Fig. 2.84 describe a true initiation event to which J_I might best relate, or some final fracture event subsequent to a certain amount of slow crack growth, with transition to cleavage failure in those cases where constraint is maintained. The latter circumstance is quite consistent with the general viewpoints of fracture mechanics, whereas the former would imply the need for a fully three-dimensional treatment of elastic–plastic problems and the loss of much small-scale test capability.

If the data of Figs. 2.47 and 2.48 are examined it is found that to reach substantial plasticity yet not complete collapse (for example, by defining a position 'round the elbow' of the load-deflection diagram with a secant modulus, say two-thirds of the elastic line), then in some cases, for example, shallow-notch bending, $JE/\sigma_Y^2 a$ increases as a/W decreases and the deflection $qE/W\sigma_Y$ is surprisingly independent of a/W ratio for deep or shallow notch depths. Thus, although $JE/\sigma_Y^2 W$ decreases as a decreases at *constant W*, $JE/\sigma_Y^2 a$ increases as W increases at *constant a*. This latter circumstance might be relevant if a minimum detectable (constant) size of crack were considered irrespective of the component size. These trends reverse for shallow-crack tension where (Fig. 2.47) the admittedly rather small trends with a/W for tension are opposite to those for bending. In LEFM these trends are normalised by the shape factor Y but, as noted in Section 2.2.2.1.4, the Y factor does not eliminate these trends in plasticity so that when one term (a or W) is held constant, various size effects appear as the other term is altered.

In conclusion it may be stated that our present understanding of size effects in fracture is not commensurate with their importance in fracture toughness testing and design.

2.4.3 Three-Dimensional Problems

A three-dimensional elastic–plastic computational ability exists for finite-element [189] and finite-difference [190] methods, but results have not been published extensively, presumably because of the computational costs. A strict usage of J in the form of the present contour integral is not possible. Although it is possible to run contours in a two-dimensional section, there is a flow of energy across the 'ends' of the cylinder of which the contour is a cross-section.

The evaluation of J from dP/da is still feasible, but the interpretation is, of course, in a characterising role and not as energy. This procedure was suggested [80] for axi-symmetric problems. In the general three-dimensional case, some mean value would be implied. By advancing nodal points on a limited part of the crack periphery in a finite-element model, a more or less local value can be inferred, as proposed by Parks [191]. Applications so far appear to be in two-dimensional elastic–plastic or three-dimensional elastic problems, but extension to three-dimensional elastic–plastic studies seems implied.

If small elements or special singularity elements are used, then COD could be modelled, but this does not seem to have been the case so far. Figure 2.85 shows the extent of plastic zone in the surface and through the minor axis for a semi-elliptic crack [189]. The degree of plane strain is expressed for the elastic calculation[25] as $\sigma_{\theta\theta}/v(\sigma_{xx} + \sigma_{yy})$ and reaches > 0.9 from the minor axis round the elliptical contour for $60°$. From $30°$ around to the major axis (i.e. the face of the plate) the ratio drops from 0.9 to zero (plane stress). On a section only $5°$ in from the free face the $\sigma_{\theta\theta}$ ratio is 0.5, i.e. 'half-way to plane strain'. Unfortunately, similar data are not given for the elastic–plastic loading. The computation is carried to limit load ($\sigma = 0.99\,\sigma_Y$, non-hardening) where, as seen in Fig. 2.85, the plastic enclaves reach the far boundary. The data for various stresses still seem linear, with load and crack opening with K^2 so that it seems likely that $J \simeq G$ still and the 'elbow' of the load–deflection curve has barely been reached. Thus, although close to limit load, the overall displacements (not given) are probably still comparable in magnitude with the elastic terms. An interesting computational development is reported by Hu & Liu in Ref. [192], in which elements near the crack are treated in plane strain whilst elements remote from the crack are treated in plane stress. A transition region is fixed in Ref. [192] to match experiment, in which the stiffness of the elements is gradually relaxed. The overall load–deflection data are but little different from plane-stress computations but strains near the tip ($< B/2$ approximately, for the data in Ref. [192]) are significantly reduced (by up to about 30%) from the values computed in plane stress, in good agreement with measured surface strains.

It is not clear how these local strains would compare with the values from a full plane-strain computation. Relaxation of stress below that predicted by a plane-strain calculation is not discussed, nor are J or COD values described. The authors suggest that if the ligament $> 10B$ the outer region of plane stress will be found. It is interesting to speculate whether the effect

---

[25] $\sigma_{\theta\theta}$ is the stress tangential to the crack profile corresponding to 'through thickness' in the two-dimensional model.

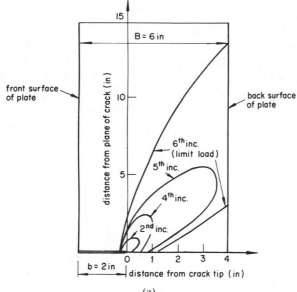

(a)

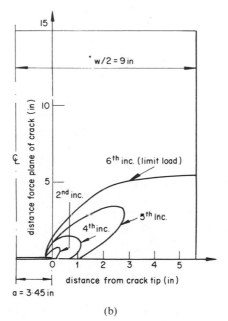

(b)

FIG. 2.85.   Extent of plastic zone in three-dimensional studies of a part-through crack: (a)
mid-plane along minor axis; (b) front surface.

is related to plastic zone size and hence some measurement in terms of COD or J, since even a square bar, unnotched, would normally be considered in plane stress. The use of mixed plane-stress/plane-strain elements has also been mentioned by Anderrson [193] who allowed the plastic zone to degenerate to plane stress whilst maintaining the remainder of the piece in plane strain, i.e. the reverse of Ref. [192]. Andersson also found that this improved the agreement between experiment and computation. In short, any relaxation from plane strain increases the agreement with overall data!

Andersson also showed that COD in three-point bend agreed well with plane-strain predictions, up to near what would be the non-hardening plane-stress collapse load, and thereafter following the plane-stress prediction. The mixed behaviour computation was not apparently continued into this regime near collapse where, by implication, the overall prediction would still have been nearer plane strain, while experiment followed plane stress.

The weight of evidence for other than thin sheet seems to support the Hu–Liu model of the tip being initially in plane strain while the overall limit load is in practice controlled by plane stress in the gross cross section, although as large geometry change occurs and plasticity becomes extensive, relaxation of plane strain near the crack tip will surely be found.

Usage of this hybrid type of computation seems a promising method for the better modelling of many structural configurations in the plastic regime without the complexity of a full three-dimensional treatment, at least for through-crack configurations.

As noted in Section 2.3.2, the BCS–Dugdale-type model has been extended to penny-shaped and semi-circular surface cracks [125], thereby making a substantial contribution to three-dimensional problems, since direct comparison with two-dimensional cases can now be made (although none has as yet come to the writer's attention). Comparison with previously published experimental results shows a very reasonable agreement (Fig. 2.86), although, as for most of the HSW data (Section 2.3.2), no distinction is made between initiation and final fracture.

Kawai et al. [194] have proposed a variant of the finite-element method for three-dimensional elastic–plastic problems. Rigid elements are connected by springs distributed over the surfaces, or, for non-linear problems, by springs and dashpots. Criteria can be used to modify the spring or dashpot characteristics to simulate crack initiation in a COD-type model. Some preliminary results for plastic zone development in V-notch bars are shown in Ref. [194] and it is suggested that considerable computing

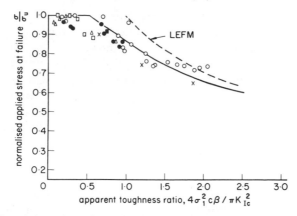

FIG. 2.86. Comparison of experimental and BCS–Dugdale-type model predictions for semi-circular surface flaws [125]. △ Titanium 6AL-4V, ● Aluminium 2014-T651, × Alloy steel 4340, ■ Aluminium 2014-T6, □ Aluminium ($-250\,^{\circ}$C) 2014-T6 (from (1969). NASA Tech. note D5340). ○ Rotor steel 1CMV (from (1973). C. A. Parsons Ltd. Tech. Memo MET 73-154).

time is saved because the stiffness matrix for the rigid-element spring system is much smaller than for conventional finite-element systems.

2.4.4 Material Selection

The choice of material is governed by many factors and there are probably few cases where fracture toughness is the most important feature. However, in preliminary assessments of a particular usage it may be desirable to have at least a rough yardstick. Just as $(K_{\mathrm{Ic}}/\sigma_{\mathrm{Y}})^2$ has been used as a parameter to assess the toughness ranking of materials, so $\delta_{\mathrm{c}}/\varepsilon_{\mathrm{Y}}$ can similarly be used in the yielding regime. This follows since, using the relationships (1.5) and (2.4), we obtain

$$\delta_{\mathrm{c}}/\varepsilon_{\mathrm{Y}} = (K_{\mathrm{Ic}}/\sigma_{\mathrm{Y}})^2 \qquad (2.139)$$

so that, for an applied stress level that is a given fraction of yield stress (but not for a given absolute value of stress), the size of a defect to which the material is tolerant is proportional to these terms. The equivalent group using J is $J_{\mathrm{c}}E/\sigma_{\mathrm{Y}}^2 a$. If it is desired to attach a numerical value to this term to 'ensure ductile behaviour' in some loosely specified application, then, from Fig. 2.48 for $\varepsilon/\varepsilon_{\mathrm{Y}} \geq 1\cdot2$, $JE/Y^2\sigma_{\mathrm{Y}}^2 a \geq 2\cdot5\varepsilon/\varepsilon_{\mathrm{Y}}$. Taking $Y^2 = 4$, then 'safe' usage with shallow cracks and a nominal applied strain of up to $2\varepsilon_{\mathrm{Y}}$ requires $J_{\mathrm{c}}E/\sigma_{\mathrm{Y}}^2 > 20a$. For $E/\sigma_{\mathrm{Y}} = 200$ (high-strength steels and light alloys) this requires $J_c/\sigma_Y > a/10$; for $E/\sigma_Y = 800$ (low-strength steels) $J_c/\sigma_Y > a/40$. Alternatively, if $J_c E$ is equated to K_{Ic}^2, then a toughness such that $(K_{Ic}/\sigma_Y)^2 > 20a$ is implied. If a is to be the order of tens of millimetres then clearly, for structural steels, behaviour on the upper shelf of a temperature transition curve is implied, for example, with K_{Ic}MNm$^{-3/2}$ numerically equal to σ_YMNm^{-2}, for $a_{crit} = 50$ mm. This is not to say that usage below the upper shelf is 'unsafe', but that the circumstances must be examined in more detail than the broad brush estimate given here, and the effect of fabrication or possible service embrittlement on the toughness must be considered.

Although such procedures are very useful for ranking materials and checking susceptibility to damage or embrittlement during fabrication or in service it will be clear from what has been said in previous sections that selection of material on an absolute basis depends on circumstances other than the choice of elastic–plastic parameter, and indeed on aspects other than fracture resistance. Major factors that affect a fracture assessment are: the choice of J (or δ) at initiation or at some later point allowing for stable crack growth; for many structural steels acceptance of toughness values from parent plates or from weldments; where a change of micro-mode seems possible, the still-disputed question of full-thickness or small-scale testing and use of static or dynamic results; and the appropriate treatment for residual and thermal stresses and indeed of many three-dimensional cases.

Indeed McClintock's argument [195] based on slip-line solutions, that the different degrees of constraint associated with various in-plane configurations would prevent the satisfactory development of a one-parameter theory of yielding fracture mechanics, has still not been fully resolved. As discussed in Section 2.2.2.1.3 (v), differences seem to emerge after very extensive yield whilst a one-parameter description, J or δ, is adequate for less extensive yield.

Thus the relation between the degree of ductility adequate for one situation and that for another cannot yet be stated in a fully quantitative manner for a general problem with arbitrary change of circumstance. There has been a very real improvement in the understanding of structural requirements and test techniques that is still continuing, but the interaction of configuration, methods of fabrication, service loading, and of course environmental variables that have not been discussed here at all, can still only be treated confidently yet not too conservatively if the relevant circumstances are carefully matched in the laboratory assessment of properties.

REFERENCES

1. Wells, A. A. (1961). Unstable crack propagation in metals: cleavage and fast fracture, *Symp. Crack Propagation*, College of Aeronautics, Cranfield, Paper B4.
2. Cottrell, A. H. (1961). Theoretical aspects of radiation damage and brittle fracture in steel pressure vessels, Iron Steel Inst. Spec. Rep. No. 69, pp. 281–96.
3. Barenblatt, G. I. (1962). The mathematical theory of equilibrium cracks in brittle fracture, *Advances in applied mechanics*, **7**, Academic Press, New York.
4. Wells, A. A. (1963). Application of fracture mechanics at and beyond general yielding, *Brit. Weld. J.*, **10**, pp. 563–70.
5. Dugdale, D. S. (1960). Yielding of steel sheets containing slits, *J. Mech. Phys. Solids*, **8**, pp. 100–4.
6. Burdekin, F. M. & Stone, D. E. W. (1966). The crack opening displacement approach fracture mechanics in yielding materials, *J. Strain Anal.*, **1**(2), pp. 145–53.
7. Bilby, B. *et al.* (1964). Plastic yielding from sharp notches, *Proc. Roy. Soc. (London)*, Ser. A, **279**, pp. 1–9.
8. British Standards Institution (1972). Draft for Development on Methods for Crack Opening Displacement (COD) Testing, DD19.
9. Ingham, T. *et al.* (1971). The effect of geometry on the interpretation of COD test data, *Practical application of fracture mechanics to pressure vessel technology*, Inst. Mech. Engrs., London, pp. 200–8.
10. IIW (Comm. X) UK Briefing Group (1976). Some proposals for dynamic toughness measurement, *Dynamic fracture toughness*, Weld. Inst./Am. Soc. Met., pp. 127–46.
11. Robinson, J. N. & Tetelman, A. S. (1974). Measurement of K_{Ic} on small specimens using critical crack tip opening displacement, *Fracture toughness and slow-stable cracking*, ASTM–STP 559, p. 139.
12. Robinson, J. N. (1976). An experimental investigation of the effect of specimen type on the crack tip opening displacement and *J*-integral fracture criteria, *Int. J. Fract.*, **12**, pp. 723–39.
13. Krafft, J. M. *et al.* (1961). Effects of dimensions on fast fracture instability in notched sheets, *Symp. Crack Propagation*, College of Aeronautics, Cranfield, Paper A1.
14. Nichols, R. W. *et al.* (1969). The use of critical crack opening displacement techniques for the selection of fracture-resistant materials, *Practical fracture mechanics for structural steel*, Proc. Conf. Culcheth, UKAEA/Chapman & Hall, London, Section F.
15. Smith, R. F. & Knott, J. F. (1971). Crack opening displacement and fibrous fracture in mild steel. In Ref. 9, pp. 65–75.
16. Harrison, T. C. & Fearnehough, G. D. (1969). The influence of specimen dimensions on measurements of the ductile crack opening displacement, *Int. J. Fract. Mech.*, **5**, pp. 348–9.
17. Burdekin, F. M. (1967). Initiation of brittle fracture in structural steels, *Brit. Weld. J.*, **14**, pp. 649–59.

18. Burdekin, F. M. & Taylor, T. E. (1969). Fracture in spherical pressure vessels, J. Mech. Engng Sci., 11(5), pp. 486–97.
19. Cowan, A. & Kirby, N. (1969). The application of COD measurement to large-scale test behaviour. In Ref. 14, Section D.
20. Kanazawa, T. et al. (1973). Study on evaluation of fracture toughness of structural steels using COD bend test, Int. Inst. Weld., Rep. IIW Doc X-702-73.
21. Urbensky, J. & Müncer, L. (1971). The evaluation of resistance towards brittle fracture initiation of welded joints of low-temperature steel. In Ref. 9, pp. 246–50.
22. Hayes, D. J. & Turner, C. E. (1974). Application of finite element techniques to post yield analysis of proposed standard three-point bend fracture test pieces, Int. J. Fract. Mech., 10, pp. 17–32.
23. Sumpter, J. D. G. (1973). Elastic–plastic fracture analysis and design using the finite-element method. Ph.D. Thesis, University of London.
24. Srawley, J. E. et al. (1970). On the sharpness of cracks compared with Wells' COD, Int. J. Fract. Mech., 6, pp. 441–4.
25. Wells, A. A. & Burdekin, F. M. (1971). Discussion of 'On the sharpness of cracks compared with Wells' COD' and Response by Srawley et al. (1971). Int. J. Fract. Mech., 7, pp. 242–6.
26. Rice, J. R. & Johnson, M. A. (1970). The role of large crack tip geometry change in plane strain fracture, Inelastic behaviour of solids, M. F. Kanninen et al., eds. McGraw-Hill, New York, pp. 641–72.
27. Harrison, J. D. et al. (1968). A proposed acceptance standard for welded defects based upon suitability for source, 2nd Conf. Significance of Defects in Welded Structures. Weld. Inst., London.
28. Burdekin, F. M. & Dawes, M. G. (1971). Practical use of linear elastic and yielding fracture mechanics with particular reference to pressure vessels. In Ref. 9, pp. 28–37.
29. Dawes, M. G. (1974). Fracture control in high yield strength weldments, Weld. J. Res. Suppl., 53, pp. 369s–79s.
30. Archer, G. L. (1975). The relationship between notch tip and notch mouth opening displacements in SENB fracture toughness specimens, Weld. Inst. Res. Rep. E63/75.
31. Dawes, M. G. (1976). The application of fracture mechanics to brittle fracture in steel weld metals. Ph.D. Thesis, CNAA, Weld. Inst.
32. Dolby, R. E. (1976). Factors controlling weld toughness—the present position. Weld. Inst. Rep. 14/1976/M.
33. Welding low-temperature containment plant. Weld. Inst., Nov., 1973.
34. Welding in off-shore constructions. Weld. Inst., Feb., 1974.
35. Dawes, M. G. (1970). Designing to avoid brittle fracture in weld metal, Mat. Const. and Brit. Weld. J., 2, pp. 55–9.
36. Masubuchi, K. et al. (1966). Interpretive report on weld metal toughness, Weld. Res. Council Bull., No. 111.
37. Dolby, R. E. & Saunders, G. G. (1972). Subcritical HAZ fracture toughness of C:Mn steels, Mat. Const. and Brit. Weld. J., 4, pp. 185–90.
38. Kihara, H. et al. (1969). Effects of notch size, angular distortion and residual stress on brittle fracture initiation characteristics of welded joints for high-strength steels, Int. Inst. Weld. Rep. IIW Doc X-508-69.

39. Hall, W. J. *et al.* (1967). *Brittle fracture of welded plate*, Prentice-Hall, Englewood Cliffs.
40. Egan, G. R. (1972). Application of yielding fracture mechanics to the design of welded structures. Ph.D. Thesis, University of London.
41. Barr, R. R. & Terry, P. (1975). Application of fracture mechanics to the brittle fracture of structural steels, *J. Strain Anal.*, **10**(4), pp. 233–42.
42. Barr, R. R. & Burdekin, F. M. (1976). Design against brittle fracture, Rosenhain Centenary Conf. *The Contribution of Physical Metallurgy to Engineering Practice*. Roy. Soc., London, pp. 149–66.
43. Assessment of size of defects, Int. Inst. Weld. Comm. X. Doc. XWGSD-13 (1974).
44. BSI (1976). Draft standard rules for the derivation of acceptance levels for defects in fusion welded joints. Draft for Public Comment, Feb.
45. Turner, C. E. & Burdekin, F. M. (1974). Review of current status of yielding fracture mechanics, *Atomic Energy Review*, **12**(3), pp. 439–503.
46. Wells, A. A. (1968–70). Crack opening displacements from elastic–plastic analyses of externally notched tension bars, *Engng Fract. Mech.*, **1**(3), pp. 399–410.
47. Turner, C. E. (1972). A unification of the *J* contour integral and COD criterion for fracture by use of in-plane constraint factors, CODA Panel of Navy Dept. Adv. Comm. on Struc. Steel, Jan.
48. Prager, W. & Hodge, P. G. (1951). *Theory of perfectly plastic solids*, Wiley, New York.
49. Hayes, D. J. & Williams, J. G. (1972). Practical method for determining Dugdale model solutions for cracked bodies of arbitrary shape, *Int. J. Fract. Mech.*, **8**, pp. 239–56.
50. Chell, G. G. (1976). Bilby, Cottrell and Swinden model solutions for centre and edge cracked plates subject to arbitrary model loading, *Int. J. Fract.*, **12**, pp. 135–47.
51. Eshelby, J. D. (1956). A continuum theory of lattice defects. In *Progress in solid state physics*, **3**, Academic Press, New York, p. 79.
52. Cherepynov, G. P. (1967). Crack propagation in continuous media, *Appl. Math. Mech.*, **31**, pp. 503–12.
53. Rice, J. R. (1968). A path-independent integral and the approximate analysis of strain concentration by notches and cracks, *J. Appl. Mech.*, **35**, pp. 379–86.
54. Rice, J. R. (1968). Mathematical analysis in the mechanics of fracture. *Fracture: an advanced treatise*, **2**, H. Liebowitz, ed. Academic Press, New York, Chapter 3.
55. Bergkvist, H. & Huong, G. L. L. (1977). Energy release rates in cases of axial symmetry, *Int. Conf. Fract. Mech. and Techn.*, Hong Kong, March. (See also *Int. J. Fract.*, **13**, p. 556.)
56. Hutchinson, J. W. (1968). Singular behaviour at end of tensile crack in hardening material, *J. Mech. Phys. Solids*, **16**, pp. 13–31.
57. Rice, J. R. & Rosengren, G. F. (1968). Plane strain deformation near crack tip in power-law hardening material, *J. Mech. Phys. Solids*, **16**, pp. 1–12.
58. McClintock, F. (1968). Plasticity aspects of fracture. *Fracture: an advanced treatise*, **3**, H. Liebowitz, ed. Academic Press, New York, Chapter 2.
59. Hill, R. (1950). *Mathematical theory of plasticity*, Oxford University Press, London.

60. Rice, J. R. (1965). An examination of the fracture mechanics energy balance from the point of view of continuum mechanics, ICF-1, Sendai, A309–40.
61. Begley, J. A. & Landes, J. D. (1972). The J integral as a fracture criterion in fracture toughness testing. *Fracture toughness*, ASTM–STP 514, pp. 1–23.
62. Landes, J. D. & Begley, J. A. (1972). The effect of specimen geometry on J_{Ic}, *Fracture toughness*, ASTM–STP 514, pp. 24–39.
63. Egan, G. R. (1973). Compatibility of linear elastic K_{Ic} and general yielding (COD) fracture mechanics, *Engng Fract. Mech.*, **5**, pp. 167–85.
64. Kanazawa, T. *et al.* (1975). A preliminary study on the J integral fracture criterion, Int. Inst. Weld. Rep. IIW Doc. X-779-75.
65. Griffis, C. A. (1975). Elastic–plastic fracture toughness: a comparison of J-integral and crack opening displacement characterisations, *J. Press. Vess. Tech.*, *ASME J*, **97**(4), pp. 278–83.
66. Hayes, D. J. (1970). Some applications of elastic–plastic analysis to fracture mechanics. Ph.D. Thesis, University of London.
67. Swedlow, J. L. *et al.* (1965). Elastic–plastic stresses and strains in cracked plates, ICF-1, Sendai, pp. A229–82.
68. Marcal, P. V. & King, I. P. (1969). Elastic–plastic analysis of two-dimensional stress systems by the finite element method, *Int. J. Mech. Sci.*, **9**, pp. 143–55.
69. Boyle, E. F. (1972). Calculation of elastic and plastic crack extension forces. Ph.D. Thesis, The Queen's University, Belfast.
70. Riccardella, P. C. & Swedlow, J. L. (1974). A combined analytical–experimental fracture study. *Fracture analysis*, ASTM–STP 560, pp. 134–54.
71. Lyall-Saunders, J. (1960). On the Griffith–Irwin fracture theory, *J. Appl. Mech.*, **27**, pp. 352–3.
72. Bucci, R. J. *et al.* (1972). J integral estimation procedures, *Fracture toughness*, ASTM–STP 514, pp. 40–69.
73. Sumpter, J. D. G. & Turner, C. E. (1976). Use of the J contour integral in elastic–plastic fracture studies by finite-element methods, *J. Mech. Engng Sci.*, **18**, pp. 97–112.
74. Andersson, H. (1974). Finite element treatment of a uniformly moving elastic–plastic crack tip, *J. Mech. Phys. Solids*, **22**, pp. 285–308.
75. Kobayashi, A. S. *et al.* (1973). A numerical and experimental investigation on the use of the J-integral, *Engng Fract. Mech.*, **5**, pp. 293–305.
76. Light, M. F. *et al.* (1976). Some further results on slow crack growth predictions by a finite element method, *Int. J. Fract.*, **12**, pp. 503–4.
77. Kfouri, A. P. & Miller, K. J. (1976). Crack separation energy rates in elastic–plastic fracture mechanics, *Proc. Inst. Mech. Engrs*, **190** (48/76), pp. 571–84.
78. Luxmoore, A. R. & Morgan, K. (1977). A re-examination of the fracture mechanics energy balance for elastic–perfectly plastic materials, *Int. J. Fract.*, **13**, pp. 553–5.
79. Rice, J. R. (1976). Elastic–plastic fracture mechanics. In *The mechanics of fracture*, F. Erdogan, ed. ASME, pp. 23–54.
80. Rice, J. R. *et al.* (1973). Some further results of J integral analysis and estimates. *Progress in flaw growth and toughness testing*, ASTM–STP 536, pp. 231–45.

81. Underwood, J. H. (1976). J_{Ic} test results from two steels. *Cracks and fracture*, ASTM–STP 601, pp. 312–29.

82. Turner, C. E. (1973). Fracture toughness and specific fracture energy: a re-analysis of results, *Mat. Sci. Engng*, **11**, pp. 275–82.

83. Srawley, J. (1976). On the relation of J_I to work done per unit uncracked area, *Int. J. Fract.*, **12**, pp. 470–4.

84. Merkle, J. G. & Corten, H. T. (1974). A *J*-integral analysis for the compact tension specimen considering axial force as well as bending effects, *J. Press. Vess. Tech.*, *ASME J*, **96**(4), pp. 286–92.

85. Sumpter, J. D. G. & Turner, C. E. (1976). Method for laboratory determination of J_c, ASTM–STP 601, pp. 3–18.

86. Landes, J. D. & Begley, J. A. (1974). Test results from *J*-integral studies: an attempt to establish a J_{Ic} testing procedure, *Fracture analysis* ASTM–STP 560, pp. 170–86.

87. Wou, C. K. S. (1972). Fracture mechanics and the relation between *J* and strain energy in three-point bending. M.Sc. Thesis, University of London.

88. Dawes, M. G. (1975). *J* estimation procedures for weldments, Weld. Inst. Res. Rep. E65/75.

89. Chipperfield, C. G. (1976). A method for determining dynamic J_Q and δ_i values and its application to ductile steels. In Ref. 10, pp. 169–80.

90. Rice, J. R. (1974). Limitations to the small-scale yielding approximation for crack tip plasticity, *J. Mech. Phys. Solids*, **22**, pp. 17–26.

91. Ewing, D. F. & Hill, R. (1967). The plastic constraint of V-notched tension bars, *J. Mech. Phys. Solids*, **15**, pp. 115–25.

92. Ewing, D. J. F. & Richards, C. E. (1974). The yield-point loads of singly notched pin-loaded tensile strips, *J. Mech. Phys. Solids*, **22**, pp. 27–36.

93. Green, A. P. & Hundy, B. G. (1956). Initial plastic yielding in notch bend tests, *J. Mech. Phys. Solids*, **4**, pp. 128–45.

94. Alexander, J. M. & Komoly, T. J. (1962). On yielding of rigid/plastic bar with Izod notch, *J. Mech. Phys. Solids*, **10**, pp. 265–75.

95. Hilton, P. D. & Sih, G. C. (1973). Application of the finite element method to the stress intensity factors, *Mechanics of fracture*, **1**, *Methods of analysis and solutions of crack problems*. G. C. Sih, ed. Noordhoff Int. Publ., Leyden, p. 426.

96. Miller, K. J. & Kfouri, A. P. (1974). Elastic–plastic finite element analysis of crack tip fields under biaxial loading conditions, *Int. J. Fract. Mech.*, **10**, pp. 393–404.

97. Turner, C. E. (1978). An analysis of the fracture implications of some elastic–plastic finite-element studies, *Numerical methods in fracture mechanics*. A. R. Luxmoore & M. J. Owen, eds. University College of Swansea, pp. 569–80.

98. McMeeking, R. M. & Parks, D. M. (1977). A criterion for *J*-dominance of crack tip fields in large scale yielding, *ASTM Symposium on Elastic–Plastic Fracture*, Atlanta, Nov.

99. Harrison, J. D. (1977). A comparison between four elasto-plastic fracture mechanics parameters, *Int. J. Pres. Ves. & Piping*, **5**, pp. 261–74.

100. Sumpter, J. D. G. & Turner, C. E. (1973). Fracture analysis in areas of high nominal strain, *Proc. 2nd Int. Conf. Pressure Vessel Technology*, **2**, ASME, New York, pp. 1095–104.

101. Sumpter, J. D. G. & Turner, C. E. (1976). Design using elastic–plastic fracture mechanics, *Int. J. Fract.*, **12**, pp. 861–73.
102. van de Ruijtenbeek, M. G. & Broekhoven, M. J. G. (1977). An application of *J*-integral concept to cracks emanating from a hole, SMIRT-4 (San Francisco), Paper G3/4.
103. Liebowitz, H. & Eftis, J. (1971). On non-linear effects in fracture mechanics, *Engng Fract. Mech.*, **3**, pp. 267–81.
104. Eftis, J. *et al.* (1975). On fracture toughness in the non-linear range, *Engng Fract. Mech.*, **7**, pp. 491–503.
105. Shih, C. F. (1974). Small-scale yielding analysis of mixed node plane-strain crack problems, *Fracture analysis*, part II, ASTM–STP 560, pp. 187–210.
106. Muscati, A. & Turner, C. E. (1977). Post-yield fracture behaviour of shallow-notched alloy steel bars in three point bending. *Fracture mechanics in engineering practice*. P. Stanley, ed. Applied Science, London, pp. 289–304.
107. Chell, G. G. & Spink, G. M. (1977). A post-yield fracture mechanics analysis of three-point bend specimens and its implications to fracture toughness testing, *Engng Fract. Mech.*, **9**, pp. 101–23.
108. Milne, I. & Worthington, P. J. (1976). The fracture toughness of a low alloy pressure vessel steel in the post yield regime, *Mat. Sci. and Engng*, **26**(2), pp. 185–95.
109. Begley, J. A. *et al.* (1974). An estimation model for the application of the *J* integral. In Ref. 70, pp. 155–69.
110. Dawes, M. G. (1977). Elastic–plastic fracture toughness based on the COD and *J* integral concepts, *ASTM Symposium on Elastic–Plastic Fracture*, Atlanta, Nov.
111. Shabbits, W. O. *et al.* (1971). Fracture toughness of ASTM A533 Grade B Class 1 heavy section submerged arc weldments, *J. Bas. Engng, ASME D*, **93**, pp. 231–6.
112. Blackburn, W. S. (1972). Path independent integrals to predict onset of crack instability in an elastic–plastic material, *Int. J. Fract. Mech.*, **8**, pp. 343–6 and Correction, **9**, (1973), p. 122.
113. Hellen, T. K. (1976). Finite-element energy methods in fracture mechanics. Central Elec. Gen. Board, Berkeley Nuclear Labs., Mar.
114. Hellen, T. K. (1976). Use of the J^* integral for non-linear fracture mechanics. Central Elec. Gen. Board, Berkeley Nuclear Labs., RD/B/N3770, Sept.
115. Blackburn, W. S. *et al.* (1977). An integral associated with the state of a crack tip in a non-elastic material, *Int. J. Fract. Mech.*, **13**, pp. 183–200.
116. Hellen, T. K. & Blackburn, W. S. (1977). The use of a path-independent integral in non-linear fracture mechanics. SMIRT-4 (San Francisco), Paper G3/3.
117. Rice, J. R. (1975). Discussion: 'The path dependence of the *J*-contour integral', Chell, G. G. and Heald, P. T., *Int. J. Fract.*, **11**, pp. 352–3.
118. Ainsworth, R. A. *et al.* (1978). Fracture behaviour in the presence of thermal strains, *Tolerance of flaws in pressurised components*, Inst. Mech. Engrs., London, May.
119. Bui, H. D. (1974). Dual path–independent integrals in the boundary-value problems of cracks, *Engng Fract. Mech.*, **6**, pp. 287–96.
120. Harrison, J. D. *et al.* (1977). The COD approach and its application to welded structures, *ASTM Symposium on Elastic–Plastic Fracture*, Atlanta, Nov.

121. Chell, G. G. & Ewing, D. J. F. (1977). The role of thermal and residual stresses in linear and post-yield fracture mechanics, *Int. J. Fract.*, **13**, pp. 467–80.
122. Heald, P. T. *et al.* (1972). Post-yield fracture mechanics, *Mat. Sci. and Engng*, **10**, pp. 129–38.
123. Chell, G. G. (1976). The stress intensity factors and crack profiles for centre and edge cracks in plates subject to arbitrary stresses, *Int. J. Fract.*, **12**, pp. 33–46.
124. Chell, G. G., Milne, I. & Kirby, J. H. (1975). Practical fracture mechanics in the post-yield regime, *Metals Technology*, **2**(12), pp. 549–53.
125. Chell, G. G. (1977). The application of post-yield fracture mechanics to penny-shaped and semi-circular cracks, *Engng Fract. Mech.*, **9**, pp. 55–64.
126. Chell, G. G. & Davidson, A. (1976). A post-yield fracture mechanics analysis of single-edge notched tension specimens, *Mat. Sci. and Engng*, **24**, pp. 45–52.
127. Chell, G. G. (1977). Post-yield fracture mechanics theory and its application to pressure vessels, *Int. J. Pres. Ves. & Piping*, **5**, pp. 123–49.
128. Newman, J. C. (1973). Fracture analysis of surface- and through-cracked sheets and plates, *Eng. Fract. Mech.*, **5**, pp. 667–89.
129. Neuber, H. (1961). Theory of stress concentration for shear-strained prismatical bodies with arbitrary non-linear stress–strain law, *Trans. ASME*, pp. 544–50.
130. Newman, J. C. (1976). Fracture analysis of various cracked configurations in sheet and plate materials, *Properties related to fracture toughness*, ASTM–STP 605, pp. 104–23.
131. Gowda, C. V. B. & Topper, T. H. (1970). On the relation between stress and strain-concentration factors in notched members in plane stress, *J. Appl. Mech.*, March, pp. 77–84.
132. Buekner, H. F. (1958). Propagation of cracks and the energy of elastic deformation, *Trans. ASME*, **80**, pp. 1225–41.
133. Randall, P. N. & Merkle, J. G. (1971). Gross strain crack tolerance of steels, *Nuclear Engng and Design*, **17**, pp. 46–63.
134. Randall, P. N. (1969). Gross strain measure of fracture toughness of steels. HSSTP TR-3, TRW Systems Group, Redondo Beach, Calif., Nov.
135. Randall, P. N. (1971). Gross strain crack tolerance of A533B steel. HSSTP TR-14, TRW Systems Group, Redondo Beach, Calif., May.
136. Randall, P. N. (1972). Effects of strain gradients on the gross strain crack tolerance of A533B steel. HSSTP TR-19, TRW Systems Group, Redondo Beach, Calif., May.
137. Parker, E. R. (1957). *Brittle behaviour of engineering structures*, Wiley, New York.
138. Stone, D. E. W. & Turner, C. E. (1965). Brittle behaviour in laboratory scale mechanical testing, *Proc. Roy. Soc. (London)*, Ser. A., **285**, pp. 83.
139. Randall, P. N. & Merkle, J. G. (1973). Effects of strain gradients on the gross strain crack tolerance of A533B steel. In Ref. 80, pp. 404–22.
140. Randall, P. N. & Merkle, J. G. (1972). Effects of crack size on the gross-strain crack tolerance of A533B steel, *J. Engng Ind.*, *ASME B*, **94**(1), pp. 935–41.
141. Merkle, J. G. (1971). Fracture safety analysis concepts for nuclear pressure vessels, considering the effects of irradiation, *J. Basic. Engng, ASME D*, **93**(2), pp. 265–73.

142. Merkle, J. G. (1975). Tests of 6-inch-thick pressure vessels. Series 2, ORNL Report, p. 5059.
143. Irwin, G. R. et al. (1967). Basic aspects of crack growth and fracture. NRL Rep. 6598, US Naval Research Lab.
144. Holt, A. B. (1977). A critical evaluation of the tangent modulus method of elasto-plastic fracture analysis, SMIRT-4 (San Francisco), Paper G 3/6.
145. Witt, F. J. (1971). The application of the equivalent energy procedure for predicting fracture in thick pressure vessels. In Ref. 9, pp. 163–7.
146. Witt, F. J. & Mager, T. R. (1971). Fracture toughness K_{Icd} values at temperatures up to 550 °F for ASTM A-533 grade B, class 1 steel, Nuclear Engng and Design, 17, pp. 91–103.
147. Begley, J. A. & Landes, J. D. (1973). A comparison of the J-integral fracture criterion with the equivalent energy concept. In Ref. 80, pp. 246–63.
148. Merkle, J. G. (1973). Analytical applications of the J-integral. In Ref. 80, pp. 264–80.
149. Lee, J. D. & Liebowitz, H. (1977). The non-linear and bi-axial effects on energy release rate, J-integral and stress intensity factor, Engng Fract. Mech., 9, pp. 765–80.
150. Dowling, A. R. & Townley, C. H. A. (1975). Effect of defects on structural failure: a two-criteria approach, Int. J. Pres. Ves. & Piping, 3, pp. 77–107.
151. Townley, C. H. A. (1976). The integrity of cracked structures under thermal loading, Int. J. Pres. Ves. & Piping, 4, pp. 207–21.
152. Harrison, R. P. et al. (1976). Assessment of the integrity of structures containing defects. Central Elec. Gen. Board R/H/R6 (Revised Apr. 1977.)
153. Darlaston, B. J. L. et al. (1977). A UK proposal for the assessment of the significance of flaws in pressurised components, SMIRT-4 (San Francisco), Paper G 2/1.
154. Irvine, W. H. & Quirk, A. (1977). An elastic–plastic theory of fracture mechanics and its application to the assessment of the integrity of pressure vessels. In Ref. 9, pp. 76–84.
155. Irvine, W. H. (1978). An elasto-plastic theory of fracture and its application to the analysis of experimental data. Tolerance of flaws in pressurised components. Inst. Mech. Engrs, London, pp. 13–26.
156. Soete, W. (1977). An experimental approach to fracture initiation in structural steels, Fracture 1977, ICF-4, 1, Taplin, D. M. R., ed. Univ. Waterloo, pp. 775–804.
157. Soete, W. & Denys, R. (1977) Fracture toughness testing of welds, Rev. de la Soudure, 1, pp. 1–8.
158. Clausing, D. P. (1970). Effect of plastic strain state on ductility and toughness, Int. J. Fract. Mech., 6, pp. 71–86.
159. Fracture toughness evaluation by R-curve methods, ASTM–STP 527 (1973).
160. McCabe, I. E. & Heyer, R. H. (1973). R-curve determination using a crack-line wedge loaded (CLWL) specimen. In Ref. 159, pp. 17–35.
161. Judy, R. W. & Goode, R. J. (1973). Fracture extension resistance (R-curve) characteristics from three high-tensile steels. In Ref. 159, pp. 48–61.
162. Neale, B. K. & Townley, C. H. A. (1977). Comparison of elastic–plastic fracture mechanics criterion, Int. J. Pres. Ves. & Piping, 5, pp. 207–39.

163. Green, G. & Knott, J. F. (1975). Effects of side grooves on initiation and propagation of ductile fracture, *Metals Tech.*, **2**, pp. 422–7.
164. Green, G. & Knott, J. (1975). On effects of thickness on ductile crack growth in mild steel, *J. Mech. Phys. Solids*, **23**, pp. 167–83.
165. Tanaka, K. and Harrison, J. D. (1976). An *R*-curve approach to COD and *J* for an austenitic steel, Weld. Inst. Rep., 7/1976/E.
166. Garwood, S. J. (1976). The measurement of crack growth resistance using yielding fracture mechanics. Ph.D. Thesis, University of London, Dec.
167. Garwood, S. J. *et al.* (1975). The measurement of crack growth resistance curves (*R*-curves) using the *J* integral, *Int. J. Fract.*, **11**, pp. 528–31.
168. Garwood, S. J. & Turner, C. E. (1977). The use of the *J* integral to measure the resistance of mild steel to slow-stable crack growth, *Fracture* 1977, ICF-4, **2**, Taplin, D. M. R., ed. Univ. Waterloo, pp. 279–84.
169. Logsdon, W. A. (1976). Elastic–plastic (J_{Ic}) fracture toughness values: their experimental determinations and comparison with conventional linear elastic (K_{Ic}) fracture toughness values for five materials, *Mechanics of crack growth*, ASTM–STP 590, pp. 43–60.
170. Mai, Y. W., Atkins, A. G. & Caddell, R. M. (1976). Determination of valid *R*-curves for materials with large fracture toughness-to-yield-strength ratio, *Int. J. Fract.*, **12**, pp. 391–408.
171. de Koning, A. U. *et al.* (1978). Energy dissipation during stable crack growth in aluminium alloy 2024-T3. In Ref. 97, pp. 525–36.
172. Wnuk, M. P. (1973). Prior-to-failure extension of flaws in a rate-sensitive Tresca solid. In Ref. 80, pp. 64–75.
173. Wnuk, M. P. (1977). Initial stages of crack extension in time-dependent and/or ductile solids, *Fracture* 1977, ICF-4, **3**, Taplin, D. M. R., ed. Univ. Waterloo, pp. 54–62.
174. Wnuk, M. P. (1977). An assessment of fracture toughness near or after general yield (Part A), Final Progress Report under NSF Grant, South Dakota State University, Mar.
175. Wilhelm, D. P. *et al.* (1977). A *J*-integral approach to crack resistance for aluminum, steel and titanium alloys. *J. Engng Mat. Sci. & Technol. (Trans. ASME.* **98**, Series 8, Part 2, pp. 97–104).
176. Rice, J. R. Some computational problems in elastic–plastic micro-crack mechanics. In Ref. 97, pp. 434–49.
177. Garwood, S. J. & Turner, C. E. (1978). Slow stable crack growth in structural steel. *Int. J. Fract.*, **14**, 195–8.
178. Willoughby, T. *et al.* The effect of specimen orientation on the *R*-curve. *Int. J. Fract.*, **14**, 249–51.
179. Broberg, K. B. (1974). The importance of stable crack extension in linear and non-linear fracture mechanics, *Prospects in fracture mechanics*. Sih, G. C. *et al.*, eds. Noordhoff Int. Publ., Leyden, pp. 125–38.
180. Paris, P. C. *et al.* (1977). A treatment of the subject of tearing instability, Washington University, St. Louis, for US Nuclear Regulatory Commission, NUREG-0311. Aug.
181. Turner, C. E. (1979). Description of stable and unstable crack growth in the elastic–plastic regime in terms of J_r resistance curves. Presented at the *Eleventh National Symposium on Fracture*, Virginia. *Fracture mechanics*, ASTM-677.

182. Knott, J. (1973). *Fundamentals of fracture mechanics*, Butterworth. London.
183. Neuber, H. (1946). *Theory of notch stresses*, Edwards, Ann Arbor, Mich.
184. Krafft, J. M. (1964). Correlation of plane strain crack toughness with strain hardening characteristics of low, medium and high strength steel, *Appl. Mats. Res.*, **3**, pp. 88–101.
185. Ritchie, R. O. *et al.* (1973). On the relationship between critical tensile stress and fracture toughness in mild steel, *J. Mech. Phys. Solids*, **21**, pp. 395–410.
186. Sumpter, J. D. G. (1976). The prediction of K_{Ic} using J and COD from small specimen tests, *Metal Sci.*, **10**, pp. 354–7.
187. Landes, J. D. & Begley, J. A. (1974). Test results from J-integral studies: an attempt to establish a J_{Ic} testing procedure. In Ref. 70, pp. 170–86.
188. Milne, I. & Chell, G. G. (1977). Effect of size on the J fracture criterion. *ASTM Symposium on Elastic–Plastic Fracture Mechanics*, Atlanta, Nov.
189. Levy, N. *et al.* (1971). Progress in three-dimensional elastic–plastic stress analysis for fracture mechanics, *Nucl. Engng and Design*, **17**, pp. 64–75.
190. Ayres, D. J. (1970). A numerical procedure for calculating stress and deformation near a slit in a three-dimensional elastic–plastic solid, *Engng Fract. Mech.*, **2**, pp. 87–106.
191. Parks, D. M. (1978). Virtual crack extension in a general finite element technique for J-integral evaluation. In Ref. 97, pp. 464–78.
192. Hu, W. & Liu, H. W. (1976). Crack tip strains—a comparison of finite element method calculations and Moiré measurements. In Ref. 81, pp. 522–34.
193. Andersson, H. (1972). Finite element analysis of a fracture toughness test specimen in the non-linear range, *J. Mech. Phys. Solids*, **20**, pp. 33–51.
194. Kawai, T. *et al.* (1978). A new discrete model for analysis of solid mechanics problems. In Ref. 97, pp. 26–37.
195. McClintock, F. A. (1965). Effects of root radius, stress, crack growth and rate on fracture instability, *Proc. Roy. Soc.* (London), Ser. A, **285**, pp. 58–72.

3

Experimental Methods for Elastic–Plastic and Post-yield Fracture Toughness Measurements

J. D. Landes

Westinghouse Electric Corporation, Pittsburgh, USA

and

J. A. Begley

The Ohio State University, Columbus, USA

3.1 INTRODUCTION

In Chapter 2 the COD and J integral concepts have been emphasised as forming the mainstream of yielding fracture mechanics. This chapter will review the state of the art of experimental methods for measuring critical COD and J toughness parameters. Historical details are treated briefly with more attention devoted to attempts to develop standardised test practices such as the British Standards Institution DD19:1972 COD Test Methods [1] and the recent ASTM E.24.01.09 Cooperative J_{Ic} Test Program. Typical test data are included and single specimen J_{Ic} test techniques are reviewed.

In Chapter 2, notably in Section 2.4, it was pointed out that fracture testing poses two problems of choosing the proper toughness measurement point in view of slow stable crack growth and of observing the proper size limitations. Inappropriate treatment of either area leads to erroneous conclusions regarding the validity of fracture theories, proper fracture toughness values and the flaw tolerance of structures.

Following the approach emerging from Chapter 2 to relate the critical COD (δ_i) and J (J_{Ic}) values to the onset of crack growth, the task at hand in this chapter is to describe how and under what limitations J_{Ic} and δ_i values may be experimentally determined.

3.2 MULTIPLE SPECIMEN TESTING

This section describes the development of J_{Ic} and COD testing. General techniques are included along with procedures specific to bend bar and CTS specimen geometries. Later sections are devoted to advanced single specimen J_{Ic} tests, typical toughness data and a recommended practice for J_{Ic} testing.

3.2.1 Compliance *J* Calibration

As outlined in Section 2.2.2.1.3 sub (iv), the general compliance relationship (2.29) enables us to obtain *J* from the change in potential

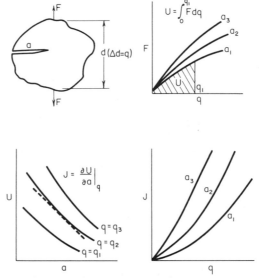

FIG. 3.1.   Schematic of steps in a compliance *J* determination.

energy *U* associated with a change in crack length *a*. The potential energy is equal to the area under load–load point deflection curve. At a given deflection the area under load–load point deflection record may be found for bodies with different crack lengths. The slope of a plot of *U* versus *a* at a given deflection is $(\partial U/\partial a)|_q$ and hence *J* may be determined. This process is shown schematically in Fig. 3.1. The final plot is *J* versus the applied deflection. In general this curve depends on the crack length *a*. If the applied deflection is found for the initiation of crack growth the critical *J* value, J_{Ic}, may be found.

Using the above procedure Landes & Begley [2] determined J_{Ic} values for a ASTM A471 Ni–Cr–Mo–V rotor steel at 250 °F. The values for bend bar and centre-cracked panel specimens were in good agreement, 0·172 MN/m versus 0·187 MN/m. A maximum load measurement point was utilised.

Metallography of similar specimens indicated a small but presumed negligible amount of growth at the maximum load point. The amount of stable crack growth was later found to be significant. Refined tests [3] which corrected a numerical error in the original calculations and clearly identified the initiation of a crack growth led to the agreement of test results at lower values of 0·12 MN/m and 0·14 MN/m. Compensating errors in early work resulted in fictitiously high toughness values but a valid conclusion that *J* could be utilised as a geometry independent post-yield fracture criterion.

3.2.2 The Rice–Paris–Merkle *J* Formula

A compliance determination of *J*, while possessing the advantage of generality, is tedious and expensive in both time and material. A significant advance in fracture toughness testing was provided by the development of the Rice–Paris–Merkle *J* formula for deeply cracked bars subjected to bending [4]. The basis of derivation of this formula depends on the fact that the angle of bend, θ, for deeply cracked bars, is only a function of the applied moment, *M*, and the square of the remaining ligament, *b*.

$$\theta = (M/b^2) \tag{3.1}$$

Using the potential energy definition of *J*,

$$J = -\left.\frac{\partial U}{\partial a}\right|_q = \left.\frac{\partial U}{\partial a}\right|_F \tag{3.2}$$

and the fact that d*a* is equal to $-$d*b*, *J* can be written as

$$J = \int_0^\theta \left.\frac{\partial \theta}{\partial a}\right|_M dM = -\int_0^\theta \left.\frac{\partial \theta}{\partial b}\right|_M dM \tag{3.3}$$

This is illustrated schematically in Fig. 3.2. From eqn. (3.1)

$$\left.\frac{\partial \theta}{\partial b}\right|_M = -f'(M/b^2)\left.\frac{\partial(M/b^2)}{\partial b}\right|_M = \frac{2M}{b^2}f'(M/b^2) \tag{3.4}$$

Similarly

$$\left.\frac{\partial \theta}{\partial M}\right|_b = \frac{f'(M/b^2)}{b^2} \tag{3.5}$$

and thus

$$\left.\frac{\partial \theta}{\partial a}\right|_{M_n} = -\left.\frac{\partial \theta}{\partial b}\right|_{M_n} = \frac{2M}{b}\left.\frac{\partial \theta}{\partial M}\right|_b \tag{3.6}$$

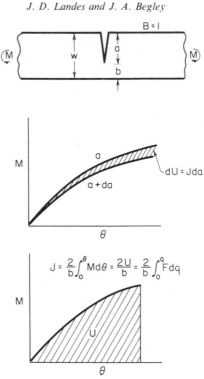

FIG. 3.2.   Schematic showing Rice–Paris–Merkle J formula for bend-type specimens [4].

Substituting eqn. (3.6) in eqn. (3.3)

$$J = \frac{2}{b} \int_0^M M \left. \frac{\partial \theta}{\partial M} \right|_a \, \mathrm{d}M = \frac{2}{b} \int_0^\theta M \, \mathrm{d}\theta \tag{3.7}$$

yields the result that J is simply related to the work done on the specimen. Again the above derivation holds true for deeply cracked bars in bending such that θ is only a function of M/b^2 and the angle of bend of an uncracked bar is negligible compared to the angle of bend of a cracked bar.

   Corrections to eqn. (3.7) have been proposed by Merkle & Corten [5] and also by Sumpter & Turner [6]. These corrections account for the tensile component of loading in compact toughness specimens and resolve the question of the significant uncracked deflection of bend specimens. For CTS specimens with $0.5 \leq a/W \leq 0.7$ eqn. (3.7) underestimates the applied J by 10 to 15 %. For three-point bend specimens of span to width ratio 4·0 and $0.5 \leq a/W \leq 0.7$ the work of Sumpter & Turner, [6] Srawley [7] and compliance calibrations [8, 9] supports the view that eqn. (3.7) provides J values to an accuracy of better than 5 % without subtracting the energy absorbed by an uncracked bar at the load of interest. Later sections list the Merkle–Corten [5] approach to CTS specimens and review the work of Sumpter & Turner [6] in providing a unified approach to J and COD testing.

3.2.3 The J Resistance Curve

The use of formulae relating J to the load–deflection curve of a cracked body obviates the need for tedious J compliance calibrations and permits attention to be focused on the detection of the onset of crack growth. As discussed earlier, the choice of the onset of crack growth as a toughness measurement point is consistent with the limitations of a critical J fracture criterion. A quantitative, large-scale plasticity treatment of the extent of stable crack growth and final instability must await further theoretical and experimental developments.

For materials exhibiting stable crack growth there is, in general, no easily recognisable feature of the load–deflection record which indicates the start of crack growth. A simple direct approach to determine the initiation of crack growth is to load a number of nearly identical samples to various deflections. After unloading the extent of crack growth can be marked by oxidation, chemical staining or fatigue cracking. In ferritic steels subsequent fracture at low temperatures leads to a cleavage fracture surface which clearly delineates the prior region of ductile tearing. A plot of the extent of crack growth versus the applied deflection can then be extrapolated to zero crack growth. The applied deflection at fracture initiation and the load–deflection record then permit the computation of a J_{Ic} value.

A variation of the above procedure was first used by Landes & Begley [10] in the development of a J_{Ic} test technique. The requirement of nearly identical test specimens may be relaxed somewhat if the applied J is plotted versus crack growth, Δa. Also, extrapolation to the onset of crack growth then leads directly to a J_{Ic} value.

As evident from Section 2.4.1, caution should be exerted in referring to the plot of J versus Δa as a J resistance curve, in view of the objections against the use of J after initiation, when unloading takes place. The reader is referred to the work of Garwood *et al.* [11] described in the aforementioned section for an attempt to justify the use of J after crack growth has been initiated.

Finally, while the J_{Ic} point is not geometry dependent, the slope of the J

resistance curve can be markedly affected by the specimen geometry. This point is illustrated in a later paragraph.

A schematic of the steps involved in the multiple specimen *J* resistance curve technique is shown in Fig. 3.3. A number of specimens are first loaded to various deflections to provide different amounts of crack growth. The specimens are unloaded and the extent of crack growth is marked by heat

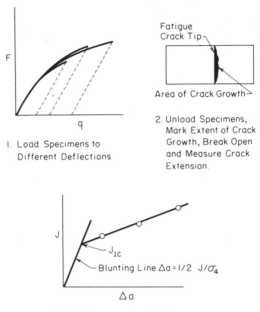

FIG. 3.3.   Multiple specimen J_{Ic} test technique.

tinting or other techniques. The specimens are broken open and the extent of crack growth is measured optically. Finally the applied *J* value for each specimen is plotted versus the amount of crack extension.

A J_{Ic} value is determined by extrapolating the *J* versus Δa curve to the onset of crack growth. The curve is not extrapolated to a zero Δa value. Figure 3.4, derived from data such as are shown on Fig. 2.80, illustrates how crack tip blunting leads to apparent crack growth when using the heat tint method and visual observation. Blunting changes the surface roughness and this change in texture is included in the crack length measurement. Assuming a circular blunted crack tip the change in surface texture will be about one half of the COD value. Taking COD equal to J/σ_Y results in a

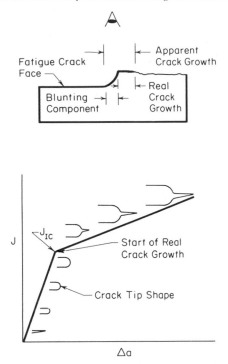

FIG. 3.4. Schematic of crack tip blunting component of crack growth.

formalism of the apparent crack growth due to crack tip blunting being given by

$$\Delta a = 1/2 J/\sigma_Y \qquad (3.8)$$

This is the equation of the blunting line. The onset of crack extension is taken as the point where the J versus Δa curve meets the blunting line [10]. The value at this point is termed J_{Ic}.

The blunting line is, admittedly, a formalism. The shape of a blunted crack varies from one material to another as does the relationship of J/σ_Y to COD [12]. However, the general features of the process of blunting and then crack growth of Fig. 3.4 have been verified by the work of Robinson [12] and Fields & Miller [13]. The blunting line does become important for low strength high toughness materials where extrapolation to a zero Δa can lead to fictitiously low J_{Ic} values.

Sample curves [10] illustrating the multiple specimen J_{Ic} test technique are shown in Fig. 3.5. Note that the J_{Ic} values are the same for both compact

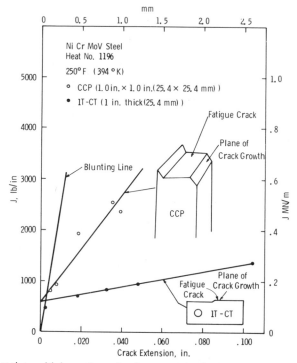

Fɪɢ. 3.5.   Heat tint multiple specimen *J* resistance curves for CTS and CCP Specimens of a Ni–Cr–Mo–V steel [3].

toughness and centre-cracked panels. Similar results have been obtained by Robinson [12] who observed that while J_{Ic} and δ_i values were geometry independent the relationship of δ_i to J/σ_Y did depend on the material. The slopes of the *J* resistance curves in Fig. 3.5 are quite different, indicating that resistance to stable crack growth may be quite geometry sensitive.

Obviously the preceding J_{Ic} multiple specimen technique is predicated upon some degree of stable crack growth. For fully cleavage fractures the onset of crack growth and final instability are often simultaneous. In this case the onset of crack growth should be clearly discernable by an abrupt load drop. With this indication of a measurement point the *J* resistance curve technique need not and, in fact, cannot be applied. Care should be taken to ensure that no stable growth does occur or the toughness value will be erroneously high.

A recommended J_{Ic} test procedure is fully described in Section 3.5. Details of specimen preparation, testing and data analysis are included.

The multiple specimen J resistance curve technique is the basis of the test procedure. Single specimen techniques as discussed in Section 3.3 may also be utilised. The recommended J_{Ic} test procedure is based largely on a co-operative J_{Ic} test program conducted by the ASTM E.24.01.09 Task Group on 'Elastic Plastic Fracture'. Official Task Group recommendations are still in preparation.

3.2.4 J Estimation Procedures for Test Specimens

Using the potential energy definition of J from eqn. (3.2), estimates of cracked body load–deflection records can be used to approximate applied J values. This technique, as advanced by Bucci *et al.* [14], can be applied to structural components as well as test specimens.

Figure 3.6a shows the relationship between J and the area, dU, between load–deflection records of cracked bodies differing only in crack length by the amount, da. If the load–deflection records are approximated by an elastic slope and a rigid–plastic limit load, F_Y, the area between curves and thus J may be computed. In Fig. 3.6b the elastic contribution to J is simply

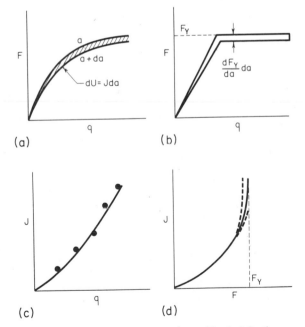

FIG. 3.6.   Estimating J from approximations of load–deflection records.

the first triangular area. This may be computed from LEFM solutions and is given by

$$J_{el} = G = K^2(1 - v^2)/E \qquad (3.9)$$

where K is evaluated at the load F_Y. The plastic contribution to J is represented by the rectangular area and is given by

$$J_{pl} = \frac{dF_Y}{da}(q - q_e) \qquad (3.10)$$

Hence the total J is

$$J = J_{el} + J_{pl} = \frac{K^2(1 - v^2)}{E} + \frac{dF_Y}{da}(q - q_e) \qquad (3.11)$$

Equation (3.11) shows how stress intensity solutions and rigid–plastic limit load theory may be combined to provide J estimates.

The above procedure has been refined by Bucci *et al.* [14] to account for the transition from elastic to fully plastic loading. A plastic zone correction can be added to the original crack size. Elastic compliance solutions then permit a deflection to be calculated for a given load and effective crack size. The increased deflection due to the effective crack size leads to a calculated load–deflection record with a smoother transition from elastic to fully plastic loading. Figure 3.6c shows the results of such a calculation for a centre-cracked panel. Calculated and compliance measured J values agree quite well.

The shape of the J versus deflection curve of Fig. 3.6c should be noted. The early curved portion can be computed from LEFM. In the plastic region J versus q can be approximated by a straight line. From rigid–plastic limit load theory the slope of this line is

$$\left. \frac{\partial J}{\partial q} \right|_a = \frac{dF_Y}{da} \qquad (3.12)$$

If J is plotted against the applied load, F, a curve similar to that of Fig. 3.6d is obtained. The applied J becomes unbounded at the limit load, F_Y. A small error in the estimated limit load or a variation in the shape of the curve near the limit load can lead to gross errors in the estimated J. For fully plastic situations estimates of J versus deflection are clearly subject to much less error than estimates of J versus load.

3.2.5 COD Testing
A fully developed approach to COD testing is provided in British Standards

Institution document DD19:1972, *Methods for Crack Opening Displacement (COD) Testing* [1]. While this publication is of a provisional nature and due for revision it is backed by substantial testing and affords a rational, consistent approach to COD testing. Salient features of COD testing of precracked bend bars are described following this text.

Early efforts to measure COD values utilised a paddle-wheel COD meter. A thin blade was inserted in the starter notch close to the fatigue crack tip. Rotation of the spring-loaded blade was measured during loading and thus, as the crack opened, a COD value could be computed. Deformation of the blade and an inability to locate the blade at the sharp crack tip complicated COD computations. This technique has been abandoned in favour of theoretical [15] and experimental [16–18] studies of rotation points of bend bars.

If an effective rigid-body rotation point can be defined then measurements of the crack opening at the face of the specimen can be used to calculate the crack opening displacement, δ, at the crack tip. Figure 3.7 shows a sketch of a deformed test specimen. The apparent centre of rotation, $r(W - a)$, is some distance below the crack tip. The location of this rotation point has been defined as a function of deformation. The crack tip δ value can be computed from measurements of the surface crack opening, d, using the following equations. The theoretical equations in BS DD19 [1] based on the work of Wells [15] are as follows for three-point bend specimens of span to width ratio, 4·0.

$$\delta = \frac{0{\cdot}45(W - a)}{0{\cdot}45W + 0{\cdot}55a + z}\left(d - \frac{\gamma\sigma_Y W(1 - v^2)}{E}\right) \tag{3.13}$$

for

$$d \geq \frac{2\gamma\sigma_Y W(1 - v^2)}{E}$$

and

$$\delta = \frac{0{\cdot}45(W - a)}{0{\cdot}45W + 0{\cdot}55a + z}\left(\frac{d^2 E}{4\gamma\sigma_Y W(1 - v^2)}\right) \tag{3.14}$$

for

$$d < \frac{2\gamma\sigma_Y W(1 - v^2)}{E}$$

The term is a non-dimensional limiting value of the elastic clip gauge displacement and z is defined in Fig. 3.7.

$$\gamma = \frac{d_{el} E}{\sigma_Y W(1 - v^2)} \tag{3.15}$$

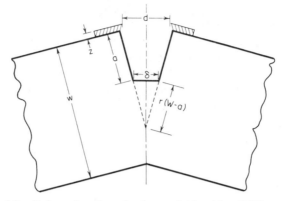

FIG. 3.7. Deformed section of a three-point bend bar COD specimen.

The parameter d_{el} is the maximum elastic component of the clip gauge displacement, d. Values of γ are tabulated below as a function of a/W.

a/W	0·2	0·3	0·4	0·5	0·6
γ	0·70	1·03	1·35	1·54	1·72

An experimental COD calibration [1, 17] leads to the formula

$$\delta = \frac{(W - a)d}{W + 2a + 3z} \qquad (3.16)$$

This formula is equivalent to assuming a constant rotation point located at a distance of $\frac{1}{3}(W - a)$ below the crack tip. A constant centre of rotation is physically unreasonable in covering the full range of elastic to fully plastic loading. Nonetheless, eqn. (3.16) has been shown to provide good approximations to COD values in the range of 0·06 to 0·60 mm.

Using the above equations, δ may be computed from the clip gauge displacement. The fracture toughness then depends upon the choice of a suitable measurement point. As discussed in Section 2.2.1.2, the onset of crack growth is the most reasonable choice. Following the draft practice DD19 the onset of crack growth may be determined by:

1.   An abrupt load drop with no prior indication of crack extension (as expected for fully cleavage fractures).
2.   The start of a section of falling or constant load on the load–deflection curve when cracking is verified by an audible acoustic emission, electric potential measurements or other techniques.

3.   The onset of crack extension as measured by electrical potential or other techniques.

The above measurement point criteria provide a critical value termed, δ_i, for the onset of crack extension. This approach is consistent with the J_{Ic} concept.

There is, however, a provision in DD19 to report a maximum load δ_m value in the absence of techniques to measure the start of crack extension.

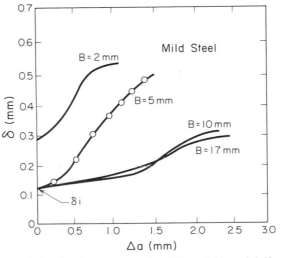

Fig. 3.8. δ resistance curve after Chipperfield *et al.* [19].

This provision is specified as a means of material comparison. If stable crack growth does occur, maximum load δ and J values do not in general provide a geometry independent toughness parameter and may seriously overestimate the fracture toughness.

The resistance curve technique, not covered in Ref. [1], can be applied to δ_i as well J_{Ic} testing. Figure 3.8 shows the results of work performed by Chipperfield *et al.* [19]. The value of applied δ is plotted versus crack growth, Δa, for a series of specimens. In these mild steel specimens, cleavage fracture at low temperature clearly revealed the extent of previous ductile crack growth at the test temperature. Extrapolation to zero crack growth provides the fracture toughness, δ_i. The multiple specimen resistance curve technique, while expensive in terms of the number of test specimens, eliminates the need for sophisticated techniques and equipment to monitor the onset of crack extension. Hence, determination of the onset of crack growth and thus proper toughness values is within the capabilities of any mechanical testing laboratory.

Figure 3.8 brings an essential point into consideration. A high δ_i value is noted for the thinnest test piece, whereas for a thickness equal to or greater than 5 mm the fracture toughness is independent of test specimen thickness. This finding is in agreement with the argument of Begley & Landes [20] that the limiting dimensions of a cracked body can influence the dominance of the crack tip plastic singularity and hence lead to size limitations in the applicability of a J_{Ic} fracture criterion. If the δ_i approach also rests on the dominance of a crack tip plastic singularity, size limitations will also be encountered.

For a general argument on the size effect the reader is referred back to Section 2.4.2. The quantitative limits briefly mentioned there have been generalised by Landes & Begley [10] so as to include crack depth and remaining ligament size restrictions such that

$$a, \ W - a, \ B \geq \alpha J_{Ic}/\sigma_Y \qquad (3.17)$$

Experimental studies of size effects in J_{Ic} testing, discussed in Section 3.4, indicate that α should be of the order 25. The δ_i data in Fig. 3.8 are consistent with the proposed size limitation of eqn. (3.17) when using a value of $m = 2$ in the general relationship (2.43), based on Robinson's experimental data for mild steel [12]. Thus a thickness effect on the δ_i values in Fig. 3.8 is noted somewhere between $\alpha = 20$ and $\alpha = 8$.

The above data, as well as those in later sections, reinforce the thematic approach in this chapter; that is, J_{Ic} and δ_i values are interchangeable, geometry independent fracture toughness parameters providing the limiting dimensions of the cracked body exceeds some multiple of δ_i or J_{Ic}/σ_Y. The size restriction of eqn. (3.17) for $\alpha = 25$ permits fracture toughness tests on specimens one to two orders of magnitude smaller than those required for valid K_{Ic} tests.

For additional details of COD testing, such as fatigue precracking and loading fixtures, the reader is referred to Ref. [1].

3.2.6 A Unified Approach to J_{Ic} and δ_i Testing

The work of Robinson [12] clearly demonstrates that J_{Ic} and δ_i are compatible and in fact interchangeable toughness parameters. It would be advantageous to obtain both J_{Ic} and δ_i values from the same test. A step in this direction has been provided by Sumpter & Turner [6]. They have developed a procedure to compute J values using the standard crack face clip gauge data utilised in COD tests of three-point bend bars.

Using the shorthand notation of eqn. (2.34), i.e. $J = J_{el} + J_{pl}$, the elastic component is simply equal to linear elastic G value

$$J_{el} = G = \frac{k^2(1 - v^2)}{E} \qquad (3.18)$$

where the maximum load during the test is used to calculate K. The plastic component is related to the plastic work done on the specimen.

$$J_{pl} = \frac{2U_{pl}}{B(W - a)} \qquad (3.19)$$

This plastic work, U_{pl}, is the product of the fully plastic limit load, F_Y, obtained from the test record and the total plastic deflection of the test specimen. For fully plastic loading a constant centre of rotation is appropriate. Thus the plastic component of the clip gauge opening can be related to the plastic component of ram travel. For an a/W of 0·5 the fully plastic centre of rotation is chosen to be $0·4(W - a)$.

Thus,

$$U_{pl} = \frac{F_Y(d - d_{el})S}{a + 0·4(W - a) + z} \qquad (3.20)$$

where S is the span of the bend bar, d is the clip gauge opening, d_{el} is the maximum elastic clip gauge opening and z is shown in Fig. 3.7. Summing the components to J leads to

$$J = \frac{K^2}{(1 - v^2)E} + \frac{2}{B(W - a)} \frac{F_Y(d - d_{el})S}{(a + 0·4(W - a) + z)} \qquad (3.21)$$

As discussed previously, δ values can be computed from the clip gauge data. Equation (3.21) provides a good estimate for J. Detection of the onset of crack extension then permits determination of both J_{Ic} and δ_i values.

3.3   SINGLE SPECIMEN TESTING

The method of using several specimens to determine J_{Ic} has some drawbacks. In addition to the cost factor involved in preparing four to six specimens there is sometimes not enough material available to prepare specimens. Also, the development of a J versus Δa resistance curve with appropriately spaced joints is sometimes difficult with the multiple specimen technique. A simpler and more economical technique would be the use of a single specimen to determine the entire curve of J versus Δa and hence the J_{Ic} value. Since J can be determined for each point along a load versus load point displacement curve, all that is needed to complete the resistance curve is an accurate measurement of crack length as a function of displacement.

To date, several methods have been proposed to measure crack length along the load–displacement curve from a single specimen. Presently, the most widely used single specimen technique is the unloading compliance method initially described by Clarke *et al.* [22]. Simply stated, the unloading compliance method uses the unloading slope of the load–displacement curve to determine crack length from the compliance of the specimen at any point of load and displacement during the test. Although the deformation theory of plasticity, which is used to develop the J integral, does not allow for unloading, it would appear that relatively small amounts of unloading, i.e. 10 % of maximum load, do not affect the overall load–displacement record of the test, Therefore, by partially unloading the specimen at specific intervals during the test, the progress of crack extension can be measured by calculating the crack length from the measured unloading compliance values. The simplest means of determining the crack length from the compliance is to use the empirically developed compliance versus crack length curves for the particular specimen being used.

A load versus displacement curve, where the unloading compliance method has been used, is shown schematically in Fig. 3.9. Each unloading of the specimen produces a linear curve whose slope reflects the crack length. Very often the slopes taken from such a curve cannot be determined with enough precision to measure crack lengths or changes in crack lengths with sufficient accuracy to develop a resistance curve. A way to attain more accuracy is to greatly amplify the curve in Fig. 3.9. One method for conveniently doing this is to electrically subtract out elastic displacements which correspond to the original crack length and replot this curve with high amplification. An example is shown schematically in Fig. 3.10. Elastic slopes corresponding to the original crack length are straight vertical lines. Subsequent crack growth produces elastic unloading lines which deviate from the vertical and allow easy determination of crack growth. Although use of this procedure is not necessary to use the unloading compliance method of crack measurement, some amplification of the curve in Fig. 3.9 is generally required.

With the amplification of the load versus displacement curve, small non-linearities in the mechanical or electrical system become apparent during

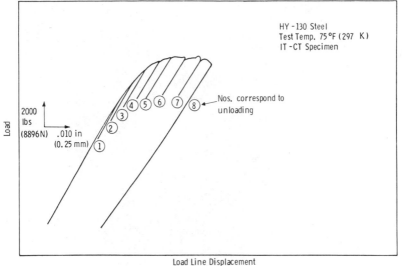

Load Line Displacement

FIG. 3.9. Record of load versus load line displacement for HY-130 steel at 75 °F (297 K) [28].

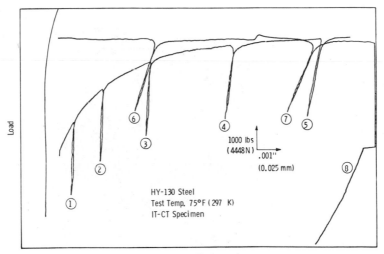

Load Line Displacement

FIG. 3.10. Record of load versus inelastic displacement for HY-130 steel at 75 °F (297 K) with amplified inputs [28]. Numbers correspond to unloading from Fig. 3.9.

FIG. 3.11.   CTS-type specimens modified for J_{Ic} testing showing (A) roller bearings for reduced pin friction and (B) razor-blade knife edges.

reversed loading. Often, rather than producing linear curves from which a slope can be determined, a loop is produced which reflects hysteresis in the mechanical and electrical systems. Accurate electrical equipment is commercially available; therefore, the most obvious way to improve the test results is to improve the mechanical system. The two things which produce the most friction in the mechanical system are the specimen loading contacts and the displacement gauge bearing surfaces. An example of how the mechanical friction can be reduced is shown in Fig. 3.11 [23]. For a CTS specimen, roller bearings were used to reduce friction in the loading pins and razor-blade knife edges were used to reduce friction in the clip gauge contact. An alternative way to reduce friction in the loading system is to use flat-bottomed clevis holes such as are used for K_{Ic} testing in the ASTM Standard E399.

Results from a single specimen of 2024–T351 aluminium where crack growth was determined from the unloading compliance method are shown in Fig. 3.12. As can be seen, this technique allows for the determination of many points on a resistance curve.

A second method for determining crack length during a J_{Ic} test is the electrical potential method. The electrical potential method for measuring crack advance has been used successfully for measuring crack propagation rates under linear elastic conditions [24, 25]. This method appears attractive for measuring crack advance for a J_{Ic} test in that a total R curve could be developed from a single specimen. While several investigators have

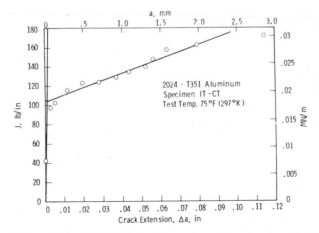

FIG. 3.12. J versus Δa for 2024–T351 aluminium alloy using unloading compliance method [28].

tried this method [26, 27], it has some major drawbacks. The amount of resolution on crack advance required in developing a resistance curve demands a very accurate system. This is further complicated by the fact that loading a specimen in the elastic–plastic regime causes a change in electrical output due solely to plasticity. This plasticity component in the signal must be subtracted from the total signal in order to determine the component solely due to crack extension.

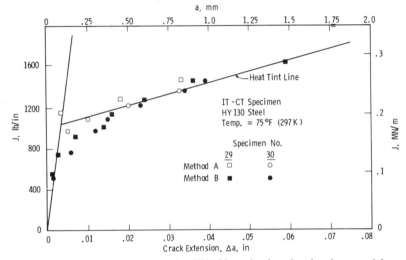

FIG. 3.13. J versus crack extension for HY-130 steel using the electric potential crack monitor [28].

Marandet & Sanz [26]. have used an AC electrical system with a dummy specimen as a reference and found that the point of first crack extension can be defined. Although this would define the J_{Ic} point, it is currently recommended that a complete R curve be generated for a J_{Ic} measurement. Methods have been suggested for eliminating the electrical signal change due solely to plasticity [28]. Results using the electrical potential method to determine crack growth in an HY–130 steel are shown in Fig. 3.13.

Other methods used to measure crack growth during a single specimen J_{Ic} test include use of an ultrasonic transducer [29], use of changes in resonant frequency of the specimens [30] and visual measurement. The former, suggested by Underwood et al. [29], involves using an ultrasonic transducer on the back of the specimen to transmit parallel to the crack plane, Fig. 3.14. The blunted crack tip which develops during the J_{Ic} test makes a good target which can be monitored for crack growth by the ultrasonic transducer.

These new methods for measuring crack advance represent a distinct advantage in J_{Ic} testing in that a single specimen can be used to generate a whole curve of J versus Δa. There are some disadvantages with these methods. They generally require sophisticated electronic equipment not always available in test laboratories. The amount of crack advance

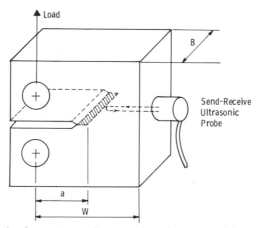

FIG. 3.14. Schematic of compact toughness specimen instrumented for end-on ultrasonic crack length measurements [29].

measured in these tests is usually very small so that the equipment must be capable of a high degree of precision in resolving these small crack advances.

Whenever possible, these methods should be used in J_{Ic} testing; however, some cautions should be exercised. Curves of J versus crack advance should be generated and J_{Ic} determined from these curves, rather than attempting to determine the point of first crack advance directly from an electrical output. A small indication of crack advance due to the crack tip blunting or even due to some electronic instability could easily be misconstrued as the point of first crack advance. Whenever a new method is tried, the resultant curve of J versus Δa should be compared with a similar curve generated by the standard method described in the previous section. Also, since some materials exhibit a degree of scatter from one specimen to another, a curve generated from a single specimen may not be completely representative of the material. A sufficient number of specimens must be tested to establish some reasonable limits on the degree of scatter.

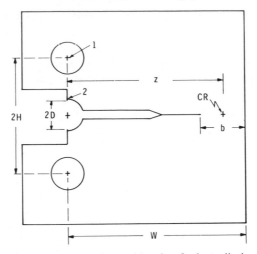

FIG. 3.15. Schematic of compact specimen with points for large displacement corrections.
(1) Load point; (2) displacement point; (3) centre of rotation.

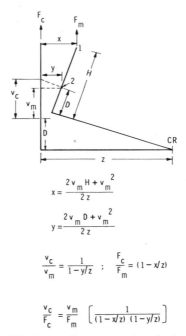

$$x = \frac{2 v_m H + v_m^2}{2 z}$$

$$y = \frac{2 v_m D + v_m^2}{2 z}$$

$$\frac{v_c}{v_m} = \frac{1}{1 - y/z} \quad ; \quad \frac{F_c}{F_m} = (1 - x/z)$$

$$\frac{v_c}{F_c} = \frac{v_m}{F_m} \left[\frac{1}{(1 - x/z)\,(1 - y/z)} \right]$$

FIG. 3.16. Schematic of elastic compliance corrections for large displacements [28].

An additional factor which may become important in a single specimen J_{Ic} test, particularly when the unloading compliance method is used, is the change in geometry due to large displacements. This change in geometry can result in a change in the linear elastic compliance slope without an accompanying change in specimen crack length. A method for correcting this problem in the CTS specimen has been suggested by Paris [23] and further developed by Donald & Schmidt [31].

The CTS specimen is illustrated in Fig. 3.15, where a constant centre of rotation point is assumed and the two points affected by large geometry changes are labelled (1) the centre of load application and (2) the point of displacement measurement.

Figure 3.16 illustrates the correction for large displacements. In the schematic v_c and v_m are elastically calculated and measured displacements and F_c and F_m are elastically calculated and measured loads. v_c/F_c and v_m/F_m are the elastically calculated compliance, from which the crack length is determined, and the measured compliance from the unloading slope, respectively. H is the distance from the crack line to the point of load application and D is the distance to the point of load-line displacement measurement.

3.4 EXAMPLES OF RESULTS

J_{Ic} tests have been successfully conducted on a number of metal alloy systems. Some of the results will be presented here to demonstrate the success of the method, illustrate the types of results and point out some of the problems and limitations inherent in the method.

Ferritic steels provide a good illustration of the use of the J_{Ic} test since these alloys have a transition from a low toughness brittle fracture mode at low temperatures to a high toughness ductile fracture mode at higher temperatures. This is illustrated in Fig. 3.17 for an ASTM A217 2 1/4 Cr–1 Mo cast steel where fracture toughness is plotted as a function of test temperature [32]. In the low temperature brittle fracture regime, fracture toughness can be measured by linear elastic techniques using the ASTM E-399, K_{Ic} test. In the transition range as toughness increases, the size of the K_{Ic} specimen must also increase. At some point the specimen size necessary for a K_{Ic} test becomes prohibitively large. At this point the J_{Ic} test can be used to measure toughness with a small specimen. The J_{Ic} values were converted to K_{Ic} by

$$K_{Ic} = \sqrt{J_{Ic}E/(1 - v^2)} \tag{3.22}$$

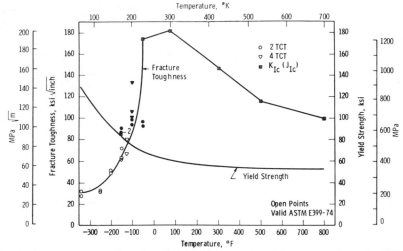

FIG. 3.17.   Temperature dependence of the yield strength and fracture toughness for ASTM
A217 2 1/4Cr–1Mo cast steel [32].

for the sake of comparison in Fig. 3.17. As the test temperature is increased
and the fracture mode becomes purely ductile, a maximum 'upper shelf'
toughness value is reached. Toughness then decreases with increasing
temperature.

Resistance curves for two steels, an AISI 403 modified 12 Cr stainless
rotor steel and an ASTM A470 Cr–Mo–V rotor steel, which are used to

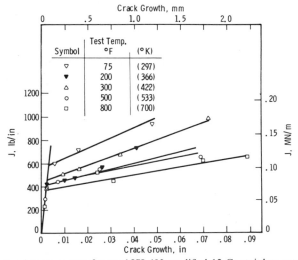

FIG. 3.18.   J resistance curves for an AISI 403 modified 12 Cr stainless rotor steel [32].

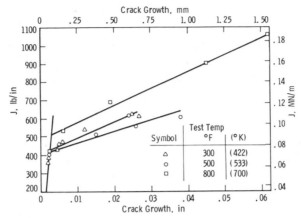

FIG. 3.19.   *J* resistance curves for an ASTM A470 Cr–Mo–V rotor steel [32].

determine J_{Ic} values in the ductile fracture regime, are shown in Figs. 3.18 and 3.19. The curves of toughness versus test temperature are shown in Figs. 3.20 and 3.21 for these two steels [32]. Again, fracture toughness is determined by the K_{Ic} test in the lower temperature brittle fracture regime and into the transition region. As the specimens size needed for a K_{Ic} test become larger, smaller specimen J_{Ic} tests are used to determine toughness and the values of J_{Ic} converted to K_{Ic}.

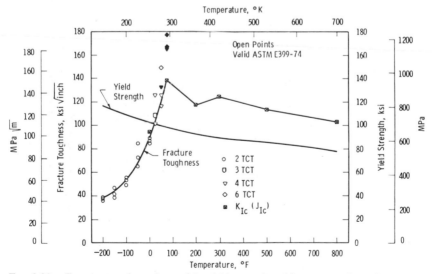

FIG. 3.20.   Temperature dependence of the yield strength and fracture toughness for an AISI 403 modified 12 Cr stainless rotor steel [32].

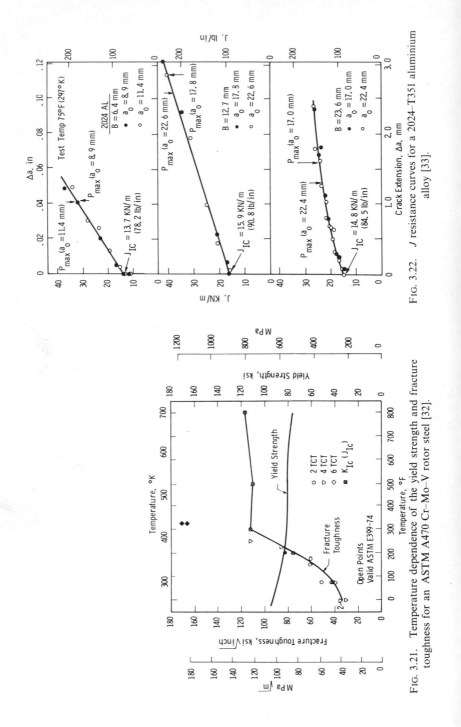

Fig. 3.22. J resistance curves for a 2024-T351 aluminium alloy [33].

Fig. 3.21. Temperature dependence of the yield strength and fracture toughness for an ASTM A470 Cr–Mo–V rotor steel [32].

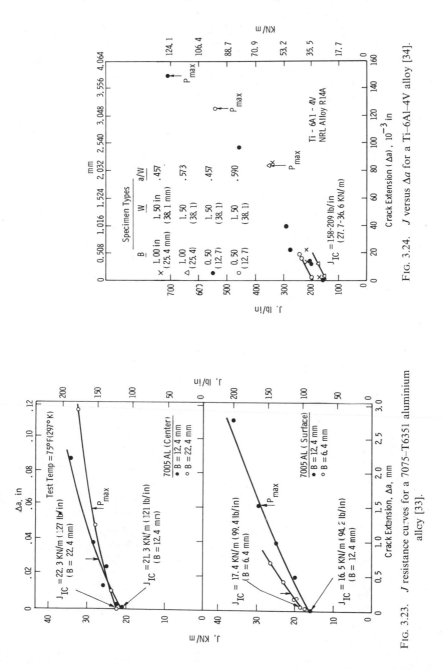

FIG. 3.24. *J* versus Δ*a* for a Ti–6Al–4V alloy [34].

FIG. 3.23. *J* resistance curves for a 7075–T6351 aluminium alloy [33].

J. D. Landes and J. A. Begley

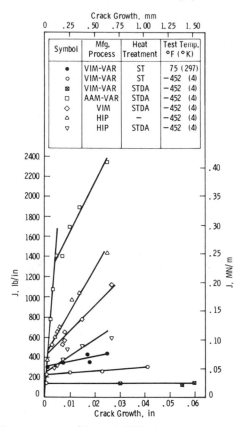

FIG. 3.25. *J* resistance curves of Inconel X750 for various manufacturing process/heat treatment combinations [35]. VIM–VAR: vacuum induction melted followed by vacuum arc remelt. AAM–VAR: air arc melted followed by vacuum arc remelt. VIM: vacuum induction melted. HIP: hot isostatic pressed. ST: solution treated. STDA: solution treated and double aged.

Resistance curves are shown for a 2024–T351 aluminium alloy in Fig. 3.22 [33]. J_{Ic} values have been determined for three different specimen thicknesses at two crack lengths and found to be reasonably consistent. Similar data for a 7005–T6351 aluminium alloy is shown in Fig. 3.23 [33]. Data from a Ti–6Al–4V alloy are shown in the form of a resistance curve in Fig. 3.24 [34]. Resistance curves for an Inconel X750 alloy with various combinations of manufacturing process and heat treatment are given in Fig. 3.25 [35].

An A286 stainless steel alloy does not exhibit cleavage fracture, even at

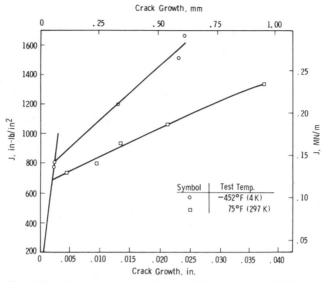

FIG. 3.26.   *J* resistance curves for an A286 stainless steel [36].

very low temperatures. Therefore, *J* versus Δa resistance curves must be developed to determine J_{Ic}. Resistance curves taken at room temperature and a very low temperature are shown in Fig. 3.26 [36]. Resistance curves for two different thicknesses of a 2 1/4 Cr–1 Mo steel are shown in Fig. 3.27 [37].

J_{Ic} values can be determined from a single specimen when crack advance occurs by a brittle unstable mechanism, such as steels fracturing by a cleavage mode. For this case J_{Ic} can be measured at the point of maximum load as illustrated in Fig. 3.28. Care must be taken to ensure that no stable crack growth occurred before the instability point. This can be done by several methods, the easiest being by visual examination of the fracture surface. If the instability point was the point of first crack advance, the fracture surface will show a negligible region of ductile fracture between the fatigue crack area and the cleavage fracture area. An example of J_{Ic} values determined for cleavage fracture is given in Fig. 3.29 [38]. J_{Ic} was determined with specimens of various sizes for an A533 Gr. B, Cl. 1 steel. As specimen size is decreased the J_{Ic} values show an increasing amount of scatter. These data lend support to a weakest link statistical model for fracture in the cleavage range previously discussed by Landes & Begley [10].

The J_{Ic} concept is applied to fracture from a sharp crack; however, the *J*

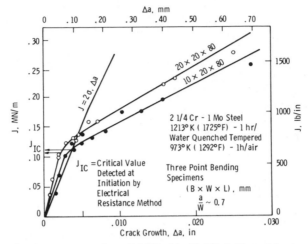

FIG. 3.27. *J* resistance curves for a 2 1/4Cr–Mo 10 CD 9–10 steel for two sizes of bend
specimens [37].

integral is defined for a blunt notch as well as a crack. The *J* integral concept
was applied to crack initiation and growth from a blunt notch in the ductile
fracture regime [39]. A critical value of *J* was defined in the same manner as
for a sharp crack, i.e. the point of first crack advance. Figure 3.30 shows
resistance-type curves for crack advance from a notch in an ASTM A471
steel. The critical value of *J* was taken at the intercept of the fitted line with

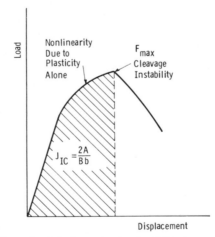

FIG. 3.28. Schematic of $J_{\rm Ic}$ determination for fully cleavage fractures in steels.

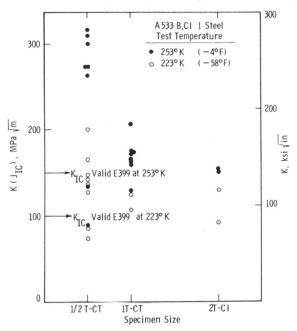

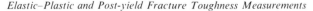

FIG. 3.29. K_{Ic} values from J_{Ic} tests in the cleavage range versus CTS specimen size, A533B Cl 1 steel [38].

the ordinate since a blunting line would not have any significance. The critical value of J for crack advance is plotted as a function of notch root radius for this steel and a 6061–T651 aluminium alloy in Fig. 3.31. The critical value of J is seen to be a linear function of notch root radius which extrapolated to J_{Ic} for a zero notch radius.

J_{Ic} is a fracture criterion which can be defined and applied independently from linear elastic considerations. However, since K_{Ic} has been the standard measure of fracture toughness since the early development of fracture mechanics methodology, a comparison between J_{Ic} and K_{Ic} is often of interest. To compare these two fracture parameters on an equal basis would require that they be compared at a consistent measurement point. Since J_{Ic} is defined by E399 at 2 % crack advance they are not taken at a consistent measurement point. However, in some cases they can be compared on a nearly equal basis. Figure 3.32 illustrates schematically fracture characterisation by both J_{Ic} and K_{Ic}. The two points are coincident when the fracture occurs by a cleavage mechanism and are nearly coincident when the R curve has a very shallow slope. There can be a significant

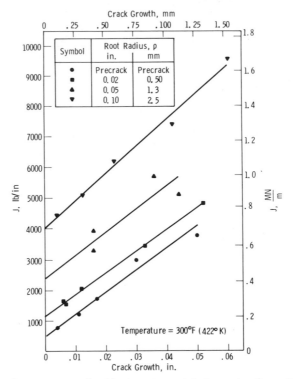

FIG. 3.30. *J* resistance curves for blunt centre notched panels of an ASTM A471 Ni–Cr–Mo–V rotor steel [39].

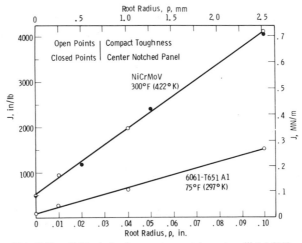

FIG. 3.31. Critical *J* values versus notch root radii (ρ) [39].

difference between the two points where the *R* curve has a fairly steep slope, as is often the case for a ductile fracture mechanism. In this case J_{Ic} would be lower than the K_{Ic} giving a conservative lower bound for fracture toughness.

Examples of comparisons between J_{Ic} and K_{Ic} are given first for a cleavage mode of fracture where the two measurement points should be

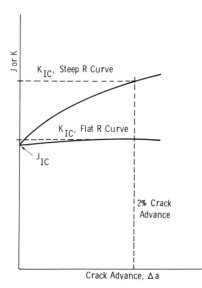

FIG. 3.32.   Schematic of *R* curve showing the differences between the J_{Ic} and K_{Ic} measurement points.

nearly coincident. Figure 3.33 [42] shows a comparison of valid K_{Ic} values with J_{Ic} values for various steels which failed by cleavage. In both cases the K_{Ic} and J_{Ic} values agree within experimental error. Data previously shown in Fig. 3.29 [37] for an A533B steel fracturing in the transition region by cleavage gives a comparison of a valid K_{Ic} value with J_{Ic} values from specimens of various sizes. Although the J_{Ic} values show increasing scatter with decreasing specimen size, the lower bound of the J_{Ic} values compare with K_{Ic} as would be expected from a weakest link statistical model [10].

In the ductile, 'upper shelf' regime of fracture for steels a comparison is difficult since valid K_{Ic} tests generally require very large specimens and such data are not often available. A comparison of K_{Ic} with J_{Ic} is given for an ASTM A471 Ni–Cr–Mo–V rotor in Fig. 3.34 [32]. Fracture toughness from 1T–CT specimens is determined by the J_{Ic} method and converted to

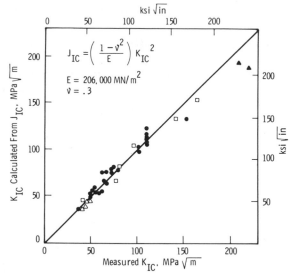

FIG. 3.33.   Correlation between K_{1c} values calculated from J_{1c} and actual K_{1c} values [37]. (●) Marendet & Sanz [37]; (▲) Begley & Landes [20]; □ Logsdon [42]; △ Crensot Loire [43].

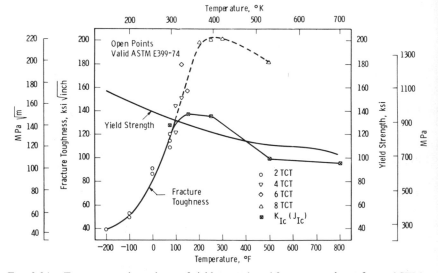

FIG. 3.34.   Temperature dependence of yield strength and fracture toughness for an ASTM A471 Ni–Cr–Mo–V rotor steel [32].

K_{Ic} to compare with toughness values from valid K_{Ic} tests on 8T–CT specimens. As would be expected the J_{Ic} values are lower than the K_{Ic} values.

An R curve determined by linear elastic methods on a large specimen of 0·25 in (6·35 mm) thick 2024–T351 aluminium is used for comparison with similar values determined on a smaller specimen by J integral techniques, Fig. 3.35 [41]. The elastic–plastic values on the R curve compare well with

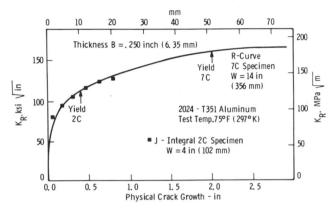

FIG. 3.35. Resistance curve for a 2024–T351 aluminium alloy from large and small CLWL specimen tests; comparison of K and J analysis [41].

the linear values; however, the range of crack growth measurement capacity is reduced.

In Section 3.2.5 the value $\alpha = 25$ was proposed for the general size limitation defined by eqn. (3.17). Examples of data generated to determine values for α are given in Fig. 3.36, for an ASTM A216 cast steel [40] which fractured by a ductile mechanism, and Fig. 3.37, for an A533B steel which fractured by cleavage [38]. For both cases a value of 25 for α is acceptable.

Geometry effects are important in the determination of J_{Ic} because, like K, it must characterise fracture independently of geometrical considerations if it is truly a crack tip parameter. Figure 3.5 shows J_{Ic} values determined for Ni–Cr–Mo–V rotor steel using both a CTS specimen (with predominantly a bending type of load) and a centre-crack panel, CCP (with a purely tension loading) [3]. As is shown, the J_{Ic} values are consistent for the two geometries; however, the subsequent R curves as well as the fracture paths were different for the two specimens.

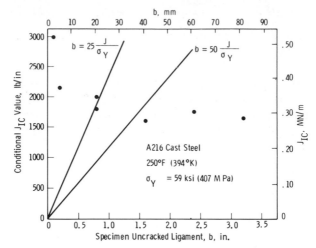

FIG. 3.36. Conditional J_{lc} versus specimen uncracked ligament size for evaluation of J_{lc} specimen size limitations [40].

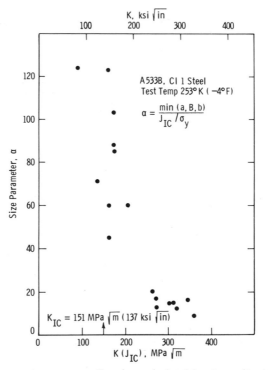

FIG. 3.37. Size parameter, α, versus K_{lc} values calculated for J_{lc} results; A533B Cl 1 steel [38].

3.5 PRESENT RECOMMENDED J_{Ic} TEST PRACTICE

A recommended practice for J_{Ic} testing is presently in the development stage through ASTM Committee E–24. The purpose of this recommended practice is to establish a common method for determining J_{Ic} through the development of J versus Δa resistance curves so that results from various laboratories can be compared. Although this recommended practice is still in the development stage and may not be completed for about a year, most of the details of the method have been agreed upon by a steering committee working through E–24. A summary of the recommended procedure in its present form is presented here.

J_{Ic} is the value of the J integral at incipient crack growth. It is measured on the J versus Δa resistance curve as the point where a fitted line of crack extension intersects the blunting line. Although the procedure is primarily described for the method of crack front marking where several specimens are required, it can also be applied to single specimen techniques where an average amount of crack extension is measured continuously or at multiple discrete points. Presently, this technique applies only to the compact specimen and the three-point bend specimen with a span to width ratio, S/W, equal to 4. Further analysis is to be completed on the four-point bend and centre-cracked panel specimens before recommendations as to the testing procedure can be made. A step by step procedure is as follows:

1. Compact or three-point bend specimens should be used with a minimum a/W ratio of 0·5.
2. Specimens should be precracked in fatigue with maximum loads less than one half of the nominal crack tip yield; see ASTM Standard E399–74 for appropriate nominal stress expressions.
3. Load each specimen to different displacement control if possible, Fig. 3.38. The actual values of displacement will be characterised by the criteria necessary to define the J versus Δa R curve as given below. Note: all displacements should be load line displacements with measurements taken from the specimen itself. If this is not possible, corrections must be made to the load versus displacement record to account for measurement location.
4. Load versus load line displacements should be recorded by an x–y recorder.
5. Unload each specimen and mark the crack. Heat tinting is an easy way to mark the crack for most steels. For other materials, further fatigue cycling to extend the crack will adequately mark the crack front.

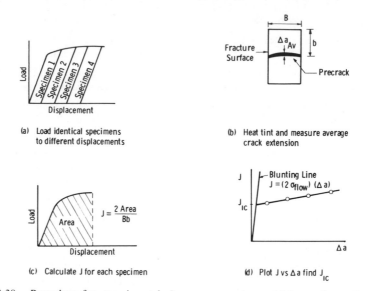

(a) Load identical specimens to different displacements

(b) Heat tint and measure average crack extension

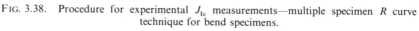

(c) Calculate J for each specimen

(d) Plot J vs Δa find J_{IC}

FIG. 3.38. Procedure for experimental J_{1c} measurements—multiple specimen R curve technique for bend specimens.

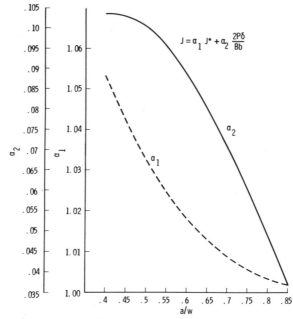

FIG. 3.39. Non-dimensional coefficients for Merkle–Corten J corrections of a compact specimen.

6. Pull the specimen apart and measure the crack extension. The crack extension should be taken to include all crack extension from the fatigue precrack to the end of the heat tinting mark, Fig. 3.38b. Recent experience has shown that it may be necessary to use an average of nine measurement points across the full crack front for an accurate measure of Δa.

7. For compact specimens the following estimation procedure should be used to calculate J from the load versus load point displacement record.

$$J = (\alpha_1 + \alpha_2)\frac{2A}{Bb} \qquad (3.23)$$

where α_1 and α_2 are coefficients developed by Merkle & Corten [5] to account for the tension component in the compact specimen. The values of α_1 and α_2 are plotted against a/W ratios in Fig. 3.39. The remaining symbols are as follows, A is the area under the load versus load point displacement curve, B is the thickness of the specimen, b is the remaining ligament of the specimen, P is the final load value and V is the final load point displacement. For the three-point bend specimen with $S/W = 4$, the value of J is found by

$$J = \frac{2A}{Bb} \qquad (3.24)$$

8. The following criteria should be used to develop the J versus Δa R curve, as shown in Fig. 3.40.

    a. The blunting line should be drawn according to the equation $J = 2\sigma_Y \Delta a$, where σ_Y is the effective yield strength, the average of the yield strength and ultimate strength from a uniaxial tensile test.

    b. At least four points should be used to develop the R-curve line. (Note, the first specimen should be loaded to maximum load in order to determine the displacements necessary for the remaining specimens).

    c. The maximum crack extension allowable, Δa_2, providing that the specimen size criterion is not invalidated during the test, is given by the maximum of the two following criteria: (i) crack extensions up to 1·5 mm past the value of crack extension at the intersection of the R-curve line and the blunting line, Δa_0; (ii) crack extensions up to $\Delta a = (5 \times J_{Ic}/\sigma_Y) + \Delta a_0$.

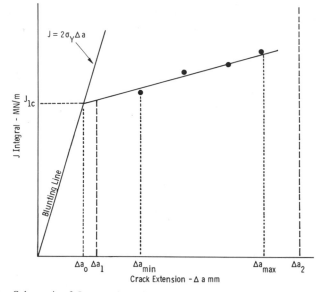

FIG. 3.40.   Schematic of J versus Δa resistance curve showing the recommended procedure for J_{Ic} determination.

    d. The minimum amount of crack extension, Δa_{min}, that may be used to define the R curve are those values greater than $0·1$ mm past Δa_0, i.e. Δa_1. (Values to the left of this line cannot be sufficiently well distinguished from the blunting line to be used for the R-curve determination.)

    e. The total range of data required during the test should be such that the distance between the maximum, Δa_{max}, and minimum, Δa_{min}, crack extensions will not be less than two-thirds of the distance between the maximum crack extension, Δa_{max}, and the intersection of the R-curve line with the blunting line, Δa_0.

    f. All specimens must adhere to the size requirements of B, $b \geq 25\,J/\sigma_Y$, for all values of J obtained during the test.

    g. A linear fit line, via least squares on J, should be fitted through all of the valid points, as defined above.

    h. If the above criteria are met then the intersection of the blunting line and the R-curve line is the critical value of J at the onset of crack extension, or J_{Ic}.

  9.   For tests in which unstable crack propagation occurs during loading, the maximum load may be considered as the point at which

the critical value of J may be measured, J_{Ic}, providing it is possible to ensure that stable crack growth was less than twice the value of crack extension as defined by the blunting line, $(\Delta a < J/\sigma_Y)$. If not, an R curve should be developed to determine J_{Ic}.

REFERENCES

1. British Standards Institution (1972). *Methods for Crack Opening Displacement (COD) Testing*, Draft for Development 19.
2. Landes, J. D. & Begley, J. A. (1972). The effect of specimen geometry on J_{Ic}, ASTM–STP 514, Part II, pp. 25–39.
3. Begley, J. A. & Landes, J. D. (1976). Serendipity and the J integral, *Int. J. Fract.*, **12**(3), pp. 764–6.
4. Rice, J. R., Paris, P. C. & Merkle, J. G. (1973). Some further results on J integral analysis and estimates, ASTM–STP 536, pp. 231–45.
5. Merkle, J. G. & Corten, H. C. (1974). A J integral analysis for the compact specimen considering axial force as well as bending effects, *J. Press. Vess. Tech., Trans. ASME*, Series J 96, **4**, pp. 286–92.
6. Sumpter, J. D. G. & Turner, C. E. (1976). Method for laboratory determination of J_c, ASTM–STP 601, pp. 3–18.
7. Srawley, J. E. (1976). On the relation of J_I to work done per unit uncracked area: (total or component (due to crack)), *Int. J. Fract.*, **12**, pp. 470–4.
8. Chipperfield, C. G. (1976). Detection and toughness characterisation in 316 stainless steels, *Int. J. Fract.*, **12**(6), pp. 873–6.
9. Begley, J. A., unpublished data.
10. Landes, J. D. & Begley, J. A. (1974). Test results from J integral studies: an attempt to establish a J_{Ic} testing procedure, ASTM–STP 560, pp. 170–86.
11. Garwood, S. J., Robinson, J. N. & Turner, C. E. (1975). The measurement of crack growth resistance curves (R-curves) using the J integral, *Int. J. Fract.*, **11**(3), pp. 528–30.
12. Robinson, J. N. (1976). An experimental investigation of the effect of specimen type on the crack opening tip displacement and J integral fracture criteria, *Int. J. Fract.*, **12**(5), pp. 723–37.
13. Fields, B. A. & Miller, K. J. (1977). A study of COD and crack initiation by a replication technique, *Eng. Fract. Mech.*, **9**, pp. 137–46.
14. Bucci, R. J., Paris, P. C., Landes, J. D. & Rice, J. R. (1972). J integral estimation procedures, ASTM–STP 514, Part II, pp. 40–69.
15. Wells, A. A. (1971). The status of COD in fracture mechanics, 3rd Canadian Congress of Applied Mechanics, Calgary.
16. Robinson, J. N. & Tetelman, A. S. (1975). The relationship between crack tip opening, displacement, local strain and specimen geometry, *Int. J. Fract.*, **11**(3), pp. 453–68.
17. Ingham, T., Egan, G. R., Elliot, D. & Harrison, T. C. (1971). The effect of geometry on the interpretation of COD test data, in *Practical Application of fracture mechanics to pressure vessel technology*, Inst. of Mech. Engs., London, p. 200.

18. Elliot, D. & Mary, M. J. (1969). The effect of position of measurement on COD, BISRA, Open Report MG/C/47/69.
19. Chipperfield, C. G., Knott, J. F. & Smith, R. F. (1973). Critical crack opening displacement in low strength steels, 3rd International Conference on Fracture, Munich.
20. Begley, J. A. & Landes, J. D. (1972). The *J* integral as a fracture criterion, ASTM-STP 514, Part II, pp. 1–23.
21. Paris, P. C. (1972). Discussion to Ref. 20, p. 21.
22. Clarke, G. A., Andrews, W. R., Paris, P. C. & Schmidt, D. W. (1976). Single specimen tests for J_{Ic} determination, *Mechanics of Crack Growth*, ASTM-STP 590, pp. 27–42.
23. Paris, P. C. & Herman, L. (1975). Improved compliance techniques for observing crack growth with reference to *J* testing, presented at the 9th National Symposium on Fracture Mechanics, Pittsburgh, Pennsylvania, Aug.
24. Landes, J. D. & Wei, R. P. (1973). Kinetics of subcritical crack growth under sustained loading, *Int. J. Fract.*, **9**(3), pp. 277–93.
25. Johnson, H. H. & Wilner, A. M. (1965). Moisture and stable crack growth in a high strength steel, *Appl. Mat. Research*, **4**, Jan., pp. 34–40.
26. Mandaret, B. & Sanz, G. (1976). Characterisation of the fracture toughness of steels by the measurement with a single specimen of J_{Ic} and the parameter K_{Bd}, presented at the 10th National Symposium on Fracture Mechanics, Philadelphia, Pennsylvania, Aug.
27. Harrison, J. D. (1976). The potential drop method in J_{Ic} testing, presented at the ASTM E-24 Committee Meetings, ASTM Committee Week, Lake Buena Vista, Florida, Mar.
28. Clarke, G. A. & Landes, J. D. (1977). Toughness testing of materials by *J* integral techniques, presented at the 106th AIME Annual Meeting, Atlanta, Georgia, Mar.
29. Underwood, J. H., Winters, D. C. & Kendall, D. P. (1976). End on ultrasonic crack measurements in steel fracture toughness specimens and thick-wall cylinders, in *The detection and measurements of cracks*, The Welding Institute, Cambridge, England, pp. 31–9.
30. Hickerson, J. P. (1976). Use of the resonant frequency technique in J_{Ic} testing, presented at the ASTM E-24 Committee Meetings, ASTM Committee Week, Lake Buena Vista, Florida, Mar.
31. Donald, K. & Schmidt, D. (1977). Rotational effects on compact specimens, presented at the ASTM E-24 Meetings, ASTM Committee Week, Norfolk, Virginia, Mar.
32. Logsdon, W. A. & Begley, J. A. (1977). Upper shelf temperature dependence of fracture toughness for four low to intermediate strength ferritic steels, *Eng. Fract. Mech.* **9**, pp. 461–70.
33. Griffis, C. A. & Yoder, G. R. (1976). Initial crack extension in two intermediate strength aluminium alloys, *Trans. ASME, J. Eng. Mat. & Technol.*, **98**, pp. 152–8.
34. Yoder, G. R. & Griffis, D. A. (1976). Application of the *J* integral to the initiation of crack extension in a titanium 6A1-4V alloy, *Mechanics of crack growth*, ASTM-STP 590, pp. 61–81.
35. Logsdon, W. A. (1977). Cryogenic fracture mechanics properties of several

manufacturing process/heat treatment combinations of Inconel X750, in *Advances in cryogenic engineering*, K. D. Timmerhaus, R. P. Reed and A. F. Clark, eds, vol. 22. Plenum Publ. Co., London.

36. Logsdon, W. A., Wells, H. M. & Kossowsky, R. (1976). Fracture mechanical properties of austenitic stainless steel for advanced cryogenic application, *Proceedings 2nd International Conference on Mechanical Behaviour of Materials*, Boston, Massachusetts, Aug.

37. Marendet, B. & Sanz, G. (1976). Characterisation of the fracture toughness of steels by the measurement with a single specimen of J_{Ic} and the parameter K_{Bd}, presented at the 10th National Symposium on Fracture Mechanics, Philadelphia, Pennsylvania, Aug.

38. Sunamoto, D., Satoh, M., Funada, T. & Tomimatsu, M. (1977). Specimen size effect on *J*-integral fracture toughness, *Proceedings of the Fourth International Conference on Fracture*, **3**, ICF4, pp. 267–72, Waterloo, Canada, June.

39. Begley, J. A., Logsdon, W. A. & Landes, J. D. (1976). A ductile rupture blunt notch fracture theory, presented at the 10th National Symposium on Fracture Mechanics, Philadelphia, Pennsylvania, Aug.

40. Landes, J. D. & Begley, J. A. (1977). Recent developments in J_{Ic} testing, *Recent Developments in Fracture Mechanics Test Method Standardization*, ASTM–STP 632.

41. McCabe, D. E. (1976). Determination of *R* curves in structural materials using nonlinear mechanics methods, presented at the 10th National Symposium on Fracture Mechanics, Philadelphia, Pennsylvania, Aug.

42. Logsdon, W. A. (1974). Elastic–plastic (J_{Ic}) fracture toughness values: their experimental determination and comparison with conventional linear elastic (K_{Ic}) fracture toughness values for five materials, Westinghouse Scientific Paper, 74-1E7-FM-PWR-P1.

43. Creusot Loire & Ponsot, J. Results obtained during the contract no. 72-7-0602.00.221.75.01 of the French Délégation à la Recherche Scientifique et Technique.

4

Methods for Computing Contour Integrals

T. K. Hellen

CEGB, Berkeley, UK

4.1 INTRODUCTION

In this paper we consider methods for evaluating crack tip parameters for practical applications. In view of the complex nature of the equations of state which exist even in the most simple test pieces, a desirable approach is to compute the displacements, stresses and strains at required points in the structure being analysed by suitable numerical techniques. Crack tip parameters may then be evaluated such as the J integral defined in Chapter 2. In this case, by the path independence discussed in Section 2.2.2.1.3, the contour of integration may be chosen to pass through areas of structure away from the tip, where stress and strain gradients are relatively mild. This in turn enhances the accuracy of the numerical integration in techniques such as the finite element method. It also reduces the importance of the actual crack tip modelling, such as the effects of blunting with increased local plasticity.

From an experimental point of view, determination of J and of its critical value J_c has been discussed in Chapter 3 for test-piece geometries. For structures of complex shape containing cracks, J values for a given loading system can be predicted in two different ways. The first is analytical, using assumptions such as the Dugdale model or power hardening laws. Edmunds & Willis [1] have obtained J values for mode III behaviour in elastic–perfectly plastic materials using matched asymptotic expansions. The common factors with analytical techniques are that varying degrees of approximation are required and the classes of structure and loading are limited. Only after considerable mathematical labour and user experience can the accuracies and breadth of application of each analytical technique be assessed.

The second approach pertains to numerical techniques. Here, the use of computers is required and permits the practical application of such proven techniques as the finite element method and boundary integral equations. The former, being at the present time the most highly developed and widely used method available, is utilised exclusively throughout this chapter. A brief summary of the method and of the matrix and vector notations going with it is given in Appendix 6 and 5, respectively. Since the method is based upon assumptions for displacements (and/or stresses), described by polynomials over elements of finite size, it cannot exactly represent the singularity existing at the crack tip. Appendix 6 therefore includes a brief section on the various approximations used for crack tip modelling.

By thus abstracting the general aspects of the finite element method as applied to fracture mechanics in appendices, the main text of this chapter is able to concentrate on those aspects typical of the occurrence of plasticity and on some applications to cracked plate structures.

There are two important interrelated consequences to the occurrence of plasticity. The first is the need for applying incremental theory, which in finite element terms usually means that the applied load has to be broken down into a series of incremental steps with the displacements, stresses and strains resulting from each step being stored on a permanent data file. A post-processor program may then be executed to read this file and calculate J values for a variety of contours and load levels. Although incremental theory is used, the J results will, theoretically, be path dependent, although according to Section 2.2.2.1.3 reasonable independence has been found in practice. Moreover, J formally ceases to be path independent in other cases too, such as for thermal or creep strains, non-homogeneous materials or three-dimensional structures. In such cases the J^* integral discussed in Section 2.2.2.2 may be considered. To further familiarise the reader with this alternative contour integral, J^* as well as J computations will be included in this chapter.

The second consequence of leaving the realm of linear elasticity is the need for constitutive equations adequately describing the material's plastic behaviour. This aspect will be dealt with in Sections 4.2.3 and 4.2.4.

4.2   FINITE ELEMENT FORMULATION OF ELASTIC–PLASTIC ANALYSIS

4.2.1 Basis of Method
The solution of non-linear problems such as deformation behaviour in the plastic range may be attempted by one of three basic techniques: incremental, iterative or a mixture of these two. For a concise description

and comparative evaluation of these techniques the reader is referred to such textbooks as Ref. [2].

The most generally applicable of these is the incremental approach, which shall be followed here. In this approach the applied loads (here taken to include any combination of mechanical or thermal loads, prescribed displacements or strains) are defined as a set of increments after that load which causes first yield. During the application of each increment linear behaviour is assumed, *inter alia* the stiffness matrix is kept constant at the value obtained at the end of the previous increment.

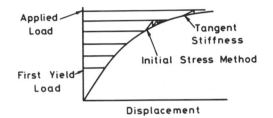

FIG. 4.1. Solution procedure with load applied in small increments.

Figure 4.1, in addition to showing schematically the load–deflection curve for a typical point in the structure, indicates two methods frequently used for implementing the incremental approach. In the tangent stiffness method the tangent moduli are used to compute the stiffness matrix at every step. In the initial stress method a number of iterations are performed for each load step until the stresses equilibrate the loads. Each iteration represents a pseudo-elastic solution of the basic equation (A6.12) in which the difference between the correct stress and the computed value from the preceding iteration step is treated as the 'initial stress'. Iteration terminates when the residual stresses required to satisfy equilibrium are sufficiently small. Another method combines these two approaches by modifying the stiffness matrix for each load increment, but is held constant for the iterations within the increment.

The implementation of both the tangent stiffness and the initial stress method is outlined in Section 4.2.5. Because of the finite number of points of reference necessary to the finite element method, the first yield level is defined as when the highest equivalent stress over these finite number of points attains the yield value. The number of load increments required can only be established by experience. In practice, however, such experience is rapidly attained and suitable numbers are given in the subsequent examples. Results at each load step are stored and used for subsequent J or J^* calculations.

4.2.2 Input Requirements

From an input point of view, the majority of the data which have to be supplied by the user is the same as for elastic analyses. This includes mesh topology and geometry, loadings, fixings, etc. The extra data for the non-linear part can be very little for straightforward problems and fall into two categories. Firstly, various control parameters are required to state the

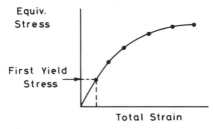

FIG. 4.2.   Non-linear stress–strain data definition.

number of load steps, maximum number of iterations allowed per load step, convergence tolerances, and other similar data depending on the actual program used. Secondly, non-linear stress–strain data are required. This can be specified in equation form or, preferably, as a sequence of points defining segments, thereby allowing general-shape curves, as shown in Fig. 4.2. Different curves can be specified for different materials and temperatures.

Further refinements in the non-linear input data include options on the method of solution (initial stress, tangent stiffness or a mixture), automatic control of load increment sizes or a specification of non-equal sizes, control parameters for selective printing and storage of results on a permanent data file. The load to first yield is usually determined automatically by comparing the highest stress over the reference points of the structure with the given first yield stress. The resulting ratio is used to scale-down all results and loads to give the elastic results corresponding to the first yield load. The load increment sizes refer to the remaining load.

For creep problems, although the methods of solution are not identical to those for elastic–plasticity, the extra non-linear data requirements are very similar.

4.2.3 Yield Criteria and Flow Rules

For isotropic, isothermal conditions a function of the stress components can be defined which describes the limits of elastic behaviour and hence is called a *yield criterion*.

The most commonly used yield criterion is the von Mises criterion, defined as

$$F(\{\sigma\}, h) = \sqrt{3}\bar{\sigma} - Y(h) = 0 \tag{4.1}$$

in which $\bar{\sigma}$ equals the second stress invariant, see for example Ref. [3], h is a hardening parameter and in three-dimensional space is given by

$$\bar{\sigma}^2 = \tfrac{1}{6}[(s_x - s_y)^2 + (s_y - s_z)^2 + (s_z - s_x)^2] + \tau_{xy}^2 + \tau_{yz}^2 + \tau_{zx}^2 \tag{4.2}$$

where

$$s_{ij} = \sigma_{ij} - \tfrac{1}{3}\sigma_{\kappa\kappa}\delta_{ij} \tag{4.3}$$

The s_{ij} are termed *deviatoric* stresses, and plastic deformation is assumed to be independent of the hydrostatic stress component, $\sigma_{\kappa\kappa}$ (using summation convention), although in eqn. (4.2) σ_x can be used instead of s_x, etc.

It is straightforward to show that

$$\bar{\sigma}^2 = \tfrac{1}{2}(s_x^2 + s_y^2 + s_z^2) + \tau_{xy}^2 + \tau_{yz}^2 + \tau_{zx}^2$$

or

$$\bar{\sigma}^2 = \tfrac{1}{2}s_{ij}s_{ij} \tag{4.4}$$

again using summation convention, so that the equivalent stress required by eqn. (4.1) is

$$Y(h) = \sigma^* = (\tfrac{3}{2}s_{ij}s_{ij})^{1/2} \tag{4.5}$$

Note that when using the summation convention, the interpretation of the shear strains is mathematical rather than engineering.

Analogous to the equivalent stress is the equivalent plastic strain, defined by

$$\varepsilon_p^* = \int d\varepsilon_p^* \, dt \tag{4.6}$$

over the strain history. Due to the requirement of incompressibility, $(d\varepsilon_p)_{\kappa\kappa} = 0$ and the plastic strain increments are deviatoric:

$$(d\varepsilon_p^*)^2 = \tfrac{2}{9}[(d\varepsilon_{px} - d\varepsilon_{py})^2 + (d\varepsilon_{py} - d\varepsilon_{pz})^2 + (d\varepsilon_{pz} - d\varepsilon_{px})^2]$$
$$+ \tfrac{1}{3}[d\gamma_{pxy}^2 + d\gamma_{pyz}^2 + d\gamma_{pzx}^2] \tag{4.7}$$

Once yielding has occurred additional information, in the form of a constitutive relation between strain and stress increments, is required to describe the plastic behaviour of a work-hardening (or -softening) material. The constitutive equation or *flow rule* to be used here is the Prandtl–Reuss flow rule, stating that the plastic strain increments and deviatoric stresses are both directed normal to the yield surface.

Written in vector form, this flow rule reads

$$d\{\varepsilon_p\} = d\lambda \frac{\partial F(\{\sigma\}, \{\varepsilon_p\}, h)}{\partial \{\sigma\}} \tag{4.8}$$

where $d\lambda$ is a constant of proportionality, which can be determined from uniaxial test data, and h a hardening parameter. If the term $\{\varepsilon_p\}$ is absent, isotropic hardening occurs, i.e. during plastic flow the loading surface expands uniformly about the origin in stress space, maintaining the shape and orientation of the yield surface. As an example Fig. 4.3 shows the von Mises yield surface in section in σ_1, σ_2 space and the direction of the increment of plastic strain, $d\{\varepsilon\}_p$. Two yield surfaces are shown,

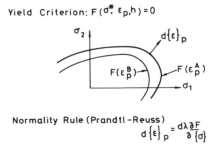

Yield Criterion: $F(\sigma^*, \varepsilon_p, h) = 0$

Normality Rule (Prandtl–Reuss)
$$d\{\varepsilon\}_p = d\lambda \frac{\partial F}{\partial \{\sigma\}}$$

FIG. 4.3.   Two curves corresponding to two plastic strains ε_p^A and ε_p^B (isotropic hardening).

corresponding to two different plastic strains ε_p^A and ε_p^B to show the surface change for isotropic hardening.

If the term h is absent, kinematic hardening exists, i.e. during plastic flow the loading surface translates as a rigid body in stress space, maintaining the size as well as the shape and orientation of the yield surface. If both the $\{\varepsilon_p\}$ and h terms are present, a combination of isotropic and kinematic hardening will occur. For a discussion of the various hardening models the reader is referred to Armen et al. [4] or to Besseling [5]. In the present work only isotropic hardening will be discussed—in addition to elastic–perfectly plastic behaviour—hence no dependence on $\{\varepsilon_p\}$ is considered.

The hardening parameter dh equals the plastic work done during an increment of plastic deformation:

$$dh = \bar{\sigma} \, d\bar{\varepsilon}_p \qquad (4.9)$$

Hence

$$\frac{\partial F}{\partial h} = -\frac{dY}{dh}$$

from eqn. (4.1), i.e.

$$\frac{\partial F}{\partial h} = -\frac{d\bar{\sigma}}{dh} = -\frac{d\bar{\sigma}}{d\bar{\varepsilon}_p \bar{\sigma}} \frac{1}{\bar{\sigma}} = -\frac{H}{\bar{\sigma}} \qquad (4.10)$$

As shown by Zienkiewicz *et al.* [6], the stress–strain curve has an instantaneous slope H given by

$$H \, d\lambda = -\frac{\partial F}{\partial h} \, dh \qquad (4.11)$$

For an elastic–perfectly plastic material, $H = 0$. Combining eqns. (4.9), (4.10) and (4.11):

$$d\lambda = d\bar{\varepsilon}_p \qquad (4.12)$$

Note that a similar analysis may be performed based on the Tresca yield criterion. Experimental evidence indicates that the von Mises criterion is generally the more realistic of the two. Hill [3] presents results due to Taylor and Quinney to demonstrate this.

4.2.4 The Relationship Between Displacement, Plastic Strain and Stress

The total strain increment at any point in the structure equals the sum of the elastic, plastic, creep and thermal (or initial) strains. Written in vector form, this gives

$$d\{\varepsilon\} = d\{\varepsilon_e\} + d\{\varepsilon_p\} + d\{\varepsilon_c\} + d\{\varepsilon_0\} \qquad (4.13)$$

and the stress increment $d\{\sigma\}$ is given by

$$d\{\sigma\} = [D] \, d\{\varepsilon_e\} \qquad (4.14)$$

where the vector notation and [D] are described in Appendices 5 and 6, respectively.

For the present analysis the vectors $d\{\varepsilon_c\}$ and $d\{\varepsilon_0\}$ may be omitted. Their readmission in the final formulae is straightforward; they simply have to be subtracted from $d\{\varepsilon\}$ whenever this vector appears.

For any required criterion with a hardening parameter h, if the stress vector $\{\sigma\}$ at any point in the given structure is sufficient to cause yielding, then

$$F(\{\sigma\}, \{\varepsilon_p\}, h) = 0 \qquad (4.15)$$

During plastic deformation, $dF = 0$ and so

$$\left(\frac{\partial F}{\partial \{\sigma\}}\right)^T d\{\sigma\} + \frac{\partial F}{\partial h} dh = 0 \qquad (4.16)$$

From eqns. (4.13), (4.14) and (4.8), this gives

$$\left(\frac{\partial F}{\partial \{\sigma\}}\right)^T [D] d\{\varepsilon\} = d\lambda \left(\frac{\partial F}{\partial \{\sigma\}}\right)^T [D] \frac{\partial F}{\partial \{\sigma\}} - \frac{\partial F}{\partial h} dh \qquad (4.17)$$

From eqns. (4.11) and (4.17),

$$\left(\frac{\partial F}{\partial \{\sigma\}}\right)^T [D] d\{\varepsilon\} = d\lambda \left[H + \left(\frac{\partial F}{\partial \{\sigma\}}\right)^T [D] \frac{\partial F}{\partial \{\sigma\}} \right] \qquad (4.18)$$

from which $d\lambda$ may be determined. Thus

$$d\{\sigma\} = [D] d\{\varepsilon\} - [D] d\{\varepsilon_p\}$$

$$= [D] d\{\varepsilon\} - d\lambda [D] \frac{\partial F}{\partial \{\sigma\}}$$

i.e.

$$d\{\sigma\} = ([D] - [D_p]) d\{\varepsilon\} \qquad (4.19)$$

$[D_p]$ may be considered as a plastic modulus matrix, given by eqns. (4.8) and (4.19) as

$$[D_p] = [D] \frac{\partial F}{\partial \{\sigma\}} \left(\frac{\partial F}{\partial \{\sigma\}}\right)^T [D] \div \left[H + \left(\frac{\partial F}{\partial \{\sigma\}}\right)^T [D] \frac{\partial F}{\partial \{\sigma\}} \right] \qquad (4.20)$$

Defining the plastic stress increment by $d\{\sigma_p\}$,

$$d\{\sigma\} = d\{\sigma_e\} - d\{\sigma_p\}$$

$$d\{\sigma_p\} = [D] d\{\varepsilon_p\}$$

and

$$d\{\sigma_p\} = [D_p] d\{\varepsilon\} \qquad (4.21)$$

Thus, the plastic stress increment can be determined from the total strain, which in turn is derived from the current incremental displacements and the plastic modulus matrix.

4.2.5 Solution Methods

The maintenance of equilibrium can be expressed using the notation of Appendix 6. Based on the minimisation principle used there, eqn. (A6.12) may be rewritten

$$\int_V [B]^T \{\sigma\} \, dV = \{R\} \tag{4.22}$$

where the integration is over the structural volume and $\{R\}$ is the external load vector, including thermal and other non-mechanical effects.

An increase in load of $d\{R\}$ will effect a stress increase of $d\{\sigma\}$, resulting in a new equilibrium condition:

$$\int_V [B]^T \, d\{\sigma\} \, dV = d\{R\} \tag{4.23}$$

The attainment of equilibrium is ensured by iterating to remove the quantity I:

$$I = \int_V [B]^T \, d\{\sigma\} \, dV - d\{R\} \tag{4.24}$$

Different solution methods arise by representing the integral in different forms. For instance, from eqns. (4.19) and (A6.3)

$$d\{\sigma\} = [D_d][B] \, d\{u\}$$

where

$$[D_d] = [D] - [D_p]$$

In eqn. (4.23), this gives

$$\int_V [B]^T [D_d][B] \, dV \, d\{u\} = d\{R\}$$

or

$$[K_T] \, d\{u\} = d\{R\} \tag{4.25}$$

which gives the *tangent stiffness method* [7–9]. As already indicated in Section 4.2.1, this method repeats the elastic accumulation of the modulus matrix at every load increment with substitution of the current constitutive properties instead of the initial elastic ones. From a computational point of view, this process is lengthy and expensive, although the load–deflection curve is followed along tangents in the form of a Newton–Raphson method. Load increments have to be small and within each increment convergence is obtained when the displacement or load residuals are suitably small. The method is necessary when material properties strongly depend on temperature or when large strains are considered.

The *initial stress method* [6, 10] is based on the relation

$$d\{\sigma\} = d\{\sigma_e\} - d\{\sigma_p\}$$

which gives in eqn. (4.25)

$$\int_V [B]^T d\{\sigma_e\} dV - \int_V [B]^T d\{\sigma_p\} dV = d\{R\}$$

The first integral is the elastic stiffness matrix $[K_0]$ and

$$[K_0] d\{u\} = d\{R\} + d\{S\} \tag{4.26}$$

Here, at any stage of iteration within a load increment, the current state of plastic stress increment is used to assess the equivalent nodal force vector, $d\{S\}$, from which a re-solution gives an improved set of displacements. The principal advantage of this method over the tangent stiffness method is that the reformulation and reinversion of the stiffness matrix is not required, an important consideration particularly in three-dimensional structures. Only a reduction of the vector on the right-hand side is required at each iteration; consequently far more iterations are possible compared to a tangent stiffness approach in the same time. Its disadvantage is that convergence slows down as plasticity increases.

In order to optimise on the attainment of adequate convergence whilst minimising the computational time and cost it is worthwhile combining the above methods, using initial stress with an occasional stiffness reformulation and reinversion from the tangent stiffness method.

4.3 NUMERICAL EVALUATION OF CONTOUR INTEGRALS

4.3.1 Outline of the Method
In Section 2.2.2.1.3 attention was paid to the results of numerical evaluations of the J integral. In the following the procedure for such an evaluation will be described in some detail. This description pertains to a

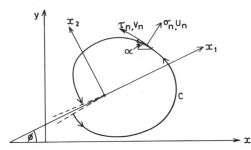

FIG. 4.4. Arbritrary crack and integral contour configuration.

computer program written by Neale [11] for arbitrary inclined cracks in two-dimensional structures. Initially, we consider linear elastic behaviour only. With reference to a cartesian global system of coordinates, suppose the local crack coordinates (x_1, x_2) are such that x_1 is along the crack and x_2 is normal to it (Fig. 4.4). This avoids unnecessary restrictions on the mesh required for the finite element program. For the present case,

$$J = \int_C \left\{ Z \, dx_2 - [\sigma_n, \tau_n] \begin{bmatrix} \dfrac{\partial u_n}{\partial \xi} \\[2mm] \dfrac{\partial v_n}{\partial \xi} \end{bmatrix} \right\} ds \qquad (4.27)$$

using vector notation (Appendix 5) and s is the distance along the contour C. The strain energy density Z for the linear elastic case is

$$Z = \frac{1}{2E}(\sigma_x + \sigma_y + \sigma_z)^2 + \frac{(1 + 2v)}{E}(\sigma_{xy}^2 - \sigma_x \sigma_y - \sigma_y \sigma_z - \sigma_z \sigma_x)$$

and

$$\sigma_n = \sigma_x \cos^2 \alpha + \sigma_y \sin^2 \alpha + \sigma_{xy} \sin \alpha \cos \alpha$$

$$\tau_n = (\sigma_y - \sigma_x) \sin \alpha \cos \alpha + \sigma_{xy}(\cos^2 \alpha - \sin^2 \alpha)$$

$$u_n = u \cos \alpha + v \sin \alpha$$

$$v_n = -u \sin \alpha + v \cos \alpha$$

For plane strain

$$\sigma_z = v(\sigma_x + \sigma_y)$$

and for plane stress

$$\sigma_z = 0$$

After suitable manipulation and change of variable, eqn. (4.27) gives the J-integral in terms of the global cartesian coordinates as:

$$J = \int Z(-\sin\phi\,dx + \cos\phi\,dy) - \int \{\sigma_x \varepsilon_x \cos\phi\,dy$$
$$+ \sigma_{xy}\varepsilon_y \sin\phi\,dy - \sigma_{xy}\varepsilon_y \cos\phi\,dx$$
$$- \sigma_y\varepsilon_y \sin\phi\,dx + \sigma_{xy}\varepsilon_{xy}\cos\phi\,dy - \sigma_{xy}\varepsilon_{xy}\sin\phi\,dx + A\} \qquad (4.28)$$

where

$$A = (\sigma_x \sin\phi - \sigma_{xy}\cos\phi)\,du \qquad \text{providing } dy \neq 0$$

$$= (\sigma_y \cos\phi - \sigma_{xy}\sin\phi)\,dv \qquad \text{providing } dx \neq 0$$

Equation (4.28) is evaluated numerically by summing the values of each portion of ds between adjacent nodes lying on the contour. Thus, for linear integration between the ith and $(i + 1)$th node eqn. (4.28) becomes

$$J = \sum_{i=1}^{n} \frac{(Z_i + Z_{i+1})}{2}(-\sin\phi\,\Delta x + \cos\phi\,\Delta y)$$

$$- (\sigma_{x_i}\varepsilon_{x_i} + \sigma_{x_{i+1}}\varepsilon_{x_{i+1}})\cos\phi\frac{\Delta y}{2} - (\sigma_{xy_i}\varepsilon_{y_i} + \sigma_{xy_{i+1}}\varepsilon_{y_{i+1}})\sin\phi\frac{\Delta y}{2} - \text{etc.}$$

$$(4.29)$$

The values of σ_{x_i}, $\sigma_{x_{i+1}}$, ε_{x_i}, Δy $(= y_{i+1} - y_i)$ are obtained from the conventional output of a finite element stress analysis.

The input data requirements to this program are very simple. Given the above stresses and strains on a data file (in this case the finite element system used was BERSAFE [12]) different data inputs can be made at different times to explore the use of contours. The contour required is specified as a list of node numbers in sequence. Further data include Young's modulus, Poisson's ratio, a plane strain or plane stress indicator, crack inclination angle ϕ (Fig. 4.4) and an indicator of whether linear or quadratic integration is required along element sides.

For plastic analysis, the numerical evaluation of J requires consideration of an extra contribution to the Z term. Also, a repeat of this evaluation is required during the loading history to follow the incremental theory. Neale [14] extended his elastic computer program to include plasticity, with

$$Z = Z_e + Z_p$$

where

$$Z_e = \tfrac{1}{2}\sigma_{ij}\varepsilon_{ij} \qquad \text{(summation convention)}$$

is the elastic strain energy, and the plastic work term is

$$Z_p = \int \bar{\sigma}_p \, d\bar{\varepsilon}_p$$

$\bar{\sigma}_p$ and $\bar{\varepsilon}_p$ being equivalent stress and equivalent strain respectively and ε_p the total component of plastic strain. If linear strain-hardening of the form

$$\bar{\sigma}_p = \sigma_y + H\bar{\varepsilon}_p$$

is assumed, where H is the strain-hardening coefficient, then

$$Z_p = \frac{(\bar{\sigma}_p^2 - \sigma_y^2)}{2H}$$

The computation of Z_p along the contour is a straightforward addition to eqn. (4.29).

4.3.2 Example of Linear Elastic *J* Computation

Here we consider the problem of a centre-cracked plate under uniaxial tension, for which an exact solution was given in Chapter 1. A mesh with 50 quadratic triangular and quadrilateral isoparametric elements was constructed for half the plate, the crack lying in a plane of symmetry (Fig. 4.5). The mesh has a refinement around the tip but is of quite modest density to assess the performance of these elements. In more complicated structures such density may be enforced by computer limit or constraints. The crack length to width ratio is 1:10. Using four contours from Fig. 4.5, *J* integral values using the linear part of the above program were computed. Four elastic analyses were performed, with special crack tip elements of the Blackburn type (cf. Appendix 6) around the crack tip, and with standard elements there, using both 2×2 (NGAUS = 2) and 3×3 (NGAUS = 3) Gaussian integration rules. The results, shown in Table 4.1, compare the quantity $K_I/\sigma\sqrt{\pi a}$, where K_I is derived from J_I according to eqn. (2.27), with the theoretical value of 1·19 given in Chapter 1.

TABLE 4.1

$K_I/\sigma\sqrt{\pi a}$ FOR SEVERAL CONTOURS OF THE *J* INTEGRAL

Special elements	*NGAUS*	*Contour 1*	*Contour 2*	*Contour 3*	*Contour 5*
Yes	2	1·014	1·072	1·139	1·183
Yes	3	1·029	1·083	1·137	1·183
No	2	0·961	1·062	1·133	1·177
No	3	0·969	1·074	1·130	1·177

T. K. Hellen

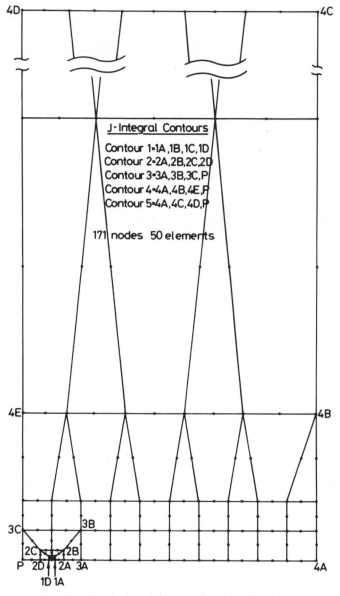

FIG. 4.5. Standard mesh for two-dimensional problems.

The outside contour gives very accurate results, within 1% of the theoretical value, being slightly better when special elements are used. As the contours approach the tip, however, deterioration in the results occurs, up to 15% and 20% low with and without special elements, respectively, on the innermost contour. These errors are due to stress and displacement perturbations in the tip vicinity affecting the numerical evaluation of J and illustrate the need to choose contours to traverse nodes remote from the crack tip.

4.3.3 Computation of the J^* Integral

The numerical evaluation of the J^* integral discussed in Section 2.2.2.2 is very similar to that of J with the additional computation of the area integral. The post-processor program, PLOPPER, to the BERSAFE system reads a data file of results at each load increment and performs both the J and J^* integral evaluations for any required contour and any required load level. The data for this program to define each individual contour in the mesh are slightly different to the previous program. Here, the contour is defined by a list of elements which lie on the outside of the contour. Their inner boundary defines the contour. The node numbers of the crack tip, the starting and finishing nodes of the contour and an initial guide to the direction of integration are also defined. The element list is null for contours traversing the outer boundary of the structure.

The program also deals with the three-dimensional problems. Here, a contour is defined over the plane through the crack tip position normal to the crack edge, but moved to pass through mid-side nodes as necessary. For heavily distorted meshes, if the layer of mid-side nodes drifts away from the plane, the surface integral is stopped and the contour is integrated over the resulting edge. When evaluating the integrals, the stresses are available at each of the four Gauss points per element in two dimensions and eight per element in three dimensions. The displacements are available at every node, and hence they and their derivatives can be interpolated at the Gauss points. For the contributions to the contour integral, values at the interior Gauss points are linearly extrapolated to two Gauss points on the contour path. In two dimensions, this path coincides with element sides and in three dimensions it joins mid-side nodes.

The area integral of eqn. (2.46) for evaluating J^* is assumed to extend over the area inside the chosen contour but not including the crack tip elements. Although the stresses may be higher in these elements, the contribution to the area integral will normally be negligible because of the small area they cover. Output from the program consists of the values J_1,

J_2, J_1^* and J_2^* in two dimensions, plus the area integral. In three dimensions, the extra term J_3^* is added. When a line of symmetry exists along the line of the crack it is necessary to double the printed J and J^* values to obtain the true results. It is advisable to take care over the choice of contour, as in certain circumstances numerical cancellations of large quantities may occur, leaving large errors in the answers. In particular, it has been found necessary to:

(a) avoid having elements with sharp angles ($< 30°$) with edges on the contour; their contribution to the contour integral may be inaccurate (their contribution to the surface integral is reasonably accurate, however),
(b) avoid elements with steep strain gradients,
(c) avoid virtually unstressed elements, especially of large area and
(d) suppress the contour integral over the crack face if this is unloaded and over the crack continuation when this is a line of symmetry such that its contribution should be zero.

Examples of some of these points are included subsequently.

4.4   SAMPLE APPLICATIONS FOR PLATE GEOMETRIES

4.4.1 General

The following examples compare results using the contour integrals and other methods. The BERSAFE finite element system has been used for the analyses. Incremental plasticity theory was modelled using the initial stress method with a von Mises yield criterion. The results were stored on magnetic tape for later evaluation of the contour integrals using the post-processor program, PLOPPER.

The element types used were the 6 and 8 node quadrilateral isoparametric elements, described in Appendix 6. Around the crack tip, special crack tip elements of the Blackburn type (Appendix 6) were used to enhance the accuracy of the elastic parts of the solution.

4.4.2 Mechanical Loading

4.4.2.1 *Centre-Cracked Plate in Tension*

A centre-cracked plate with crack length:width ratio of 1:4 was monotonically loaded, then unloaded, under plane strain elastic–plastic conditions. Both elastic–perfectly plastic and work-hardening situations were considered, the latter case being defined by $H = 0 \cdot 1E$ in the notation

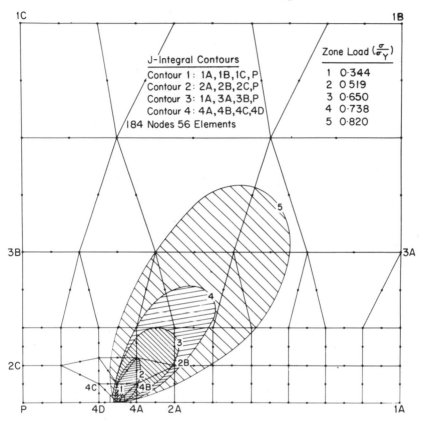

FIG. 4.6. Plastic zones for centre-cracked plate, with no work-hardening.

of Section 4.2.3. The plastic zone sizes for loads corresponding to ligament stresses $0.312 \leq \sigma/\sigma_Y \leq 0.962$ are shown for these two cases in Figs. 4.6 and 4.7 respectively, superimposed on the mesh which consisted of 56 quadratic elements and 184 nodes (for reasons of symmetry only one quarter of the plate was considered for the centre-cracked case). Higher load levels are achieved with strain-hardening than in the elastic–perfectly plastic case ($H = 0$), with more plastic material ahead of the crack. In Fig. 4.6, definitions of the contours for evaluating the J and J^* integrals are included. The numerical results for these integrals, suitably normalised, are plotted against applied stress for the two cases $H = 0$ and $H = 0.1$ in Figures 4.8 and 4.9, respectively. Both loading and unloading are shown for

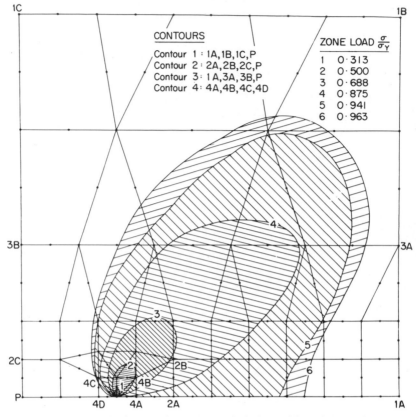

Fig. 4.7.   Plastic zones for centre-cracked plate, with work-hardening.

contours 1 and 2. Additional points show the highest values for J^* obtained by integrating along contours 3 and 4. Despite the large difference in contour lengths, the smaller ones passing through the plastic zones, very close agreement exists for J^* showing path independence in both loading and unloading situations with and without work-hardening. The J results are higher than the J^* results, indicating that a critical value will be attained at a lower applied load level. Upon unloading, J^* attains negative values whilst J returns to zero. Upon further reloading, both integrals are seen to retrace this unloading path.

The J integral results show some dependence on the contour used, particularly in the non-work-hardening case. This path dependence of J when the contours pass through areas of plasticity was also noted in some of

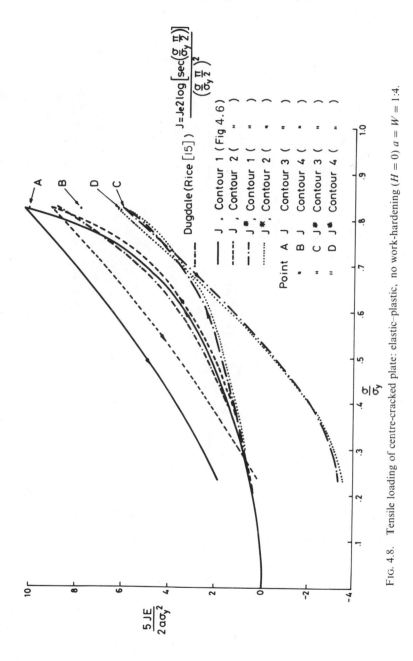

FIG. 4.8. Tensile loading of centre-cracked plate: elastic–plastic, no work-hardening ($H = 0$) $a = W = 1:4$.

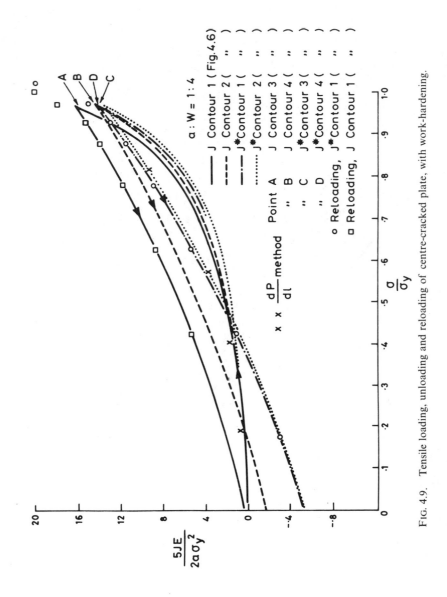

FIG. 4.9. Tensile loading, unloading and reloading of centre-cracked plate, with work-hardening.

the references quoted in Section 2.2.2.1.3. In addition to the J values computed from finite element results the approximate solution based on the Dugdale model given in Section 2.2.1.1, namely

$$J = 2J_e \log\left[\sec\left(\frac{\pi\sigma}{2\sigma_Y}\right)\right] \cdot \left(\frac{\pi\sigma}{2\sigma_Y}\right)^{-2}$$

has also been evaluated for the strain-hardening case and plotted in Fig. 4.8. The result is seen to give very good agreement with the present results, particularly those from the J integral.

Also included in Fig. 4.9 are points obtained for J and J^* using contour 1 (around the outer boundary) when reloading. In this case, a much higher final load, $\sigma/\sigma_Y = 1\cdot077$, was achieved for which $5JE/2a\sigma_Y^2$ was 29·20 and $5J^*E/2a\sigma_Y$ was 27·26. The results show that the reloading curves for both J and J^* follow the previous unloading curve. This trend shows that under cyclic loading conditions the failure stress decreases with increasing numbers of cycles. For a given load amplitude, therefore, which is below the critical load, eventual failure would occur. This is consistent with material behaviour under fatigue crack propagation.

4.4.2.2 *Edge-Cracked Plate in Tension*
Experimental results have been obtained at C. A. Parsons Ltd for a variety of different sized geometries and crack configurations for 1 CMV steels. Each was loaded in tension to determine the critical load at failure. Two particular geometries are considered here for finite element analysis and J and J^* integral determinations. In both cases, a thickness of 0·01 m and width of 0·02 m applied. One had a long crack (0·01025 m), the other a short crack (0·0012 m), corresponding to crack length to width ratios of 0·50 and 0·06, respectively. In the latter case, more crack tip yielding occurred. In both cases, a mesh similar to that in Fig. 4.5 was used with the tip refinement suitably positioned.

In both cases, results for the J and J^* integrals (suitably doubled because of the line of symmetry along the crack line) are shown in Fig. 4.10. The corresponding experimental values are also shown. For the present material, $K_{Ic}^2 = 3025 \text{MN}^2/\text{m}^3$ which implies a critical J value of $J_{Ic} = 0\cdot0131 \text{ MN/m}$. Assuming the extended Griffith criterion, from these curves, by reading off the critical stress corresponding to J_{Ic} gives the results of Table 4.2. The closeness of the results suggests that the concept of a critical J_c (equivalent to K_{Ic}) can be applied in areas of gross plastic yielding as well as in small-scale yielding.

T. K. Hellen

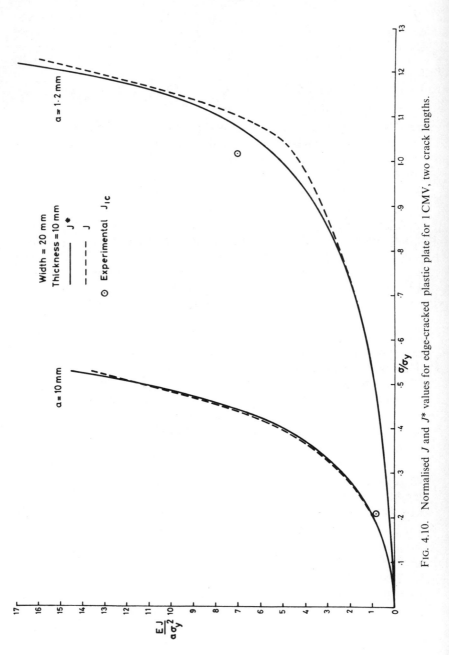

FIG. 4.10.   Normalised *J* and *J** values for edge-cracked plastic plate for 1CMV, two crack lengths.

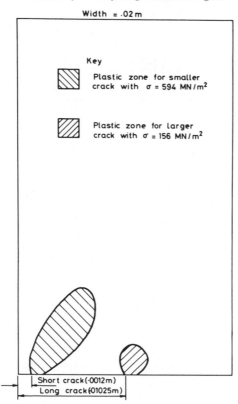

FIG. 4.11. Plastic zones at critical stress for two crack lengths.

The yield zone sizes for each crack length are shown in Fig. 4.11. The zones were obtained from the finite element output corresponding to the incremental load level nearest to the experimentally determined critical stress. Thus, for the short crack, the load level used was at an applied stress of 594 MN/m², compared with the experimental value of 580 MN/m². Six load increments were used between the first yield stress (108 MN/m²) and this value. For the longer crack, the load level was at an applied stress of 156 MN/m², compared with the experimental value of 120 MN/m². Only one load increment was required in this case above the first yield stress of 25 MN/m².

The zones indicate that the amount of plasticity spreading from the shorter crack is larger (by approximately a factor of 4) than that from the longer crack at the respective failure stresses.

TABLE 4.2
EXPERIMENTAL AND CALCULATION CRITICAL STRESS VALUES FOR TWO CRACK LENGTHS

Geometry used	Critical stress (MN/m^2)			Percentage difference	
	Experimental	Using J	Using J*	J: Experimental	J*: Experimental
Long crack	120	105	110	12%	8%
Short crack	580	625	615	8%	6%

4.4.3 Thermal Loading: Centre-Cracked Plate

4.4.3.1 Elastic Behaviour

It is possible to demonstrate the earlier assertion that the J integral becomes path dependent in two-dimensional thermoelastic cases, whilst path independence is maintained for the J^* integral. Consider a centre-cracked plate whose crack length to width to length ratio is $1:10:25$, loaded by a symmetric quadratic temperature variation in the crack direction. For an infinite strip of width b containing a central crack of length $2a$ under a temperature $\theta(x) = \theta_0 x^2/c^2$ (where $\theta(x)$ is the temperature at location x, with x measured from the centre of the crack in a direction parallel to the crack, and θ_0/c^2 is a constant defining the magnitude of the quadratic temperature variation) an estimate of the stress intensity factor may be made by using Bueckner's method [12], i.e. considering the equivalent loading on a closed crack as acting on an open crack. For cracks sufficiently small that the width may subsequently be neglected this gives a value for the stress intensity factor

$$K = E\theta\alpha(\pi a)^{1/2}(\tfrac{1}{3}b^2 - \tfrac{1}{2}a^2)/c^2 \tag{4.30}$$

where E is Young's modulus and α the coefficient of thermal expansion.

A plane mesh of quadratic isoparametric elements, similar to those above, was used with two axes of symmetry, one along the crack axis. Around the crack tip, four triangular elements were used. The values of stress intensity factor were calculated using several methods and using both standard and singularity elements, described in Appendix 6, around the crack tip.

The first method used was substitution from the values of the vertical displacement at the nodes on the crack face which are contained in a crack tip element. The effect of linear strains were eliminated [11] from these values of K_M (mid-side node) and K_V (vertex node) to give $K^* = (2 + \sqrt{2})K_M - (1 + \sqrt{2})K_V$. The ratios of K_M, K_V and K^* to K of

TABLE 4.3

STRESS INTENSITY FACTORS USING SUBSTITUTION TECHNIQUES

Dimensionless stress intensity factors	K_V/K	K_M/K	K^*/K
Without special elements	0·91	0·85	0·69
With special elements	0·95	0·96	1·00

eqn. (4.30) are shown in Table 4.3. It is seen that the large quadratic terms in the standard elements caused a deterioration in the value of K^*/K, while if special elements were used good agreement was obtained.

The virtual crack extension (VCE) method [13] was also used to calculate the stress intensity factor K_E from the energy differences of suitably small crack extensions (δa). The ratios of K_E to K are shown in Table 4.4 In this case, excellent agreement is seen for both types of crack tip element.

TABLE 4.4

STRESS INTENSITY FACTORS USING VCE METHOD

$\delta a/a$	0·01	0·001	0·0001	0·00001
Without special elements	0·98	0·98	0·98	0·98
With special elements	1·00	1·00	1·00	0·98

The J integral was evaluated around three contours corresponding to contours 2 (smallest), 3 (middle) and 5 (outer boundary) of Fig. 4.5 in terms of distance from the crack tip. Using special elements about the tip, the ratios of the integrals to K^2/E are 0·88, 0·92 and 4·52, respectively. Consequently, path dependence exists and increases with increasing contour length.

The J^* integral was also evaluated along three similar contours with both standard and special elements about the crack tip. The ratios of J^* to K^2/E

TABLE 4.5

STRESS INTENSITY FACTORS USING CONTOUR INTEGRATION

Dimensionless stress intensity factors	*Smallest contour*	*Intermediate contour*	*Outer boundary*
Without special elements	0·97	0·98	1·00
With special elements	0·98	0·99	1·00

are shown in Table 4.5. Excellent agreement is seen for all contours and both types of crack tip element.

4.4.3.2 *Elastic–Perfectly Plastic Behaviour*

Using the same mesh and quadratic temperature distribution as used in the last example, a plane stress elastic–perfectly plastic analysis was conducted. Normalised results are shown in Fig. 4.12 for various contours defined in Fig. 4.5. Some inconsistency occurs when using J^* and the outermost contours due to large elements at the right side of the plate, where large thermal strains and therefore considerable plastic yielding exists, and near the top, where considerable distortion occurs. The inner contours give better consistency and show that critical values will only be achieved slowly. The J results are very poor for the outer contours but show surprisingly good agreement for the inner contours.

To alleviate the problem of distorted elements a mesh with more elements, all square-shaped apart from the refinement about the tip, was used (Fig. 4.13). Here, the crack length was double that above. Also shown are the progressive yield zone contours during the five load increments used. It is seen that plasticity grows from the right-hand side of the plate very rapidly (compressive yielding) and dwarfs the local crack tip yielding. The normalised J and J^* results are shown in Fig. 4.14, contours 1 to 4 being roughly equally spaced progressing outwards from the tip, the latter being round the outer boundary. The J^* results show path independence for the smaller contours but, as the regions of very high plastic and thermal strain are entered, violent oscillations occur. The J results for the outer paths are very diverse. In this case, the loading was $3\frac{1}{3}$ times greater than for the previous mesh, hence the greater apparent oscillations. This demonstrates the need to avoid contours through plastic zones or areas of high thermal strain not associated with crack tip yielding.

4.4.3.3 *Creep Behaviour*

The above square-element mesh, subject to the same thermal gradient, was used to investigate the effects of secondary creep on the contour integrals. Because of the presence of both creep and thermal strains, it was anticipated that path independence would not be exhibited by J, whereas it would by J^*. For the same four contours as used in the last example, good path independence for j^* was obtained (Fig. 4.15). For the innermost contour, results occur which are 20 % higher (due to the closeness of the contour to the crack tip) whilst the outer contour gives a result slightly lower than the averaged value. The J integral results were extremely path dependent. In

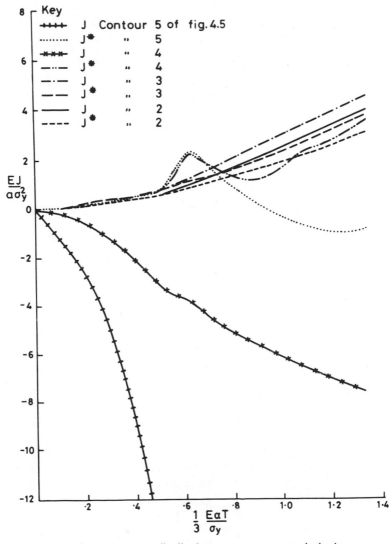

FIG. 4.12.   Quadratic temperature distribution across centre cracked plate no work-hardening, elastic–plastic, plane stress.

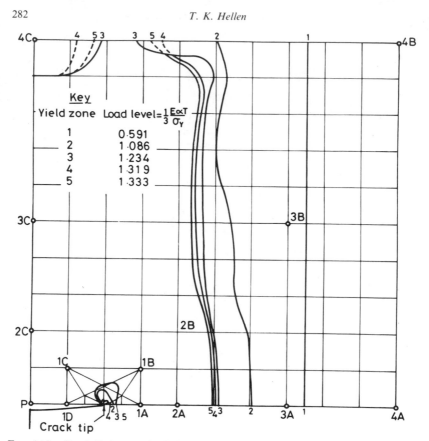

FIG. 4.13. Cracked thermo-plastic plate—yield zone contours, square element mesh. J-integral contours: contour 1—1A, 1B, 1C, 1D; contour 2—2A, 2B, 2C, P; contour 3—3A, 3B, 3C, P; contour 4—4A, 4B, 4C, P.

this example, temperatures varying quadratically over 100 °C, as before, were used. The temperatures were increased by 1000 °C whenever reference to the creep law was made. The law used was

$$\dot{\varepsilon}_c = k\sigma^n t^m e^{-(p/\theta)}$$

where θ = temperature, $k = 10$, $n = 3$, $m = 1$ and p = 30 000. Also included in Fig. 4.15 is the equivalent stress at a Gauss point near the tip. The stress is observed to decrease rapidly with time. The J^* values show only a very slight, linear, diminution over the elapsed time interval, implying that no delayed instability would occur. This does not preclude, however, rupture due to creep damage spreading from the highly strained

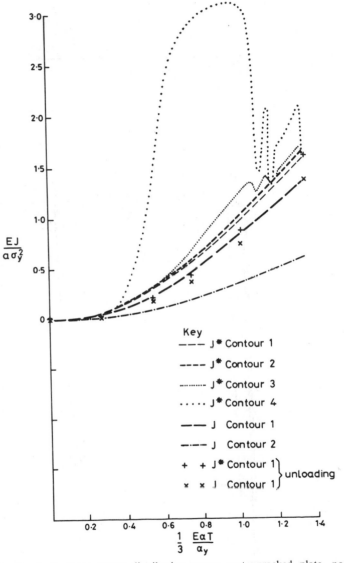

$$\frac{EJ}{a\sigma_y^2}$$

Key

——— J^* Contour 1

- - - - J^* Contour 2

·········· J^* Contour 3

· · · · · J^* Contour 4

—— J Contour 1

—·—· J Contour 2

+ + J^* Contour 1 ⎫
                    ⎬ unloading
× × J Contour 1 ⎭

$$\frac{1}{3}\frac{E\alpha T}{\alpha_y}$$

FIG. 4.14.   Quadratic temperature distribution across centre-cracked plate, no work-hardening, plane strain. $a\!:\!W = 1\!:\!5$, square element mesh.

T. K. Hellen

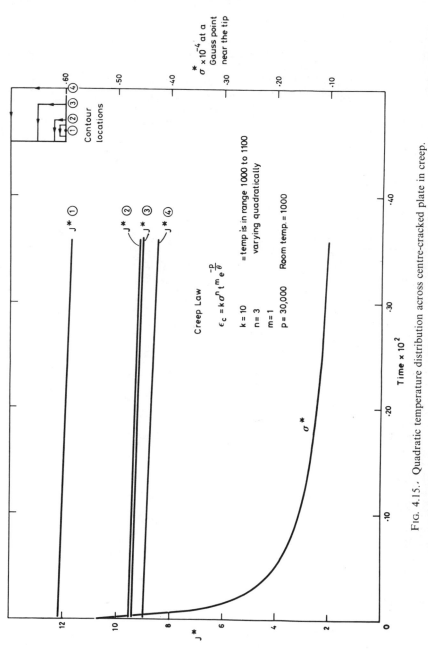

FIG. 4.15. Quadratic temperature distribution across centre-cracked plate in creep.

areas on the right-hand edge of the plate, but here normal continuous rupture predictions are valid without recourse to fracture mechanics.

4.5 CONCLUDING REMARKS

In this chapter we have considered the evaluation of contour integrals using the finite element method shown to be equivalent in the case of linear elastic fracture mechanics. For tensile loading in elastic–plastic plates, the integrals designated J and $J*$, respectively, remain roughly equal and path independence is achieved to a reasonable degree of accuracy in each case. Currently available comparisons with experimental data show good correlation, particularly for the $J*$ integral. For unloading, thermal and creep conditions it has been confirmed that the J integral ceases to be path independent. The results obtained for $J*$ also show some deviation from path independence which may, however, be due to errors that arise when evaluating the area integral when either badly distorted elements are used or large elements exist in regions of high strain gradients.

The $J*$ integral is also theoretically available in three-dimensional bodies in any of the above loading situations. Results have been obtained in mechanical loading situations and have shown satisfactory accuracy with only moderately refined meshes. For very complex structures, high computer and user involvement costs are inevitably required.

REFERENCES

1. Edmunds, T. M. & Willis, J. R. (1976). Matched asymptotic expansions in non-linear fracture mechanics—1. Longitudinal shear of an elastic–perfectly plastic specimen, *J. Mech. Phys. Solids*, **24**, pp. 205–23.
2. Desai, Ch. S. & Abel, J. F. (1972). *Introduction to the finite element method*, Van Nostrand Reinhold, New York.
3. Hill, R. (1950). *Mathematical theory of plasticity*, Clarendon Press, Oxford.
4. Armen, H., Levine, H. S. & Pifko, A. B. (1972). Plasticity: theory and finite element applications. pp. 393–438 in *Advances in computational methods in structural mechanics and design* (St. Oden, R. W. Clough and Y. Yamamoto, eds.), UAH Press.
5. Besseling, J. F. (1973). Nonlinear analysis of structures by the finite element method as a supplement to linear analysis. Report WTHD 51, Delft University of Technology.
6. Zienkiewicz, O. C., Valliapan, S. & King, I. (1969). Elasto-plastic solutions of engineering problems—the initial stress finite element approach, *Int. J. Num. Engng.*, **1**, pp. 75–100.

7. Marcal, P. V. & King, I. (1967). Elastic–plastic analysis of two-dimensional stress systems by the finite element method, *Int. J. Mech. Sci.*, **9**, pp. 143–55.
8. Pope, G. C. (1965). *Proc. Conf. Matrix Methods in Structural Mechanics*, Wright–Patterson Air Force Base, Ohio, pp. 635–54.
9. Yamada, Y., Yoshimura, N. & Sakurai, T. (1968). Plastic stress–strain matrix and its application for the solution of elastic–plastic problems by finite element method, *Int. J. Mech. Sci.*, **10**, pp. 343–54.
10. Nayak, G. C. and Zienkiewicz, O. C. (1972). Elasto–plastic stress analysis. A generalisation for various constitutive relations include strain softening, *Int. J. Num. Meth. Engng.*, **5**, pp. 113–35.
11. Neale, B. K. (1973). Finite element crack analysis using the *J*-integral method. CEGB, Report RD/B/N 2785.
12. Hellen, T. K. & Protheroe, S. J. (1974). The BERSAFE finite element system, *Computer Aided Design*, **6**(1), pp. 15–24.
13. Hellen, T. K. (1975). On the method of virtual crack extensions, *Int. J. Num. Meth. Engng.*, **9**, pp. 187–207.
14. Neale, B. K. (1975). Elastic–plastic analysis of cracked bodies using the *J*-integral method. CEGB, Report RD/B/N 3253.
15. Rice, J. R. (1968). In *Fracture: an advanced treatise*, **2**, H. Liebowitz (ed.), Academic Press, New York.

5

Status and Outlook

D. G. H. Latzko

Delft University of Technology, Delft, The Netherlands

Looking back at this point to see how far the quest for extension of fracture mechanics into the realm of elasto-plastic material behaviour has yielded a proven practical approach for failure load prediction, we should first recall the limitation of the present book to monotonic loadings and (quasi-) static crack behaviour. While the former limitation has prevented discussion of plasticity effects on crack growth under cyclic loads, it might be mentioned here that despite the additional complexity due to unloading or load reversal pointed out in Chapter 2, encouraging progress in the application notably of the *J*-integral concept to this class of problems has repeatedly been reported elsewhere (for example Ref. [1]). As to the arrest behaviour of running cracks, it should be kept in mind that prevention of crack extension (either stable or unstable) rather than arrest forms the primary aim of fracture control measures. Thus the omission of plasticity effects in crack arrest from the present book, although regrettable for certain safety studies, is hardly important for a treatise primarily concerned with fracture prevention.

Within these limitations any method eligible for engineering fracture control should be based on (a) parameter(s):

measurable by standard procedures on specimen sizes compatible with industrial laboratory practices;

predictable by computations commensurate with the skills and budgets available to the engineering services of high technology capital goods manufacturers or users and to consulting firms or licensing authorities in such fields;

representing a unique material property, i.e. of proven geometry and size independence within the range of applications envisaged;

unequivocally defining the risk of fracture; and

applicable to actual flawed structures, including those featuring complex geometries and/or subject to complex loadings of both mechanical and thermal origin.

Both COD and J integral were found to satisfy the first two requirements and have been developed to the point where standardised recommended practices are available for the determination of critical values (δ_i and J_{Ic}) indicative of a material's fracture toughness. Computational costs and effort for finding values for J or δ applied may in due course be further reduced through developments briefly to be touched upon later in this chapter.

It is perhaps worthwhile to re-emphasise geometry independence in terms of the *specimen type* as an absolute prerequisite for applicability in engineering practice. Only then will it be possible to predict, for example, the fracture behaviour of a structure loaded in bending from a specimen in tension, or vice versa. Unfortunately recent evidence on this crucial point appears somewhat conflicting. For example, extensive and well-documented tests reported by Markström [2] on both centre-cracked and double edge-cracked wide plate and CT specimens of two medium strength steels confirmed this independence, provided the onset of crack extension was determined from electrical signals, namely changes in impedance and disturbances in eddy-current fields. By contrast, similar comparisons based on computational studies reported by Luxmoore *et al.* [3] cast considerable doubt on the applicability of J_{Ic} for double edge-cracked specimen geometries. Similar discrepancies have been pointed out by, for example, Keller & Munz [4] between experimental J_{Ic} values obtained from three-point bend and CT specimens, using the blunting line approach of Fig. 3.4. Further thorough and extensive investigations are desirable to verify whether such discrepancies should be attributed to deficiencies in the determination of the onset of crack extension or to more fundamental influences acting on the crack tip, such as differences in constraint between the various geometries too large to be eliminated by crack tip blunting. Pending such investigations, care should be taken to thoroughly validate each test specimen geometry prior to recommending its use for measuring a particular fracture parameter.

The effect of *specimen size* has been thoroughly discussed in Section 2.4.2 in terms of the resulting variation in constraint on plastic deformation, while the resulting minimum thickness requirements for J testing were expressed in Section 3.4 in terms of the fraction $\alpha(J_{Ic}/\sigma_Y)$. The value(s) to be inserted for the parameter α clearly need additional study and should be based on studies of a wide range of materials. Strain-hardening effects on α

should also be evaluated. In the meantime it is interesting to note that recently published experiments by Chell & Spink [5] on a series of geometrically similar three-point bend specimens of medium strength steel with $B\sigma_Y/J_{1c}$ values varying between 125 and 35 suggest a *continuous* increase in nonlinear deformation behaviour prior to fracture with decreasing specimen size. Another size-related parameter requiring further investigation is the ratio of crack length to specimen height W, or alternatively the remaining ligament b. Further work in this area is needed for both δ_i and J_{1c}.

Only limited experimental information has been published to date on the size effect in *actual structures* failed by ductile fracture. Tests on flawed spherical vessels loaded to failure by internal pressure, reported by Lebey *et al.* [6], do indicate increasingly ductile behaviour with decreasing vessel size. In this respect it is to be regretted that the vast amount of information gleaned from the intermediate vessel tests of the Heavy Section Steel Technology (HSST) program does not include data on the size effect.

A feature inherent to the problem of post-yield fracture appearing as a weakness common to all methods discussed throughout this book is the lack of a clear-cut *fracture* criterion. Section 2.4.1 has explained this inadequacy in terms of the slow stable crack growth phenomenon providing a margin between crack extension initiation and unstable propagation. The immediate result of this phenomenon, illustrated by the R curve representation of, for example, Figs. 2.73 and 3.4, is that material having a low value of J_{1c} (or δ_i) may exhibit a high resistance to failure by ductile fracture, due to a high value for dJ/da. This aspect lends particular urgency to further investigations of the tearing modulus concept discussed in Section 2.4.1, claimed in Ref. 180 of Chapter 2 to define the onset of crack instability marking the end of the R curve, the initiation of which is marked by J_{1c}.

Applicability of the fracture control parameters discussed to complex structures and loading conditions requires both the capability for mathematical modelling of crack tip conditions and experimental verification of crack behaviour at least for some typical structural elements under conditions involving significant plasticity. In the area of computational prediction of crack tip parameters various significant developments are under way. On the one side these take the form of simplifications in the physical model describing crack tip plasticity. A case in point is the approach presented by Atkinson & Kanninen [8] for extension of the in-plane strip yield model described in Section 2.2 to the plane strain situation where the spread of plasticity occurs predominantly normal to the crack plane. By basing their mathematical derivation on the idealisation of the plastic zone as a dislocation pile-up concentrated into one single 'superdislocation' they obtain the following simple formula for COD:

$$\delta = \alpha \frac{K^2}{E\sigma_Y}$$

where α is a function solely of the angle between the mid-plane of the crack and the plane of maximum shear. This simple result becomes quite intriguing when compared to the generalised expression for J, namely $j = m\sigma_Y\delta$, indicating near equality of α and m—but for the term $(1 - p^2)$—for small amounts of plasticity.

Approximations of this kind are likewise indispensable for application of the strip yield or BCS model to part through cracks in a more satisfactory manner than the substitution of elastic data discussed in Section 2.2.1.4. Recently an example of such an application to penny-shaped and semi-circular cracks was given by Chell [9].

On the other side, developments continue towards further improved cost effectiveness of the rigorous elasto-plastic crack tip computations described in Chapter 4. These range from hybrid formulation of the finite element method as proposed, for example, by Atluri et al. [10] to the (partial) replacement of the finite element method by that based on boundary integral equations. The comparatively recent application of this latter numerical technique to fracture mechanics, described by, for example, Besuner [11], makes it difficult at present to assess its potential for problems involving significant plasticity.

The limitation to small strains mentioned in Chapter 4 and typical for numerical procedures used in practice for crack tip modelling implies that crack tip blunting, known to occur in the actual material (cf. Fig. 3.4), is disregarded in the computations. A systematic investigation into the effect of this approximation, although not particularly urgent in view of the good agreement between model and experiment reported in the majority of cases quoted throughout this book, might be helpful in explaining some of the differences mentioned earlier in this chapter between various specimen geometries differing in crack tip constraint.

In particular, the prospect of three-dimensional analysis poses the question of the meaning of the J integral in such situations. Numerical evaluations for J in various planes of a 3-D model, as carried out by Light & Luxmoore [12] may only justify their high computational costs if the existing doubts mentioned in Section 2.4.3 can be removed. Alternatively one may of course adhere to the J^* concept used in Chapter 4, provided the relation of this concept to the fracture control parameter J_{1c} is backed by appropriate experimental evidence.

The latter remark may be generalised to express some concern about the existing scarcity of experimental studies relating the fracture control concepts described in this book to data on ductile fracture pertinent to complex structural elements such as pressure vessel nozzles, branch piping, turbine bucket fixtures, etc. Such structural elements differ from the various test specimens discussed before not only by the three-dimensional geometry and loading of the flaws they may contain, but often also by the presence of thermal gradients and residual stresses, i.e. secondary stresses whose relevance for crack extension seems to require further clarification. Tests on models incorporating these aspects could be used to complete the existing experimental data obtained on less complex configurations and plotted in the transition region of Fig. 2.51. Their evaluation in terms of the various concepts described would hopefully provide the ultimate proof that post-yield fracture mechanics has fulfilled the second of its two aims mentioned in Sections 1.8 and 2.2.3, namely of predicting the fracture safety of flawed structures operating beyond the LEFM range. While mentioning the desirability of such model tests one should remain aware of the particular experimental difficulties in obtaining meaningful crack extension data for the region of slow stable crack growth under monotonic loading.

The above brief survey did not, of course, strive for completeness. Notably missing are new developments not directly connected with the concepts expounded in the present book, in particular those of a more fundamental nature as yet remote from application in fracture control concepts. Quantification of the margin between the initiation points for stable and unstable crack extension and experimental evidence pertinent to complex engineeri.ig structures emerge as the foremost goals for continuing development. For the attainment of both these goals improved insight into the fundamentals of ductile crack propagation will be instrumental.

In summary, post-yield fracture mechanics appears to have reached the age of adolescence, characterised by vigorous development, some lingering uncertainties and great promises.

REFERENCES

1. Dowling, N. E. & Begley, J. A. (1976). Fatigue crack growth during gross plasticity and the J integral, pp. 82–103 in *Mechanics of Crack Growth*, ASTM–STP 590, Philadelphia.

2. Markström, K. (1977). Experimental determination of J_c data using different types of specimen, *Engng Fract. Mech.*, **9**, pp. 637–46.
3. Luxmoore, A., Light, M. S. & Evans, W. T. (1977). A comparison of energy release rates, the J integral and crack tip displacements, *Int. J. Fract.*, **13**, pp. 257–9.
4. Keller, H. P. & Munz, D. (1977). Effect of specimen type on J integral at the onset of crack extension, *Int. J. Fract.*, **13**, pp. 260–2.
5. Chell, G. G. & Spink, G. M. (1977). A post-yield fracture mechanics analysis of three-point bend specimens and its implications to fracture toughness testing, *Engng Fract. Mech.*, **9**(1), pp. 55–64.
6. Lebey, J., Roche, R. & Brouard, D. (1977). Effect of specimen and vessels size on toughness measurements and crack propagation, *4th Int. Conf. Struct. Mech. Reactor Technol.* Paper G. 2–3.
7. Paris, P. C., *et al.* (1977). A treatment of the subject of tearing instability, NUREG–0311, Aug.
8. Atkinson, C. & Kanninen, M. F. (1977). A simple representation of crack tip plasticity: the inclined strip yield superdislocation model, *Int. J. Fract.*, **13**, 151–64.
9. Chell, G. G. (1977). The application of post-yield fracture mechanics to penny-shaped and semi-circular cracks, *Engng Fract. Mech.*, **9**, pp. 55–64.
10. Atluri, S. N., Kathiresan, K., Kobayashi, A. S. & Nakagaki, M. (1977). Inner-surface cracks in an internally pressurized cylinder analyzed by a three-dimensional displacement-hybrid finite element method, *Proc. 3rd Int. Conf. on Pressure Vessel Technol.*, pp. 527–34, ASME, New York.
11. Besuner, P. M. (1977). The application of the boundary–integral equation method to the solution of engineering stress analysis and fracture mechanics problems, *Nucl. Engng & Des.*, **43**, pp. 161–74.
12. Light, M. F. & Luxmoore, A. R. (1977). A numerical investigation of post-yield fracture, *J. Strain Anal.*, **12**, pp. 293–304.

Appendix 1

Note on Plasticity

C. E. TURNER

Imperial College of Science and Technology, London, UK

The following brief note serves to illustrate the difference between 'total' or 'deformation' and 'incremental' plasticity mentioned in Section 2.2.2. For further information on this subject the reader is referred to Ref. [1].

The Hencky stress–strain laws for 'total' plasticity are of the form

$$\varepsilon'_x = \phi\sigma'_x + (\sigma'_x/2\mu) \tag{A.1.1}$$

with similar expressions for ε'_y and ε'_z. In eqn. (A.1.1) the first term is the plastic component and the second the elastic component. Here σ' is the deviatoric stress, $\sigma' = \sigma - \sigma_M$, where σ_M is the hydrostatic (or mean) component $(\sigma_x + \sigma_y + \sigma_z)/3$. Similarly, ε' is the deviatoric strain, $\varepsilon' = \varepsilon - \varepsilon_v$, where ε_v is the volumetric strain $(\varepsilon_x + \varepsilon_y + \varepsilon_z)$. μ is the elastic shear modulus and ϕ is a plastic proportionality factor (or inverse plastic modulus) defining the work-hardening. Clearly it is a total (i.e. actual) strain that is related to the stress system although it is only the deviatoric component thereof that is directly relevant, since the mean stress does not cause plastic flow. The commonly used criteria of yielding are either the Tresca maximum shear stress criterion

$$(\sigma_1 - \sigma_2)/2 \text{ or } (\sigma_2 - \sigma_3)/2 \text{ or } (\sigma_3 - \sigma_1)/2 = \bar{\sigma}/2 \tag{A.1.2}$$

or the von Mises criterion (sometimes called shear strain energy, second stress invariant or octahedral shear stress criterion)

$$(\sigma_1 - \sigma_2)^2 + (\sigma_2 - \sigma_3)^2 + (\sigma_3 - \sigma_1)^2 = 2\bar{\sigma}^2 \tag{A.1.3}$$

where $\sigma_{1\,2\,3}$ are the principal stress components. $\bar{\sigma}$ is called the 'equivalent stress' and at yield is numerically equal to the tensile yield stress σ_Y. As work-hardening occurs so $\bar{\sigma}$ is equated to some increased value of stress again found from a tensile stress–strain curve. The two criteria do not differ by more than 15 % for any combination of stress components and in many instances may differ by only a few per cent. A direct volumetric strain,

293

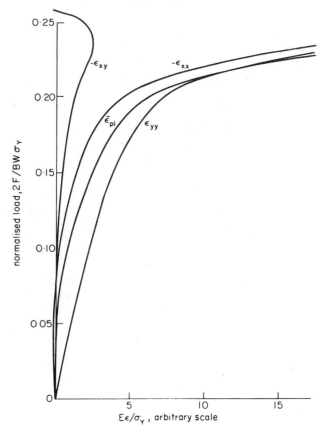

FIG. A.1.1.   Averaged strain near the crack tip. Three-point bending, $a/W = 0\cdot5$; plane strain, non-work-hardening. $\bar{\varepsilon}_{pl}$ is the equivalent plastic strain based on the von Mises criterion.

$\varepsilon_v = \sigma_M/\kappa$, where κ is the (elastic) bulk modulus, exists additionally to the above components.

Real plasticity materials behave closer to the 'incremental' stress–strain laws of Prandtl–Reuss, often stated as

$$\frac{\delta\varepsilon_x^{pl}}{\sigma_x'} = \frac{\delta\varepsilon_y^{pl}}{\sigma_y'} = \cdots \frac{\delta\gamma_{xy}^{pl}}{\tau_{xy}} = \delta\lambda \qquad (A.1.4)$$

These can be re-expressed as

$$\delta\varepsilon_x' = \delta\lambda\sigma_x' + (\delta\sigma_x'/2\mu) \qquad (A.1.5)$$

where μ is again the elastic shear modulus and $\delta\lambda$ defines the plastic work-hardening in terms of the *increment* of plastic strain (such as $\delta\varepsilon_x^{pl}$) and the deviatoric stress (such as σ_x'). The difference from eqn. (A.1.1) is that in the former case the total plastic strain is related to stress, whereas in the latter case only the increment of plastic strain is related to stress. In the former case the elastic strain can be added also in terms of stress. In the latter it is an increment of elastic strain that must be added in terms of an increment of stress. According to eqn. (A.1.5) the actual strain in any component is found by summing all the increments that may have occurred from the onset of yielding to the current time. If all increases are simple proportional increases (so-called radial stress system) then the two theories will give the same answer. However, if the ratio of the stress components alters as plastic flow occurs or, if unloading occurs, then the results will not in general be the same.

Since the plastic component of deformation occurs at constant volume, the plastic equivalent of Poisson's ratio (called the plastic contraction ratio) is numerically equal to 0·5. The actual contraction is the sum of an elastic component (with Poisson's ratio retaining its elastic value) plus the plastic component, so that for large plastic strains the effective ratio approaches 0·5 closely, Nevertheless, an elastic component always remains, associated, as noted previously, with the hydrostatic stress and (elastic) volumetric strain.

A picture of some particular strain paths found in elastic–plastic finite element computations of notched components (using constant strain triangles, von Mises criterion of yielding and the incremental laws of plasticity) is shown in Fig. A.1.1. It is seen that the ratio between the principal strains does indeed change, but not to an extent that would make the use of 'total' theory, suggested in Section 2.2.2, too unrealistic in the absence of unloading.

REFERENCE

1. Hill, R. (1950). *Mathematical theory of plasticity*, Clarendon Press, Oxford.

Appendix 2

Crack Separation Energy G^Δ

C. E. TURNER

Imperial College of Science and Technology, London, UK

A recent study by Kfouri & Miller [1] deals with three topics relevant to fracture: energy release rate, the effect of biaxial loading and the variation of toughness with temperature. An attempt will be made to present these three features separately.

First they envisage a finite region, Δa, that represents the process zone in which the actual separation occurs. They evaluate the separation energy G^Δ over the region Δa by finite element computation and to that extent the explanation of their model is closely bound to their procedures. G^Δ is defined by the sentence 'Suppose a crack is allowed to extend by an amount Δa under constant applied load by the quasi-static release of the cohesive forces holding the surfaces together at Δa and that the work absorbed at Δa is ΔU, then G^Δ will be defined as $\Delta U / \Delta a$'. It is the essence of their argument that $\Delta U / \Delta a$ must be evaluated over some finite length, so that, when there is any plastic deformation, it will differ from the limit $\partial U / \partial a$. However, for linearly elastic brittle material G^Δ degenerates to G or J, since the only dissipation, ΔU, occurs at the crack face. This use of a finite length, Δa, is consistent with the 'process zone' concept outlined in Section 2.4.2.

The crack separation energy is computed by allowing the nodal force at the crack tip node to relax progressively to zero thereby doing work as the crack face separates over the region Δa, to the next node. This process zone, Δa, is one element in extent. The size of element is arbitrary, but this is shown not to matter in the computations, the governing feature being the ratio of element size to plastic zone size. The crack tip node can be released at any desired size of plastic zone (or of applied stress for any specified value of yield stress) and the corresponding value of G^Δ found.

Kfouri & Miller used sizes of plastic zone, expressed as the ratio, R, of plastic zone to element size Δa, normalised in terms of a datum ratio, R_0, for incipient plasticity at which condition $G = G_0$. A parameter $\psi = R_F/R_0$ was then defined where R_F is the plastic zone size at fracture. ψ was taken from about unity to fifteen. For each of five or six sizes of plastic zone ratio

297

the crack was allowed to extend and G' computed. As the crack extended the plastic zone altered in size and further work was dissipated. The results were expressed in terms of ψ and a second non-dimensional function ϕ. ϕ was defined as $\phi = (\sigma_{Y0}/\sigma_Y)^2$ where σ_{Y0} is the yield stress at the brittle datum condition of toughness G_0, and did not therefore relate explicitly to G^Δ. This indirect procedure for expressing the results was necessary since to suit a real material the values have to be scaled such that the datum $G_0 = G_{Icb}$, a value of G_{Ic} at some low temperature brittle condition at which the effects of plasticity are negligible. The matter was further complicated by the arbitrary size adopted for the element Δa. The values of both G^Δ and stress depend upon element size but their ratio was shown to be independent of element size.

In this presentation G^Δ appears to enter through ψ since R_F, the plastic zone size at fracture, is determined by the critical value of G^Δ, G_c^Δ, used in the computation. In Ref. [1] this critical value is taken as G_{Icb} so that R_F is the plastic zone size at fracture for various values of yield stress but for a specific toughness, $G^\Delta = G_{Icb}$. Results are shown in Fig. A.2.1. The straight line is the Griffith condition defined by $\psi = \phi = 1$. The curve marked 'uniaxial' is the plasticity computation for uniaxial loading. The further variable of biaxial applied loading (transverse stress = $\pm$ axial stress) is also shown. These results are obtained by a complete recomputation. The value of plastic zone R_F, and hence ψ, is not unexpectedly dependent on the transverse stress component.

Another presentation is given in Ref. [2], in which ϕ is defined as $\phi = G^\Delta/G_0$ and $\psi = G/G_0$. It does not seem possible to accept both $\phi = (\sigma_{Y0}/\sigma_Y)^2$ from Ref. [1] and $\phi = G^\Delta/G_0$ from Ref. [2] as general statements, since this would imply just $G^\Delta = G_0(\sigma_{Y0}/\sigma_Y)^2$. It appears to the writer that, in Ref. [1], ϕ is more strictly a special case of the general parameter, limited to the circumstance that $G^\Delta = G_{Icb}$ for all yield stress values, whereas in Ref. [2] ϕ is another special case in which yield stress does not vary and G^Δ is not given a specific value. Thus G^Δ is now a variable expressed through ϕ and not related directly to ψ which in this definition is related not to the fracture event but to the loading condition as measured by G.

Yet another presentation by Kfouri & Rice [3] uses different variables that imply

$$\psi = \frac{G}{G_0}\left(\frac{\sigma_{Y0}}{\sigma_Y}\right)^2; \qquad \phi = \frac{G^\Delta}{G_0}\left(\frac{\sigma_{Y0}}{\sigma_Y}\right)^2 \qquad (A.2.1)$$

If yield stress does not vary then the usage of Ref. [2] is recovered.

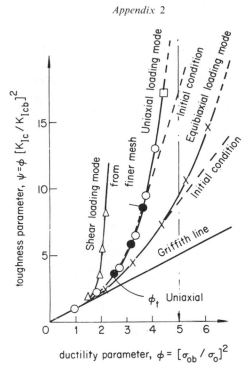

Fig. A.2.1. Effect of stress ratio on relationship between ψ and ϕ parameters in the crack separation theory; ●, uniaxial case computed with re-fined mesh [1].

If the datum G_0 is given the value G_{Icb} and this value held to be the material toughness to which G^Δ is equated at all temperatures, the usage of Ref. [1] is recovered.

What are essentially the same results as in Fig. A.2.1 are shown in Fig. A.2.2 (from Ref. [2]) together with values of J/G for the three cases of loading. Taken at constant G/G_0, or constant axial stress (i.e. constant ψ in the interpretation of Ref. [2]) clearly J/G and G^Δ/G vary, according to the transverse stress field. Conversely, taken at constant J/G or G^Δ/G then ψ (i.e. the axial stress to cause G/G_0) equally depends on the biaxiality. It is less clear whether this implies a complete lack of correlation between the three cases. If a value of J/G is translated into G^Δ/G for each biaxial case, as shown schematically by the dotted arrows, there seems to be a measure of correspondence implying that, as discussed in Section 2.2.2.1.3, J (or δ) may define at least approximately a characteristic severity of crack tip deformation even for biaxial fields. As far as the writer is aware, the J–G^Δ data have not been presented in a way to confirm or refute this viewpoint, by

plotting J against G^Δ for each loading case to see whether a unique curve is obtained, at least for the 'initial' loading condition about to be described.
A further feature marked on Fig. A.2.1 as 'initial condition' relates to the difference between the value of G^Δ for the first node that is relaxed from the initial crack that opens (corresponding to δ_i) before the nodal force is allowed to relax and the subsequent values of G^Δ for two or three further

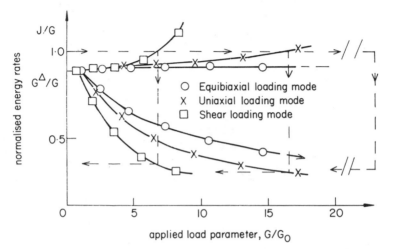

FIG. A.2.2.   Effect of various biaxial modes of loading on J and G^Δ [2].

steps. As seen, these initial values of G^Δ differ only by a few percent from the main curves which are based on the main values for the subsequent three increments of crack growth. It is suggested [1] that these effects may be due to differences in the residual stresses left by the rather different plastic zone that is associated with the initial or the growing crack.

In Ref. [1] the theory is applied to two cases of toughness as a function of temperature taken from published literature. In each case the data required to make the prediction is a value of G_{Icb} at low temperature where the behaviour is very brittle and also the yield stress, both at low temperature, σ_{Yb}, and as a function of temperature. The true toughness, G^Δ, is taken to be G_{Icb} at all temperatures and the reference condition G_0, σ_{Y0}, as G_{Icb}, σ_{Yb}. Thus, at any temperature, ϕ is known and ψ is read off the uniaxial curve, Fig. A.2.1. This gives the value G at which the separation energy rate has reached G^Δ and fracture is postulated to occur, i.e. G_c, which is a function of temperature because ϕ is, through σ_Y. In each case the comparison of G_c from G^Δ with measured toughness is good. In one case, data for A533B steel

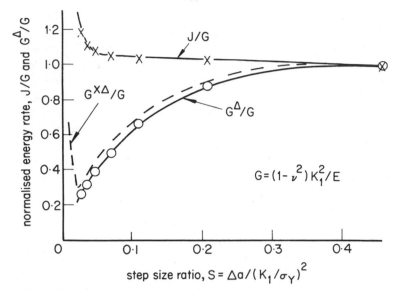

FIG. A.2.3.   Variation of *J*, *G* and the crack separation rate with the ratio of process to plastic
zone size [1].

are used and the comparison made for valid G_{Ic} tests taking the critical value of G^{Δ} as G_{Icb} at 50 K ($-223\,^{\circ}$C). The correlation extends up to about 255 K where the large-scale test data are still valid, thus implying that, even in the LEFM regime, the increase in toughness is due to a growth of plastic zone with decrease in yield stress over the 'completely brittle' conditions at 50 K. In the second case, comparison with mild steel data is made in terms of invalid LEFM results in terms of K and plasticity corrected results (the correction being made by the HSW model described in Section 2.3.2). In both cases a brittle–ductile transition temperature is predicted above which temperature ϕ increases but little for large increases in ψ. For the uniaxial case this condition is defined by $\phi = 5$ so that, for the fixed G^{Δ} variable σ_Y case of Ref. [1], $\sigma_Y = \sqrt{5}\sigma_{Yb}$ for transition to ductile behaviour.

The main novelty of the approach appears to lie in the concept of G^{Λ}. Finite element methods make it possible to compute the plasticity terms, even for the local unloading that occurs on crack growth, so that a reasonably accurate value can be found for what is in essence the rather small difference between the external work done and the plastic work absorbed remote from the immediate crack face. It is then argued that this difference term, G^{Δ} tends to zero as the crack advance, Δa, tends to zero. The evidence is shown in Fig. A.2.3, and the result is taken to be consistent with the conclusion of Rice [4] summarised in Ref. [3] as 'there is no singularity in recoverable energy density, (so that) there is no surplus of work done by external loads over stress working on the deformation of material elements', when there is a saturation of flow stress at finite values of strain. It does not seem apparent from Fig. A.2.3 that G^Δ tends to zero. Indeed it is only the ratio G^Δ/G that is decreasing. This is due entirely to the increase in G, since Δa is held constant and load increased (whereby G and plastic zone size are increased). If Fig. A.2.3 is re-dimensionalised it simply shows that G^Δ increases at constant Δa as load increases. There is no evidence whatsoever on the effect of Δa in the absolute sense. This point is discussed later in connection with more recent studies.

The concept does not appear to be directed specifically at initiation or final unstable propagation, although conventional energy balance concepts are usually directed at the latter. In so far as initial values of G^Δ are referred to, (Fig. A.2.1) clearly initiation is in question with the implication that this occurs over a finite distance, Δa, for which energy has to be available. Slow growth is discussed in more detail in Ref. [3] where it is pointed out that (for the uniaxial case) when R is small (near LEFM) the initial value of G^Δ is slightly larger than the subsequent values so that initiation and unstable propagation could occur jointly, whereas when R is large (extensive plasticity) the initial value of G^Δ is slightly lower than the subsequent values implying a measure of stability perhaps corresponding to slow stable crack growth. Note these results refer to a plane strain computation and not the growth of shear lip.

It therefore seems that the concept offers a model that contains a number of features consistent with known behaviour or other theoretical predictions, but, at this stage of development, it is indicative rather than definitive. Reference [3] also models the concept in terms of the Dugdale–BCS plasticity model (of Section 2.2.1.1) by allowing the restraining stress, f, (Fig. 2.2) to reduce linearly to zero over a region, Δa, that forms part of the plastic zone adjacent to the real crack tip. G^Δ is calculated and for the case $\varepsilon = \Delta a/(c - a) \ll 1$ (where $(c - a)$ is plastic zone extent, Fig. 2.2) a specific formula is obtained for G^Δ that, consistently with Ref. [4], tends to zero as the step size tends to zero

$$G^\Delta = J\varepsilon(\ln 1/\varepsilon + 3/2 + \ln 4)/4$$

and

$$\varepsilon = \Delta a/(c - a) = \frac{8m^2}{\pi} \frac{\sigma_Y{}^2 \Delta a}{E' J} \qquad \text{(A.2.2)}$$

where the restraining stress, f (Fig. 2.2), is written $f = m\sigma_Y$.

A best fit is found between this model and the computed results when $m = 1.91$, close to the values of 2 taken elsewhere as an average value for the constrained yield stress in crack problems. The Dugdale–BCS model does not, however, predict the difference between initial and subsequent values of G^Δ that might lead to further understanding of slow stable growth as mentioned above.

Further discussion of the model leads to an approximation that, at the ductile–brittle transition $K_{Ic(transition)} \simeq 2.7 K_{Ic(brittle)}$ if σ_Y varies with temperature but the process zone Δa and true toughness G^Δ do not. This is again based on the value $\phi = 5$ (Fig. A.2.1) for ductile behaviour. The possibility that Δa or G^Δ are themselves temperature or flow stress dependent would of course further complicate the issue. The concept is in the broadest sense compatible with a two-parameter picture (process zone and crack tip characterising terms) but the details of the model have allowed a significant advance to be made in presenting a tentative picture over a broad field and further studies of the implications of this concept seem certain. If the interpretation here put on Fig. A.2.1 is indeed confirmed, then a link will have been shown between a characterising term (J) and energetic term (G^Δ) for the elastic–plastic crack problem.

Further computations of quantities related to G^Δ have been reported recently by Atluri *et al.* [5] and Bleackley & Luxmoore [6]. Arluri *et al.* use embedded singularity elements of linear elastic or HRR type as appropriate to the degree of yield and a large-geometry-change formulation. As the crack blunts, the origin of the polar coordinate is moved from the tip to the centre of curvature of the now blunted tip so that the actual singularity is lost. For crack growth a core of singularity elements is translated by Δa and the restraining crack tip nodal forces are gradually relaxed. These are elegant procedures but may force on the solution some features not yet proven to be representative of incremental plasticity.

In Ref. [5] a term $G^{X\Delta}$ is computed in plane strain by evaluating the external, elastic and plastic work rates in the global energy balance[1] whereas, it will be recalled, G^Δ is evaluated from the local work of crack tip separation. If energy is conserved in the computations, then the terms must be identical. On Fig. A.2.3 they are indeed closely similar but Atluri *et al.* extend the results to smaller values of $\Delta a/(K/\sigma_Y)^2$, whereon $G^{X\Delta}/G$ increases markedly (dotted line in Fig. A.2.3). G^Δ/G itself is not carried into this regime but if indeed equal to $G^{X\Delta}/G$ it must also increase. In these computations by both Kfouri and Atluri *et al.*, Δa is not varied, the crack

---

[1] This is precisely $\bar{J}_i$ of Section 2.2.2.1.3 (iii), but with $\Delta a \not\to 0$.

increment ratio being altered by $(K/\sigma_Y)^2$ increasing as load increases. When Atluri *et al.* reduce Δa explicitly, the upswing is lost and the ratio $G^{X\Delta}/G$ reduces, perhaps to zero, consistent with the argument of Ref. [4]. The present writer suggests that the upswing (dotted line in Fig. A.2.3) arises as uncontained plasticity occurs, so that the pattern of a crack embedded in a plastic zone is no longer roughly self-similar. At this point, load and hence G cease to increase so that the ratios J/G or $G^{X\Delta}/G$ take a rapid upswing whatever their previous trends. If yield is contained (as for ref. [1] and the corresponding parts of the Ref. [5] data), then an appropriate self-similarity is maintained in which the decrease of $G^{X\Delta}/G$ is also maintained by G increasing. As noted, with G held fixed whilst Δa is explicitly reduced, Atluri *et al.* find $G^{X\Delta}/G$ decreases. This result is contrary to Ref. [6], where Δa is varied at constant load and a term equivalent to $G^{X\Delta}$ found to be nearly independent of step length, with a tendency to increase as Δa becomes small. The computation in Ref. [5] is for plane stress. Parts of the data cover similar ranges of $\Delta \dot{a}/a$ (1–7 % steps), although the rapid decrease of $G^{X\Delta}/G$ in Ref. [5] corresponds to yet smaller values of $\Delta a/a$ (0·1 %). It may be noted that Ref. [6] evaluates $\partial U_{el}/\partial a$ and that it increases as step size decreases. In elasticity this is G, but with plasticity it is the term called I (Section 2.2.2.1.3 (iii)), albeit now for the growing crack rather than for the monotonic load case. If $G^{X\Delta}/(\partial U_{el}/\partial a)$ is plotted from Ref. [6] it corresponds fairly well with $G^{X\Delta}/G$ of Ref. [5] for the range where $\Delta a/a$ overlaps, despite the differences in stress–strain laws and conditions of plane strain and plane stress. This may be fortuitous, but it could be asked whether in Ref. [5] the G used for normalising the constant load cases is indeed the elastic energy rate (as in LEFM) or whether it might not be the elastic energy rate computed *in the presence of plasticity*, $\partial U_{el}/\partial a$. If this were so then the apparent conflict between Refs. [4] and [5] would be resolved, but with $G^{X\Delta}$ not shown to tend to zero with step size, the exact value perhaps depending on details of the model used for crack advance.

REFERENCES

1. Kfouri, A. P. & Miller, K. J. (1976). Crack separation energy rates in elastic–plastic fracture mechanics, *Proc. Inst. Mech. Engrs*, **190** (48/76), pp. 571–84.
2. Kfouri, A. P. & Miller, K. J. (1977). The effect of load biaxiality on the fracture toughness parameters J and G^Δ, *Fracture* 1977, ICF–4, Univ. Waterloo, p. 241.
3. Kfouri, A. P. & Rice, J. P. (1977). Elastic–plastic separation energy rates for crack advance in finite growth steps, *Fracture* 1977, ICF–4, Univ. Waterloo, p. 43.

4. Rice, J. R. (1965). An examination of the fracture mechanics energy balance from the point of view of continuum mechanics, ICF–1, Sendai, pp. A309–40.
5. Atluri, S. N. *et al.* (1977). Hybrid finite element models for linear and non-linear fracture analyses, *Numerical methods in fracture mechanics*, A. R. Luxmoore & M. J. Owen, eds. Univ. Swansea, pp. 52–66.
6. Bleackley, M. H. & Luxmoore, A. R. (1977). An investigation of numerical errors in finite element elastic–plastic crack extension models, *Numerical methods in fracture mechanics*, A. R. Luxmoore & M. J. Owen, eds. Univ. Swansea, pp. 508–24.

Appendix 3

Derivation of the J Integral

C. E. TURNER

Imperial College of Science and Technology, London, UK

In a two-dimensional body of unit thickness of non-linear elastic material, the potential energy, P, measured as a change from an arbitrary datum, can be expressed as the sum of two components

$$P = \int_A Z \, da - \int_{S_T} T_i u_i \, dS \qquad (A.3.1)$$

where da is an element $dx \, dy$ of the area, A, of the component; S is circumference and S_T that part of the circumference on which forces are fixed; T_i are surface tractions and u_i surface displacements in cartesian directions and Z is strain energy density. For simplicity here (though not in Ref. [1]) a steady-state case is considered where

$$\left. \frac{\partial}{\partial a} \right|_x = - \left. \frac{\partial}{\partial x} \right|_a$$

Thus, changing to $\partial/\partial x$, differentiating under the integral sign and noting that the second integral in eqn. (A.3.1) can be taken over the whole of S (since it is automatically zero over those regions other than S_T where displacement will be fixed) then eqn. (A.3.1), when differentiated, gives

$$\frac{dP}{da} = - \int_A \frac{\partial Z}{\partial x} dx \, dy - \int_S T_i \frac{\partial u_i}{\partial x} dS \qquad (A.3.2)$$

By Green's Lemma

$$\int_A \frac{\partial Z}{\partial x} dx \, dy = \int_S Z \, dy$$

$$\therefore \quad \frac{dP}{da} = - \int_S Z \, dy - \int_S T_i \frac{\partial u_i}{\partial x} dS \qquad (A.3.3)$$

Rice showed this integral to be zero around a closed path so that on any two contours, such as Γ and Γ' (Fig. 2.17) the terms must be equal (of opposite sign when traversed in opposite directions) since the tractions along the crack face are zero. Thus the path s along which the integral is taken need not be the boundary S, but can be any arbitrary contour Γ. The right-hand side of eqn. (A.3.3) is then identical to the right-hand side of eqn. (2.18) so that J is indeed—dP/da as stated in eqn. (2.19), for a non-linear elastic material.

The stress–strain law for the material must not vary in the direction of crack growth, though variation normal to the crack plane is permissible.

REFERENCES

1. Rice, J. R. (1968). Mathematical analysis in the mechanics of fracture. In *Fracture: an advanced treatise*, **2**, H. Liebowitz, ed. Academic Press, New York, Chapter 3.
2. Eshelby, J. D. (1956). A continuum theory of lattice defects. In *Progress in solid state physics*, **3**, Academic Press, New York, p. 79.
3. Cherepynov, G. P. (1967). Crack propagation in continuous media, *J. Appl. Math. Mech.*, **31**, pp. 503–12.
4. Rice, J. R. (1968). A path-independent integral and the approximate analysis of strain concentration by notches and cracks, *J. Appl. Mech.*, **35**, pp. 379–86.

Appendix 4

The η Factor, with Particular Reference to Shallow Notches

C. E. TURNER

Imperial College of Science and Technology, London, UK

(a) η_{el}

In Ref. [1] a factor was introduced that in the present notation is called η_{el}, to relate G to the work done, $Fq/2$, in a linear elastic (triangular) load–deflection diagram taken to the deflection of interest (eqn. (2.30b)). A similar relationship was pointed out in Ref. [2]. The term was evaluated from LEFM shape factors to give

$$\eta_{el} = (W - a)Y^2 a / \int Y^2 a \, \mathrm{d}a$$

For a five-term representation of Y

$$Y = A_0 + A_1(a/W) + A_2(a/W)^2 + A_3(a/W)^3 + A_4(a/W)^4 \quad \text{(A.4.1)}$$

so that

$$Y^2 a = W[\bar{A}_0(a/W) + \cdots \bar{A}_n(a/W)^{n+1} \cdots \bar{A}_8(a/W)^9] \quad \text{(A.4.2)}$$

$$\int Y^2 a \, \mathrm{d}a = W^2 \left[\frac{\bar{A}_0(a/W)^2}{2} + \cdots \frac{A_0(a/W)^{n+2}}{n+2} \cdots \frac{A_8(a/W)^{10}}{10} + \frac{EF^2 C_0}{2W^2 B\sigma^2} \right] \quad \text{(A.4.3)}$$

where

$$\bar{A}_0 = A_0^2; \quad \bar{A}_1 = 2A_0 A_1; \quad \bar{A}_2 = 2A_0 A_2 + A_1^2; \quad \bar{A}_3 = 2A_0 A_3 + 2A_1 A_2;$$

$$\bar{A}_4 = 2A_0 A_4 + 2A_1 A_3 + A_2^2; \quad \bar{A}_5 = 2A_1 A_4 + 2A_2 A_3;$$

$$\bar{A}_6 = 2A_2 A_4 + A_3^2; \quad \bar{A}_7 = 2A_3 A_4; \quad \bar{A}_8 = A_4^2$$

In eqn. (A.4.3) C_0 is the unnotched compliance.

The usual formulae for Y [3] are not convergent series, so that a useful expression for η_{el} is not available for other than very shallow notches without the full development of eqns. (A.4.2) and (A.4.3).

For bending,

$$C_0 \simeq (S^3/48EI)[1 + (4W^2/S^2)] \qquad (A.4.4a)$$

(the shear correction term $[1 + (4W^2/S^2)]$ is for a rectangular cross-section and was omitted in Ref. [1]). Values of η_{el} are shown in Fig. 2.30c for $S/W = 4$ and $S/W = 8$. Note: $\eta_{el} \to 2$ for deep notches, particularly with $S/W = 4$. For shallow notches using only the leading terms of eqns. (A.4.2) and (A.4.3)

$$\eta_{el} \simeq 18A_0^2(a/W)[1 - (a/W)]/(S/W) \qquad (A.4.4b)$$

If shear is negligible $(S/W \geq 8$, say) $A_0 \simeq 2$, so, as $(W - a) \to W$

$$\eta_{el} \to 72a/S \qquad (A.4.4c)$$

For $S/W = 4$, $A_0 = 1\cdot93$ and shear deflection increases compliance by 25%, whence

$$\eta_{el} \simeq 46a/S \qquad (A.4.4d)$$

Thus, for shallow notches

$$\eta_{el} \propto a/S$$

In tension

$$C_0 = D/EBW \qquad (A.4.5a)$$

Values of η_{el} are shown for centre-cracked tension $(D/W = 4$ and $D/W = 8)$ in Fig. 2.30d, and approximate values for single-edge notch tension deformed such that the ends remain parallel. For larger values of gauge length η_{el} is approximately proportional to $1/D$ for all notch depth ratios. For shallow notches

$$\eta_{el} \simeq 2A_0^2(a/W)[1 - (a/W)]/(D/W) \qquad (A.4.5b)$$

so $A_0 \simeq 2$ for SEN, DEN, or $\pi^{1/2}$ for CCP, as $(W - a) \to W$.

$$\eta_{el} \to 8a/D \quad (\text{or } 2\pi a/D \text{ for CCP}) \qquad (A.4.5c)$$

An alternative definition of η_{el} can be found. Writing $FC = q$, differentiating with respect to crack length and expressing the term dC/da in terms of η_{el} from eqn. (2.32), and restricting the analysis to constant displacement (i.e. $dq/da = 0$) it is found that

$$\eta_{el} = -\frac{(W - a)}{W} \frac{1}{F_{el}} \left.\frac{\partial F_{el}}{\partial a}\right|_q \qquad (A.4.6)$$

where F_{el} is any elastic load. This formulation is directly comparable to eqn. (2.33a), which has been adopted here as the definition of η_{el}, and is a useful form in connection with Section 2.4.1.

Another relationship of interest (from eqn. (2.30b) by rearrangement) is the ratio of notched compliance, C_n, to the unnotched, C_0;

$$C_n = 2GB(W - a)/F^2\eta_{el} \qquad (A.4.7)$$

whence, expressing G in terms of K^2E and hence of Y

$$\frac{C_n}{C_0} = \frac{2Y^2a(W - a)}{DW\eta_{el}} \qquad (A.4.8)$$

where D is either tensile gauge length or span $S/9$ in three-point bending. This expression is also of use in Section 2.4.1.

(b) η_{pl}

In plasticity, usage of the term here called η originated with the relationship between J and work done (eqn. (2.30a)), with some uncertainty, as discussed in Section 2.2.2.1.3 (iv), over whether U_T or U_{pl} should be the measure of work. However, the same term arises in eqn. (2.35) for the linear slope dJ_{pl}/dq_{pl} of the J-deflection relationship in the plastic region. In both cases the term arises from the differentiation of the limit load with respect to crack length. The term also arises in eqn. (2.122a) in connection with stable crack growth. In Section 2.2.2.1.3 (iv) it was defined:

$$\eta_{pl} = -\frac{(W - a)}{W}\frac{1}{F_L}\frac{\partial F_L}{\partial(a/W)} \qquad (2.33a \text{ bis})$$

$$= N - \frac{W - a}{W}\frac{1}{L}\frac{\partial L}{\partial a/W} \qquad (2.33c \text{ bis})$$

where $F_L \propto (W - a)^N$; $N = 1$ in tension, $N = 2$ in bending.

Some of these usages have been pointed out before but it does not seem to have been noted that it is the same term that arises in each case. Ref. [4] used a constraint factor

$$L = 1 + \ln (W - a)/(W - 2a) \qquad (A.4.9)$$

from Ref. [5] for an evaluation of the slope dJ_{pl}/dq_{pl} in eqn. (2.35) for DEN pieces, and found good agreement with computed results for $2a/W = 0{\cdot}625$. A general usage of both η_{el} and also the deep-notch case

$\eta_{\mathrm{pl}} = 1(\mathrm{SEN})$ or 2 (three-point bend) in connection with collapse loads was made [6] and η_{pl} for DEN and SEN used [7]. The effect of work-hardening in combination with shallow-notch tension was noted [4, 8] on the slope $dJ_{\mathrm{pl}}/dq_{\mathrm{pl}}$ but not related to η_{pl} in the sense of the usage common to several problems.

Where an expression exists for the constraint factor L, η_{pl} can be evaluated from eqn. (2.33c). Clearly, if 'full constraint' is reached for some notch depth, L is thereafter constant until loss of constraint occurs as $a/W \to 1$, and in this deep-notch region $\eta_{\mathrm{pl}} \simeq N$. For shallow notches η_{pl} is sensitive to $\partial L/\partial(a/W)$ and adequate expressions do not seem to exist for various geometries of interest. Although the values of L may be known to within a few per cent, $\partial L/\partial(a/W)$ may differ appreciably for various approximate expressions for L. Clearly $1 \leq L \leq W/(W - a)$.

There is a lack of clarity in some uses of η_{pl} between a true η_{pl} at limit load in conditions of extensive plasticity and the plastic component of the overall η whilst still in contained yield.

For tension the constraint for shallow SEN, deformed with ends kept parallel, must be similar whilst in contained yield to that for DEN. Following eqn. (A.4.9) the writer used

$$L = 1 + \ln[1 + (a/W)] \tag{A.4.10}$$

whence

$$\eta_{\mathrm{pl}} \simeq 3a/W \tag{A.4.11}$$

$$J_{\mathrm{pl}}/q_{\mathrm{pl}} \simeq (3a/W)[1 + (a/W)]\sigma_{\mathrm{Y}} \tag{A.4.12}$$

Preliminary computations for shallow-notch CCP with work-hardening give $J_{\mathrm{pl}}/q_{\mathrm{pl}}$ and hence η_{pl} similar to SEN values as noted in connection with Fig. 2.47.

In bending for $a/W < 0\cdot1$ or perhaps $0\cdot2$ the writer used

$$L = 1 + \ln[1 + (2a/W)] \tag{A.4.13}$$

whence

$$\eta_{\mathrm{pl}} \simeq 10a/W \tag{A.4.14}$$

$$J_{\mathrm{pl}}/q_{\mathrm{pl}} \simeq (10a/W)[1 + (a/W)]\sigma_{\mathrm{Y}}/(S/W) \tag{A.4.15}$$

The effect of shear is neglected, span entering only through the relation between load and moment, $M = FS/4$.

Agreement between the values of η_{pl} in eqns. (A.4.11), (A.4.14) and the computations was shown to be quite reasonable (Table 2.1), and eqns. (A.4.12) and (A.4.15) predict the $J_{\mathrm{pl}}/q_{\mathrm{pl}}$ slopes in Fig. 2.47 to within similar accuracy. In the general description of Section 2.2.2.1.3 no distinction was made between uniaxial tension and tension with the ends kept parallel. The computations refer to the latter case.

(c) η (THE OVERALL TERM, COMBINING ELASTIC AND PLASTIC EFFECTS)

There are two further complications. The first is that the term required in the elastic–plastic region is not strictly as given in eqn. (2.33a bis), but the yet more general one

$$\eta = -\frac{W-a}{W}\frac{1}{F}\frac{\partial F}{\partial(a/W)} \tag{A.4.16}$$

where F is the current load and a the current crack length. In the absence of solutions for the advancing crack problem, except in a few computer formulations, estimates of $\partial F/\partial(a/W)$ have to be accepted from the elastic or limit load cases (eqns. (2.32) and (2.33a) as repeated above), with the latter strictly relevant only after extensive plasticity has allowed the slip pattern to develop. For cases where there is little change in constraint from elastic to limit load regime this approximation seems adequate. Where there is a loss of constraint at limit load the two end-state solutions, elastic and limit load, do not seem adequate to cover the intervening elastic–plastic regime. In particular, the shallow single-edge notch and centre-cracked cases seem to behave in the plastic region well before collapse in a manner comparable to the same notch depth in double-edge notch tension, yet for the former the limit load constraint factor is unity, independent of notch-depth ratio, whereas in the latter there is a marked notch-depth dependence. For the former cases $\partial F/\partial(a/W)$ in the pre-collapse region must be appreciably different from $\partial F_L/\partial(a/W)$. No expressions are known to exist for $\partial F/\partial(a/W)$, nor indeed is the term 'constraint factor' strictly applicable except in the limit state. The second difficulty is that values of η found from the relation between J and work done (eqn. (2.30e)) are further complicated if work-hardening causes general yield of the remote ends as discussed in Section 2.2.2.1.3 (iv) (following eqn. (2.37)). Very approximate estimates of η can be made using η_{el} and η_{pl}. Since,[1] from eqn. (2.30c),

$$\eta_0 U_T = \eta_{el} U_{el} + \eta_{pl\,n} U_{pl} \tag{A.4.17}$$

[1] η_{pl} is defined in relation to the plastic work in the notched region, $_n U_{pl}$, whereas η_{el} relates to the (overall) elastic work and η to the overall (total) work.

Let $_0U_{pl} = M(_nU_{pl})$, where for shallow-notch tension $M \simeq D/D_n$, then

$$\eta = (\eta_{el}U_{el} + \eta_{pl\,n}U_{pl})/(U_{el} + M_nU_{pl}) \qquad (A.4.18)$$

For any load-deflection diagram express $_0U_T$ as a multiple of U_{el} so that

$$M(_nU_{pl}) + U_{el} = {_0U_T} = ZU_{el} \qquad (A.4.19)$$

Thus

$$\eta \simeq [\eta_{el} + \eta_{pl}(Z - 1)/M]/Z \qquad (A.4.20)$$

As an example related to Fig. 2.30b, for shallow-notch SEN, $D/W = 4$, then $M \simeq 2$ (after gross yield with slip at $45°$) and for an elastic non-work-hardening diagram $_0U_T/U_{el} \simeq (2qE/D\sigma_Y) - 1$. The value is $\eta_{el} \simeq 0.2$ for this case. In general yield, at, say, $qE/D\sigma_Y = 3$, take $\eta_{pl} = 1$. Thus $\eta \simeq [0.2 + 1(5 - 1)/2]/5 \simeq 0.45$ (cf. 0.33 in Fig. 2.30b). For $D/W = 2$, $M \simeq 1$; $\eta_{el} \simeq [0.4$; thus $\eta \simeq [0.4 + 1(5 - 1)/1] \simeq 0.85$ (cf. 0.61 in Fig. 2.30b). For estimates in contained yield, $M \simeq 1$, since there is no plasticity remote from the notch, and Z has a value such as $1 < Z \lesssim 2$, say, so that $\eta \simeq \eta_{el} \simeq \eta_{pl}$.

These estimates are hardly satisfactory, but serve to show the factors that are involved.

The discussions of η_{pl} were based on formulae for limit load constraint factors but estimates for some of the shallow-notch cases were made in a heuristic manner, which, if interpreted strictly, would imply that limit load is exceeded. Clearly, improved approximations for η are still required. The term $\partial F/\partial a$ in eqn. (A.4.16) could, of course, be found experimentally for any particular geometry.

The close relationship in value found between η_{el} and η_{pl} is probably peculiar to a few configurations, of which three-point bending $S/W = 4$ is the best known example, since η_{pl} is independent of span or length over and above that relevant to the notched deformation pattern, whereas η_{el} is a function of span or length. η is also a function or span or length, depending both on configuration and on extent of yielding. Using the definitions in eqns. (2.33a) and (A.4.6), it follows $\eta_{el} < \eta_{pl}$, in some cases by a large margin, in others by quite small amounts. For most test-piece configurations $\eta_{el} < \eta < \eta_{pl}$, but for shallow-notch, long-gauge-length pieces with general plasticity where overall work could be a large multiple of both elastic and plastic-notch component (i.e. M and Z both large) then η could be smaller than either η_{el} or η_{pl}.

REFERENCES

1. Turner, C. E. (1973). Fracture toughness and specific fracture energy: a re-analysis of results, *Mat. Sci. Engng*, **11**, pp. 275–82.
2. Feddein, G. & Macherauch, E. (1973). A new method for fracture toughness determination, *ICF*-3, Munich, **3**, II–241.
3. Plane strain crack toughness testing, ASTM–STP 410, 1966.
4. Sumpter, J. D. G. (1973). Elastic–plastic fracture analysis and design using the finite-element method. Ph. D. Thesis, University of London.
5. Ewing, D. F. & Hill, R. (1967). The plastic constraint of v-notched tension bars, *J. Mech. Phys. Solids*, **15**, pp. 115–25.
6. Neale, B. K. & Townley, C. H. A. (1977). Comparison of elastic–plastic fracture mechanics criterion, *Int. J. Pres. Ves. & Piping*, **5**, pp. 207–39.
7. Sumpter, J. D. G. *et al.* (1973). Post-yield analysis and fracture in notch tension pieces, *ICF*-3, Munich, I–433.
8. Sumpter, J. D. G. & Turner, C. E. (1976). Design using elastic–plastic fracture mechanics, *Int. J. Fract.*, **12**, pp. 861–73.

Appendix 5

On Matrices and Vectors

T. K. HELLEN

CEGB, Berkeley, UK

The matrix algebra concerns the manipulation of arrays of numbers or variables in such a way that the overall behaviour can be expressed in a short-hand manner involving only collective, matrix-type terms. Matrices are used extensively in finite element theory and so, as an introductory definition of a matrix, consider the simple elastic constitutive equations for an isotropic material in plane stress:

$$\sigma_x = \frac{E}{1 - v^2} [\varepsilon_x + v\varepsilon_y]$$

$$\sigma_y = \frac{E}{1 - v^2} [v\varepsilon_x + \varepsilon_y]$$

$$\tau_{xy} = \frac{E}{2(1 + v)} \gamma_{xy} \tag{A5.1}$$

The *vector* $\{\sigma\}$ contains the stress components σ_x, σ_y and τ_{xy} expressed in terms of cartesian components and is written as a *column vector*,

$$\{\sigma\} = \begin{bmatrix} \sigma_x \\ \sigma_y \\ \tau_{xy} \end{bmatrix} \tag{A5.2}$$

The vectors used in the finite element theories in this text are all column vectors. Similarly the strain vector may be written as

$$\{\varepsilon\} = \begin{bmatrix} \varepsilon_x \\ \varepsilon_y \\ \gamma_{xy} \end{bmatrix} \tag{A5.3}$$

It is now possible to write eqn. (A5.1) as a matrix equation via an intermediate step:

$$
\begin{bmatrix} \sigma_x \\ \sigma_y \\ \tau_{xy} \end{bmatrix} = \frac{E}{1 - v^2} \begin{bmatrix} \varepsilon_x + v\varepsilon_y + 0 \\ v\varepsilon_x + \varepsilon_y + 0 \\ 0 + 0 + \dfrac{1 - v}{2}\gamma_{xy} \end{bmatrix} \tag{A5.4}
$$

i.e. three equations linked by the square brackets, or

$$
\begin{bmatrix} \sigma_x \\ \sigma_y \\ \tau_{xy} \end{bmatrix} = \frac{E}{1 - v^2} \begin{bmatrix} 1 & v & 0 \\ v & 1 & 0 \\ 0 & 0 & \dfrac{1 - v}{2} \end{bmatrix} \begin{bmatrix} \varepsilon_x \\ \varepsilon_y \\ \gamma_{xy} \end{bmatrix} \tag{A5.5}
$$

where, for each of the three equations of (A5.4), the large square bracket contains an array of terms, or elements, not connected algebraically but, by convention, the first row's three elements $[1, v, 0]$ are multiplied by $[\varepsilon_x, \varepsilon_y, \gamma_{xy}]$ of the final strain vector, respectively, to give $\varepsilon_x + v\varepsilon_y$. Similarly for the second and third rows. This type of multiplication, involving a number of separate elements each multiplied together, is known as scalar multiplication.

The large square bracket is a matrix and the scalar quantity $E/1 - v^2$ preceding it indicates that every element in the matrix has to be multiplied by it. Defining the matrix [D] as

$$
[D] = \frac{E}{1 - v^2} \begin{bmatrix} 1 & v & 0 \\ v & 1 & 0 \\ 0 & 0 & \dfrac{1 - v}{2} \end{bmatrix} \tag{A5.6}
$$

allows us to write the constitutive relations in matrix equation form as:

$$
\{\sigma\} = [D]\{\varepsilon\} \tag{A5.7}
$$

It is easy to verify that the constitutive laws for a point in an arbitrary, isotropic three-dimensional body give rise to a matrix equation identical to eqn. (A5.7) but with a redefinition of the two vectors and the matrix as follows:

$$\{\sigma\} = \begin{bmatrix} \sigma_x \\ \sigma_y \\ \sigma_z \\ \tau_{xy} \\ \tau_{yz} \\ \tau_{zx} \end{bmatrix} \qquad \{\varepsilon\} = \begin{bmatrix} \varepsilon_x \\ \varepsilon_y \\ \varepsilon_z \\ \gamma_{xy} \\ \gamma_{yz} \\ \gamma_{zx} \end{bmatrix} \qquad (A5.8)$$

$$[D] = \frac{E}{(1+v)(1-2v)} \begin{bmatrix} 1-v & v & v & 0 & 0 & 0 \\ v & 1-v & v & 0 & 0 & 0 \\ v & v & 1-v & 0 & 0 & 0 \\ 0 & 0 & 0 & \frac{1}{2}-v & 0 & 0 \\ 0 & 0 & 0 & 0 & \frac{1}{2}-v & 0 \\ 0 & 0 & 0 & 0 & 0 & \frac{1}{2}-v \end{bmatrix}$$

$$(A5.9)$$

Consider a more general matrix [A]:

$$[A] = \begin{bmatrix} a_{11} & a_{12} & \cdots & a_{1n} \\ a_{21} & a_{22} & \cdots & a_{2n} \\ \vdots & & & \\ a_{m1} & a_{m2} & \cdots & a_{mn} \end{bmatrix}$$

where the a_{ij} are individual numbers or variables. [A] contains $m \times n$ such elements, has m rows and n columns. The size of [A] is therefore denoted by $(m \times n)$.

It is usual to write a matrix and vector inside [] and {}, respectively. When writing out all the terms, the matrix retains the square brackets, as in eqn. (A5.5), but the use of both square, curved or curly brackets for vectors is adopted by different authors.

Transposition is denoted as in vectors by superscript T, so that

$$[A]^T = \begin{bmatrix} a_{11} & a_{21} & \cdots & a_{m1} \\ a_{12} & & & \vdots \\ a_{13} & & & \vdots \\ \vdots & & & \vdots \\ a_{1n} & \cdots & \cdots & a_{mn} \end{bmatrix}$$

$[A]^T$ here is of size $(n \times m)$.

An alternative way of writing a matrix is as (a_{ij}) where i and j vary over all permissible values to cover all elements. Hence, i ranges from 1 to m and j ranges from 1 to n in $[A]$.

If $a_{ij} = a_{ji}$ throughout $[A]$, $[A]$ is symmetric. By necessity, $m = n$, so $[A]$ is square. The $[D]$ matrices of eqns. (A5.6) and (A5.9) are both symmetric. The use of non-isotropy constitutive equations generally results in non-symmetric $[D]$ matrices.

Matrix addition (or subtraction) is valid for two matrices of the same size. Thus

$$[A] + [B] = [A + B]$$

and the corresponding elements in $[A]$ and $[B]$ are added:

$$(a_{ij}) + (b_{ij}) = (a_{ij} + b_{ij}).$$

A column vector of n components can be regarded as a matrix of size $(n \times 1)$. The rules of matrix multiplication require that the number of components in the row of the matrix and the column of the following vector (or, in more general terms, the following matrix) be the same. In eqn. (A5.5) this is so since there are three components across the matrix and three in the column vector $\{\varepsilon\}$.

This law is most easily remembered as follows. Consider a valid matrix multiplication of three matrices: their sizes must be of the form shown underneath the actual equation.

$$\underset{(a \times b)}{[A]} = \underset{(a \times c)}{[B]} \underset{(c \times b)}{[C]}$$

If written as $(a \times b) = (c \times d)(e \times f)$ we would require $a = c$, $b = f$, $d = e$.
In the case of eqn. (A5.5), the sizes are

$$(3 \times 1) = (3 \times 3)(3 \times 1)$$

so that the multiplication is valid. In general, $[B][C] \neq [C][B]$; indeed the latter multiplication may not even be valid.

Transposition of matrix products can be shown to give the following result:

$$\text{If} \quad [A] = [B][C]$$
$$[A]^{\mathrm{T}} = ([B][C])^{\mathrm{T}} = [C]^{\mathrm{T}}[B]^{\mathrm{T}}$$

The $(n \times n)$ identity matrix $[I]$ is

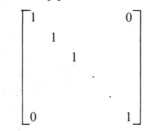

so that

$$a_{ij} = \begin{cases} 0 \text{ if } i \neq j \\ 1 \text{ if } i = j \end{cases}$$

i, j each varying over 1 to n.

If $[A]$ is $n \times n$, $[A][I] = [I][A] = [A]$

The inverse of $[A]$, $[A]^{-1}$, is defined as $[A]^{-1}[A] = [I] = [A][A]^{-1}$. Inversion cannot be defined for non-square matrices $(m \neq n)$ nor for matrices that are singular, i.e. have zero determinants.

Matrix inversion is called for whenever the relationship between two vectors $\{x\}$ and $\{y\}$ is given in the form

$$[A]\{x\} = \{y\}$$

while $\{x\}$ is unknown. Inversion yields

$$\{x\} = [A]^{-1}\{y\}$$

Integration of matrices is defined in a straightforward manner:

$$\int_V [A] \, dV - \begin{bmatrix} \int_V a_{11} dV & \int_V a_{12} dV & \cdots & \int_V a_{1n} dV \\ \int_V a_{21} dV & & & \vdots \\ \vdots & & & \vdots \\ \int_V a_{m1} dV & \cdots & \cdots & \int_V a_{mn} dV \end{bmatrix}$$

so that the usual requirements of integrability on $[A]$ are also necessary for each individual a_{ij}.

T. K. Hellen

SUMMATION CONVENTION

This is a short-hand convention where a summation is implied but no summation sign is given. Thus, if $\bar{\sigma} = \sigma_{11} + \sigma_{22} + \sigma_{33}$, we would write

$$\bar{\sigma} = \sigma_{ii}$$

it being understood that i is in the relevant range, here 1 to 3.

Usually, any subscript on the right-hand side not appearing on the left-hand side is summable. As an example, consider eqn. (4.4):

$$\bar{\sigma}^2 = \tfrac{1}{2} S_{ij} S_{ij}$$
$$= \tfrac{1}{2}(S_{11}^2 + S_{22}^2 + S_{33}^2 + 2S_{12}^2 + 2S_{23}^2 + 2S_{31}^2)$$

Appendix 6

Notes on the Finite Element Method

T. K. HELLEN

CEGB, Berkeley, UK

In view of the many excellent textbooks now available on the finite element method (e.g. Refs. [1–4]) the aims of this appendix are limited to:

briefly summarising the method's application to stress analysis, as a possible 'refresher' for readers of Chapter 4 and

motivating the selection of element type used in Chapter 4 for modelling crack tip behaviour.

The finite element method applied to stress analysis, and in particular the analysis of structures of non-linear materials, involves a representation of the given structure in a manner suitable for numerical analysis. This is effected by constructing a model of the structure made of a number of elements, within which the displacements (and possibly other variables) are assumed to vary in a fairly defined manner. This variation, together with the number of dimensions, scope of element shape and number of nodes, or points of reference, lying on it, define the element type. The size of the elements governs the accuracy of the solution: the more elements, the better the results. For each element type, a stiffness matrix is produced which is amalgamated for all elements of the structure to a total set of equations. It is the solution of these equations that necessitates the use of large computers.

Elements can be one-, two- or three-dimensional, beams, plates or shells to model structures of virtually any shape subject only to computer program limitations. The derivation of the element and structural equations follows a common course when vector and matrix notation is used. Thus, the strain vector (Appendix 5) is written in the general form $\{\varepsilon\}$, but contains a variable number and type of components depending on the form of the structure.

We proceed by deriving the basic equations of the finite element method for elastic behaviour. The form of the individual matrices are then illustrated by describing the simple constant stress triangle for two-dimensional plane stress analysis. This element has formed the basis of most finite element computer programs since their first inception and has sufficiently simple matrix derivations to illustrate well the basic theory.

For crack tip modelling the use of higher order elements is required. A discussion of the theoretical extensions for these elements, used extensively in Chapter 4, is therefore included in accordance with the second aim formulated above.

Further use of elastic finite element theory in non-linear plasticity and creep analysis is described in Chapter 4 and Appendix 7, respectively.

DERIVATION OF ELASTIC EQUATIONS

At any point in an element, the total strain vector $\{\varepsilon\}$ (see Appendix 5) corresponding to the calculated displacements is the sum of the strains due to mechanical forces, $\{\varepsilon_e\}$, and initial strains, $\{\varepsilon_0\}$, such as thermal or residual strains:

$$\{\varepsilon\} = \{\varepsilon_e\} + \{\varepsilon_0\} \tag{A6.1}$$

Plastic or creep strains, if present, would also be added to the right-hand side, but here only $\{\varepsilon_0\}$ is considered. The number of components in these vectors depends on whether the element is two or three dimensional, etc.

The stress vector is determined from the constitutive relations

$$\{\sigma\} = [D]\{\{\varepsilon\} - \{\varepsilon_0\}\} \tag{A6.2}$$

where [D] is a matrix of terms containing Young's modulus and Poisson's ratio, being symmetric under isotropic conditions. This equation has been explicitly given in Appendix 1.

The compatibility equations at a point are

$$\{\varepsilon\} = [B]\{u\}^l \tag{A6.3}$$

where $\{u\}^l$ is the total nodal displacement vector for the element containing the point, which may lie on the edge of the element, and [B] is a matrix depending on the shape functions of the point.

The point-wise strain energy density is ρ, given by

$$\rho = \tfrac{1}{2}\{\{\varepsilon\} - \{\varepsilon_0\}\}^{\mathrm{T}}[D]\{\{\varepsilon\} - \{\varepsilon_0\}\}$$

or, using eqn. (A2.3), this can be expanded to

$$\rho = \{u\}^{l\mathrm{T}}[B]^{\mathrm{T}}[D][B]\{u\}^l - \{\varepsilon_0\}^{\mathrm{T}}[D][B]\{u\}^l + \tfrac{1}{2}\{\varepsilon_0\}^{\mathrm{T}}[D]\{\varepsilon_0\} \tag{A6.4}$$

The strain energy of the element is ρ^1, given by

$$\rho^1 = \int_1 \rho \, dV$$

i.e.

$$\rho^1 = \tfrac{1}{2}\{u\}^{1T}\left(\int_1 [B]^T[D][B]\, dV\right)\{u\}^1 - \{u\}^{1T}$$

$$\times \int_1 [B]^T[D]\{\varepsilon_0\}\, dV + \tfrac{1}{2}\int_1 \{\varepsilon_0\}^T[D]\{\varepsilon_0\}\, dV \qquad (A6.5)$$

The element stiffness matrix is defined as

$$[K]^1 = \int_1 [B]^T[D][B]\, dV \qquad (A6.6)$$

The total potential energy of the element is

$$P^1 = \rho^1 - \{u\}^{1T}\{F_e\}^1 \qquad (A6.7)$$

where $\{F_e\}^1$ is the mechanical load vector, corresponding term by term with $\{u\}^1$ for the element. Thus,

$$\rho^1 = \tfrac{1}{2}\{u\}^{1T}[K]^1\{u\}^1 - \{u\}^{1T}\{F_0\}^1 + W^1 - \{u\}^{1T}\{F_e\}^1 \qquad (A6.8)$$

where

$$\{F_0\}^1 = \int_1 [B]^T[D]\{\varepsilon_0\}\, dV \qquad (A6.9)$$

is the load vector of thermal and residual strains and

$$W^1 = \tfrac{1}{2}\int_1 \{\varepsilon_0\}^T[D]\{\varepsilon_0\}\, dV \qquad (A6.10)$$

which will be termed 'thermal energy'.

By superimposing the matrices and vectors of every element but keeping the terms due to different degrees of freedom distinct in the newly accumulated matrix it is possible to write down the total potential energy of the structure. These are defined as before but without the superscript l. Then if

$$\{F_T\} = \{F_e\} + \{F_0\}$$

$$P = \tfrac{1}{2}\{u\}^T[K]\{u\} - \{u\}^T\{F_T\} + W \qquad (A6.11)$$

$[K]$ is the structural stiffness matrix and is symmetric and positive definite.

The vector $\{F_e\}$ only represents contributions from externally applied loads since, by equilibrium, internal forces cancel at internal nodes. Differentiating P with respect to each term of $\{u\}$ in turn, and equating to zero, gives the condition

$$[K]\{u\} = \{F_T\} \qquad (A6.12)$$

for a state of minimum potential energy. This represents the basic equations to be solved in the finite element method. The terms of $\{u\}$ are usually unknown, although any number of the displacements may be prescribed to represent applied constraints.

From eqns. (A6.9) and (A6.10)

$$P = -\tfrac{1}{2}\{u\}^{T}[K]\{u\} + W$$

or

$$P = W - V \qquad (A6.13)$$

where $V = \tfrac{1}{2}\{u\}^{T}[K]\{u\}$ is the strain energy when $\{F_0\}$ is null.

Because elements are only connected to a small number of other elements, the $[K]$ matrix contains large areas of zero terms. In the practical solution of eqn. (A6.12), it is important to ensure that all non-zero terms are as close to the main diagonal as possible, so that the maximum semi-bandwidth of non-zero terms in each row is minimal. This is to reduce computer core requirements and running times, both of which increase rapidly with increasing semi-bandwidth. For such equations, a very efficient solution process known as the front solution has been devised [5, 6]. The size of the semi-bandwidth is dictated by the sequencing of elements in the solution process.

Inelastic processes like plasticity and creep involve a sequence of solutions (A6.12) to follow load- or time-dependent properties. Extensions of the finite element method to these cases are given in Chapter 4 and Appendix 7.

THE SIMPLE CONSTANT STRESS TRIANGLE

This element is the original finite element used by Turner et al. [7], Clough [8] and many others since. Although it can be used for plane stress, plane strain and axisymmetric structures we restrict the present discussions to plane stress. Thus, there are two displacement degrees of freedom (u, v) at every point in the element. Referring to Fig. A6.1, consider the element with nodes 1, 2 and 3, with coordinates $(x_1, y_1), (x_2, y_2)$ and (x_3, y_3), respectively, and a uniform thickness t. The variation of displacement (known also as the

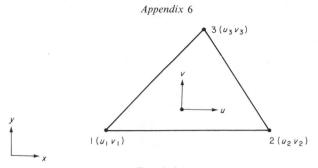

Fig. A.6.1

'shape function') is assumed to be linear in the element so that the element may be written as

$$u = \alpha_1 + \alpha_2 x + \alpha_3 y$$

$$v = \alpha_4 + \alpha_5 x + \alpha_6 y \qquad \text{(A6.14)}$$

The α_i are determined by inserting the nodal values (u_i, v_i), which are assumed unknown, to give the matrix relationship:

$$
\begin{bmatrix} u_1 \\ v_1 \\ u_2 \\ v_2 \\ u_3 \\ v_3 \end{bmatrix}
=
\begin{bmatrix}
1 & x_1 & y_1 & 0 & 0 & 0 \\
0 & 0 & 0 & 1 & x_1 & y_1 \\
1 & x_2 & y_2 & 0 & 0 & 0 \\
0 & 0 & 0 & 1 & x_2 & y_2 \\
1 & x_3 & y_3 & 0 & 0 & 0 \\
0 & 0 & 0 & 1 & x_3 & y_3
\end{bmatrix}
\begin{bmatrix} \alpha_1 \\ \alpha_2 \\ \alpha_3 \\ \alpha_4 \\ \alpha_5 \\ \alpha_6 \end{bmatrix}
\qquad \text{(A6.15)}
$$

By suitable manipulation of eqns. (A6.14) and (A6.15) it is possible to remove the α_is to give a relationship between (u, v), the nodal displacements, and geometric terms:

$$
\begin{bmatrix} u \\ v \end{bmatrix}
= \frac{1}{2A}
\begin{bmatrix}
a_1 + b_1 x + c_1 y & 0 & a_2 + b_2 x + c_2 y \\
0 & a_1 + b_1 x + c_1 y & 0
\end{bmatrix}
$$

$$
\left.
\begin{matrix}
0 & a_3 + b_3 x + c_3 y & 0 \\
a_2 + b_2 x + c_2 y & 0 & a_3 + b_3 x + c_3 y
\end{matrix}
\right]
\begin{bmatrix} u_1 \\ v_1 \\ u_2 \\ v_2 \\ u_3 \\ v_3 \end{bmatrix}
$$

$$\text{(A6.16)}$$

where

$$a_1 = x_2 y_3 - x_3 y_2 \quad a_2 = x_3 y_1 - x_1 y_3 \quad a_3 = x_1 y_2 - x_2 y_1$$

$$b_1 = y_2 - y_3 \qquad b_2 = y_3 - y_1 \qquad b_3 = y_1 - y_2$$

$$c_1 = x_3 - x_2 \qquad c_2 = x_1 - x_3 \qquad c_3 = x_2 - x_1$$

and A is the area of the triangle. It will be seen that the components of the shape function matrix are linear in x and y and that $\{u\}^1 = [u_1 v_1 u_2 v_2 u_3 v_3]^T$ (a column vector).

The strain vector is given by

$$\{\varepsilon\} = \begin{bmatrix} \varepsilon_x \\ \varepsilon_y \\ \gamma_{xy} \end{bmatrix} = \begin{bmatrix} \dfrac{\partial u}{\partial x} \\ \dfrac{\partial v}{\partial y} \\ \dfrac{\partial u}{\partial y} + \dfrac{\partial v}{\partial x} \end{bmatrix} = \frac{1}{2A} \begin{bmatrix} b_1 & 0 & b_2 & 0 & b_3 & 0 \\ 0 & c_1 & 0 & c_2 & 0 & c_3 \\ c_1 & b_1 & c_2 & b_2 & c_3 & b_3 \end{bmatrix} \begin{bmatrix} u_1 \\ v_1 \\ u_2 \\ v_2 \\ u_3 \\ v_3 \end{bmatrix}$$

$$(A6.17)$$

so that, by inspection with eqn. (A6.3),

$$[B] = \frac{1}{2A} \begin{bmatrix} b_1 & 0 & b_2 & 0 & b_3 & 0 \\ 0 & c_1 & 0 & c_2 & 0 & c_3 \\ c_1 & b_1 & c_2 & b_2 & c_3 & b_3 \end{bmatrix} \qquad (A6.18)$$

Because $[B]$ contains only constant terms over the element, $\{\varepsilon\}$ is constant also and so the element is known as a constant strain (or stress) element.

As described in Appendix 5, the elasticity matrix $[D]$ is given by

$$[D] = \frac{E}{1 - v^2} \begin{bmatrix} 1 & v & 0 \\ v & 1 & 0 \\ 0 & 0 & \dfrac{1 - v}{2} \end{bmatrix} \qquad (A6.19)$$

for plane stress. The element stiffness matrix is given by eqn. (A6.6) as

$$[K]^1 = \int_1 [B]^T [D][B] \, dV$$

and since the matrices $[B]$ and $[D]$ contain no terms which are functions of x

or y, the integrand $[B]^T[D][B]$ is constant over the element. Hence,

$$[K]^I = At[B]^T[D][B]$$

and after suitable manipulation of the product of the three matrices,

$$[K]^I = \frac{Et}{4A(1-v^2)} \begin{bmatrix} k_{11} & & & & & \\ k_{21} & k_{22} & & \text{symmetric} & & \\ k_{31} & k_{32} & k_{33} & & & \\ k_{41} & k_{42} & k_{43} & k_{44} & & \\ k_{51} & k_{52} & k_{53} & k_{54} & k_{55} & \\ k_{61} & k_{62} & k_{63} & k_{64} & k_{65} & k_{66} \end{bmatrix} \quad (A6.20)$$

where

$$k_{11} = b_1^2 + \tfrac{1}{2}(1-v)c_1^2 \qquad\qquad k_{51} = b_1 b_3 + \tfrac{1}{2}(1-v)c_1 c_3$$

$$k_{21} = \tfrac{1}{2}(1+v)b_1 c_1 \qquad\qquad k_{52} = v b_3 c_1 + \tfrac{1}{2}(1-v)b_1 c_3$$

$$k_{22} = c_1^2 + \tfrac{1}{2}(1-v)b_1^2 \qquad\qquad k_{53} = b_2 b_3 + \tfrac{1}{2}(1-v)c_2 c_3$$

$$k_{31} = b_1 b_2 + \tfrac{1}{2}(1-v)c_1 c_2 \qquad\qquad k_{54} = v b_3 c_2 + \tfrac{1}{2}(1-v)b_2 b_3$$

$$k_{32} = v b_2 c_1 + \tfrac{1}{2}(1-v)b_1 c_2 \qquad\qquad k_{55} = b_3^2 + \tfrac{1}{2}(1-v)c_3^2$$

$$k_{33} = b_2^2 + \tfrac{1}{2}(1-v)c_2^2 \qquad\qquad k_{61} = v b_1 c_3 + \tfrac{1}{2}(1-v)b_3 c_1$$

$$k_{41} = v b_1 c_2 + \tfrac{1}{2}(1-v)b_2 c_1 \qquad\qquad k_{62} = c_1 c_3 + \tfrac{1}{2}(1-v)b_1 b_3$$

$$k_{42} = c_1 c_2 + \tfrac{1}{2}(1-v)b_1 b_2 \qquad\qquad k_{63} = v b_2 c_3 + \tfrac{1}{2}(1-v)b_3 c_2$$

$$k_{43} = \tfrac{1}{2}(1+v)b_2 c_2 \qquad\qquad k_{64} = c_2 c_3 + \tfrac{1}{2}(1-v)b_2 b_3$$

$$k_{44} = c_2^2 + \tfrac{1}{2}(1-v)b_2^2 \qquad\qquad k_{65} = \tfrac{1}{2}(1+v)b_3 c_3$$

$$k_{66} = c_3^2 + \tfrac{1}{2}(1-v)b_3^2$$

The stress vector is obtained from eqn. (A6.2). Thus, ignoring for the moment any initial strains,

$$\{\sigma\} = [D]\{\varepsilon\}$$

i.e. $$\{\sigma\} = [D][B]\{u\} \quad (A6.21)$$

Hence, when the displacements at each node are known (after solution of the eqns. (A6.12)) the stress vector $\{\sigma\}$ can be obtained. Since the stress is constant over the element, it is usual to calculate it once per element and refer to it as pertaining to the element centroid. Nodal values may subsequently be obtained by straightforward averaging techniques.

For thermal problems,

$$\{\varepsilon_0\} = \begin{bmatrix} \alpha T \\ \alpha T \\ 0 \end{bmatrix}$$

where α is the coefficient of linear expansion and T is the mean element temperature above some datum. Hence the thermal load vector defined by eqn. (A6.9) is

$$\{F_0\}^1 = At[B][D]\{\varepsilon_0\}$$

This reduces, again after some matrix manipulation, to

$$\{F_0\}^1 = \frac{Et\alpha T}{2(1-v)} \begin{bmatrix} b_1 \\ c_1 \\ b_2 \\ c_2 \\ b_3 \\ c_3 \end{bmatrix} \tag{A6.22}$$

From eqn. (A6.2),

$$\{\sigma\} = [D]\{\varepsilon\} - [D]\{\varepsilon_0\}$$

so that if we write

$$\{\sigma_0\} = [D]\{\varepsilon_0\},$$

we have

$$\{\sigma_0\} = \frac{E}{1-v^2} \begin{bmatrix} 1 & v & 0 \\ v & 1 & 0 \\ 0 & 0 & \dfrac{1-v}{2} \end{bmatrix} \begin{bmatrix} \alpha T \\ \alpha T \\ 0 \end{bmatrix}$$

or

$$\{\sigma_0\} = \frac{E\alpha T}{(1-v)} \begin{bmatrix} 1 \\ 1 \\ 0 \end{bmatrix} \tag{A6.23}$$

Hence, for thermal problems, it is necessary to subtract the vector $\{\sigma_0\}$ from the calculated stress $[D]\{\varepsilon\} = [D][B]\{u\}$ in order to give the true stress.

THE ISOPARAMETRIC FAMILY OF ELEMENTS

The family of isoparametric elements includes triangles and quadrilaterals in two dimensions and wedge or brick shapes in three dimensions. The nodes are defined not only at corners but also along sides, the more nodes being used per element the higher the displacement, stress and strain variations over the element and as a result the higher the resulting accuracy of these elements compared with lower ordered ones, for a given number of structural degrees of freedom. It has generally been found more economic in terms of manpower involvement and computer costs to produce coarser meshes of the higher ordered elements than finer meshes of lower elements, such as constant stress triangles, when each mesh gives the same accuracy of results. This is particularly important in three dimensions.

The three most common isoparametric elements in two dimensions (for plane stress, plane strain or axisymmetry) are the quadrilaterals with 4 and 8 nodes, respectively, and the 6-node triangle (Fig. A6.2).

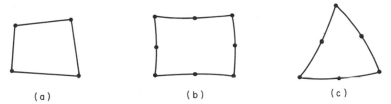

(a)   (b)   (c)

FIG. A.6.2   (a) 4-node quadrilateral; (b) 8-node quadrilateral; (c) 6-node triangle.

The 4-node quadrilateral has a linear displacement variation in each direction and approximately constant stress. This element is therefore very similar to the constant stress triangle and gives slightly better accuracy for a given number of nodes. The 8-node quadrilateral and 6-node triangle both have mid-side nodes. The displacement variation over both types of element in each direction is quadratic, with approximately linear stress variation. But now the geometry definition along each side may also be quadratic, being based on the same shape functions, and it is this property that, whatever the order of displacement variation, the side geometry is the same, that gives the name *isoparametric*. By this definition, the constant stress triangle is an isoparametric element.

In three dimensions, the two-element shapes most commonly used are wedges and bricks, such as the 15-node wedge and 20-node brick shown in Fig. A6.3, each having quadratic displacement variations in each direction.

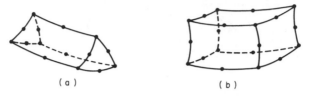

(a)　　　　　　　　　　　　　　(b)

FIG. A.6.3　(a) 15-node wedge; (b) 20-node brick.

In both dimensions, the use of cubic or higher elements does not appear to give any practical advantages and so these quadratic elements, with a better uniformity of node distribution, are generally preferred. The degree of distortion which each element can tolerate is quite high, e.g. 20-node bricks have been used to model thin cylinders, but user experience is required to avoid undue numerical errors which may arise.

Elements of triangular or quadrilateral shape may be mixed in a mesh as long as the number of nodes and the curvature along each side are the same.

The derivation of the element matrices for the isoparametric elements follows a common course, the only variation being in the shape function definition. A brief résumé, for the case of two dimensional plane stress, follows.

Suppose an element of the family has N nodes, with nodal shape functions p_i, $i = 1, \ldots, N$. At any node j, we require that $p_j = 1$ and $p_i = 0$, $i = 1, \ldots\ldots, j - 1, j + 1, \ldots, N$ by definition.

For quadrilateral-shaped elements, dimensionless coordinates ξ and η are chosen such that ξ and η vary between -1 and $+1$, and in the (ξ, η) space the quadrilateral becomes a square with side lengths of 2 units (Fig. A6.4(a)).

For triangles, dimensionless area coordinates L_1, L_2, and L_3 are chosen so that L_1, L_2 and L_3 vary between 0 and 1, and $L_1 + L_2 + L_3 = 1$. Thus, if A is the area of the triangle 123 (Fig. A6.4(b)),

$$L_1 = \frac{A_1}{A}, L_2 = \frac{A_2}{A} \quad \text{and} \quad L_3 = \frac{A_3}{A}$$

The dependence relation $L_1 + L_2 + L_3 = 1$ indicates that only two, usually

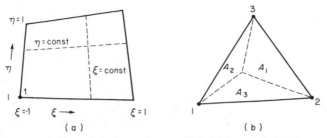

FIG. A.6.4   (a) Definition of (ξ, η) coordinates; (b) definition of $(L_1 L_2 L_3)$ coordinates.

L_1 and L_2, need be used. In the following analysis, they may be written as ξ and η.

The nodal shape functions for the 4-node quadrilateral are, with reference to the nodes numbers of Fig. A6.5(a):

$$p_1 = \tfrac{1}{4}(1 - \xi)(1 - \eta)$$
$$p_2 = \tfrac{1}{4}(1 + \xi)(1 - \eta)$$
$$p_3 = \tfrac{1}{4}(1 + \xi)(1 + \eta)$$
$$p_4 = \tfrac{1}{4}(1 - \xi)(1 + \eta)$$

For the 6-node triangle, the shape functions are, with reference to Fig. A6.5(b),

$$p_1 = L_1(2L_1 - 1)$$
$$p_2 = 4L_1 L_2$$
$$p_3 = L_2(2L_2 - 1)$$
$$p_4 = 4L_2 L_3$$
$$p_5 = L_3(2L_3 - 1)$$
$$p_6 = 4L_3 L_1$$

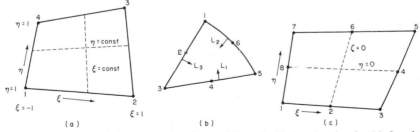

FIG. A.6.5   Nodal definitions: (a) 4-node quadrilateral; (b) 6-node triangle; (c) 8-node quadrilateral.

whilst for the 8-node quadrilateral (Fig. A6.5(c)):

$$p_1 = -\tfrac{1}{4}(1 - \xi)(1 - \eta)(\xi + \eta + 1)$$
$$p_2 = \tfrac{1}{2}(1 - \xi^2)(1 - \eta)$$
$$p_3 = \tfrac{1}{4}(1 + \xi)(1 - \eta)(\xi - \eta - 1)$$
$$p_4 = \tfrac{1}{2}(1 + \xi)(1 - \eta^2)$$
$$p_5 = \tfrac{1}{4}(1 + \xi)(1 + \eta)(\xi + \eta - 1)$$
$$p_6 = \tfrac{1}{2}(1 - \xi^2)(1 + \eta)$$
$$p_7 = \tfrac{1}{4}(1 - \xi)(1 + \eta)(- \xi + \eta - 1)$$
$$p_8 = \tfrac{1}{2}(1 - \xi)(1 - \eta^2)$$

Any field variable $\theta = \theta(\xi, \eta)$, such as displacement, temperature or geometry, is defined throughout the element by these shape functions as:

$$\theta = \sum_{i=1}^{N} p_i \theta_i$$

where N is the number of nodes on the element and θ_i are the assumed known nodal values of θ. We can thus write the matrix equation for the differentials of θ with respect to ξ and η as

$$
\begin{bmatrix} \dfrac{\partial \theta}{\partial \xi} \\[2mm] \dfrac{\partial \theta}{\partial \eta} \end{bmatrix}
=
\begin{bmatrix} \dfrac{\partial p_1}{\partial \xi} & \dfrac{\partial p_2}{\partial \xi} & \cdots & \dfrac{\partial p_N}{\partial \xi} \\[3mm] \dfrac{\partial p_1}{\partial \eta} & \dfrac{\partial p_2}{\partial \eta} & \cdots & \dfrac{\partial p_N}{\partial \eta} \end{bmatrix}
\begin{bmatrix} \theta_1 \\ \theta_2 \\ \vdots \\ \theta_N \end{bmatrix}
= [P]
\begin{bmatrix} \theta_1 \\ \theta_2 \\ \vdots \\ \theta_N \end{bmatrix}
$$

Hence

$$
[P]
\begin{bmatrix} x_1 & y_1 \\ \vdots & \vdots \\ x_N & y_N \end{bmatrix}
=
\begin{bmatrix} \dfrac{\partial x}{\partial \xi} & \dfrac{\partial y}{\partial \xi} \\[3mm] \dfrac{\partial x}{\partial \eta} & \dfrac{\partial y}{\partial \eta} \end{bmatrix}
= [J]
$$

where [J] is the Jacobian matrix relating the (x, y) and (ξ, η) coordinate systems. Writing the differential chain rule

$$\frac{\partial \theta}{\partial \xi} = \frac{\partial \theta}{\partial x}\frac{\partial x}{\partial \xi} + \frac{\partial \theta}{\partial y}\frac{\partial y}{\partial \xi} \quad \text{and} \quad \frac{\partial \theta}{\partial \eta} = \frac{\partial \theta}{\partial x}\frac{\partial x}{\partial \eta} + \frac{\partial \theta}{\partial y}\frac{\partial y}{\partial \eta}$$

in matrix form gives

$$
\begin{bmatrix} \dfrac{\partial \theta}{\partial \xi} \\[2mm] \dfrac{\partial \theta}{\partial \eta} \end{bmatrix}
= [J]
\begin{bmatrix} \dfrac{\partial \theta}{\partial x} \\[2mm] \dfrac{\partial \theta}{\partial y} \end{bmatrix}
$$

and so

$$
\begin{bmatrix} \dfrac{\partial \theta}{\partial x} \\ \dfrac{\partial \theta}{\partial y} \end{bmatrix} = [J]^{-1}[P] \begin{bmatrix} \theta_1 \\ \vdots \\ \theta_N \end{bmatrix}
$$

Replacing θ by the two components of displacement, u and v in turn, gives

$$
\begin{bmatrix} \dfrac{\partial}{\partial x} \\ \dfrac{\partial}{\partial y} \end{bmatrix} [uv] = \begin{bmatrix} \dfrac{\partial u}{\partial x} & \dfrac{\partial v}{\partial x} \\ \dfrac{\partial u}{\partial y} & \dfrac{\partial v}{\partial y} \end{bmatrix} = [J]^{-1}[P] \begin{bmatrix} u_1 & v_1 \\ \vdots & \vdots \\ u_N & v_N \end{bmatrix}
$$

We may write

$$
[J]^{-1}[P] = \begin{bmatrix} d_{11} & d_{12} & \cdots & d_{1N} \\ d_{21} & d_{22} & \cdots & d_{2N} \end{bmatrix}
$$

Referring back to eqn. (A6.17), the strain–displacement equations now become:

$$
\{\varepsilon\} = \begin{bmatrix} d_{11} & 0 & d_{12} & 0 & \cdots & d_{1N} & 0 \\ 0 & d_{21} & 0 & d_{22} & \cdots & 0 & d_{2N} \\ d_{21} & d_{11} & d_{22} & d_{12} & \cdots & d_{2N} & d_{1N} \end{bmatrix} \begin{bmatrix} u_1 \\ v_1 \\ u_2 \\ v_2 \\ \vdots \\ u_N \\ v_N \end{bmatrix} = [B] \begin{bmatrix} u_1 \\ v_1 \\ u_2 \\ v_2 \\ \vdots \\ u_N \\ v_N \end{bmatrix}
$$

thus defining the $[B]$ matrix.

The $[B]$ matrix is dependent on the d_{ij} terms, which are themselves functions of geometry. Hence, unlike the constant stress triangle case, $[B]$ here varies from point to point over the element and the stiffness matrix integration (eqn. (A6.6)) is not simple. Indeed, generally the integral is not algebraic. However, a convenient and powerful technique exists, particularly suited to computation, known as numerical integration. This technique was first used by Irons [9] in the present context, and sums the stiffness integrand calculated at strategic points, known as Gauss points, over the elements with suitable weighting functions. The number and position of Gauss points varies for the different element types and, indeed, can vary for the same element type.

Thus

$$[K]^1 = \int_1 [B]^T [D][B]\, dV$$

$$= \sum_i [B_i]^T [D][B_i] W_i$$

where i indicates values at the Gauss points and W_i is the weighting function.

The treatment of different types of loading including thermal effects is a straightforward procedure. Stresses are no longer constant over the element, in general, and so are calculated at suitable points of reference. Most users prefer values calculated at nodes. This is provided by the relationship

$$\{\sigma_i\} = [D][B_i]\{u\}$$

with $[B_i]$ derived at each node i in turn. Since the same node may appear in several elements, a simple averaging procedure of the calculated stresses obtained at the node from the different elements is necessary. Alternatively, stresses may be evaluated at the Gauss points. As shown by Barlow [10], the stresses at these points have an optimum accuracy. Indeed, in the plasticity and creep processes described in Chapter 4, stresses are referred to such points since the solution progression is based on their magnitudes.

All the above comments and derivations apply to plane strain, axisymmetric and three-dimensional versions of the elements.

ELEMENTS FOR CRACK TIP MODELLING

In LEFM, the form of the classical Westergaard equations described in Chapter 1 show that in the vicinity of the crack tip the displacements vary proportional to $\sqrt{r}$, r being the distance from the crack tip. Similarly, the stresses and strains are proportional to $1/\sqrt{r}$.

Since the finite element method as described above is based upon assumptions for displacements (and/or stresses), defined by polynomials over elements of finite size, it is not normally suitable for exactly describing the behaviour in the region of a singularity. The various possibilities for overcoming this problem in the application of the method to fracture mechanics, reviewed, *inter alia* by Gallagher [11], may be summarised as follows:

Use of standard elements, with substantial mesh refinement around the crack tip. This approach is very expensive in both preparation effort and computer time and has, therefore, been all but abandoned.

Use of special elements in the vicinity of the crack tip, either

—elements whose assumed displacements include the singularity at one of the nodes (e.g. Refs. [12] and [13]),

—mid-side node isoparametric elements with assumed displacements including the singularity at one of the nodes [15] or

—standard isoparametric elements with mid-side nodes displaced from their nominal position.

As pointed out, for example, by Barsoum [14], certain elements of the first approach suffer from the disadvantage that the elements lack the constant strain and rigid body motion modes and are therefore unsuitable for cases where thermal gradients occur near the crack tip.

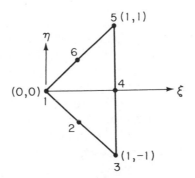

FIG. A.6.6   Dimensionless coordinates for special elements.

In view of this latter aspect, the computations described in Chapter 4 make use of elements of the second approach, namely isoparametric elements with mid-side nodes and displacement assumptions including the singularity effect.

Such an element was, *inter alia*, derived by Blackburn [15]. The shape is a mid-side node triangle within which dimensionless coordinates ξ and η are defined as shown in Fig. A6.6.

With reference to the node numbers shown (the first being at the crack tip where $\xi = \eta = 0$), then suitable shape functions are as follows:

$$p_1 = (1 - \xi^{1/2})(1 - \sqrt{2}\xi)$$
$$p_2 = (1 + \sqrt{2})(1 - \xi^{1/2})(\xi^{1/2} + \eta\xi^{-1/2})$$
$$p_3 = \tfrac{1}{2}(\xi^{1/2} + \eta\xi^{-1/2})\{(2 + \sqrt{2})(\xi^{1/2} - 1) + 1 - \xi + \eta\}$$
$$p_4 = \xi^{3/2} - \eta^2\xi^{-1/2}$$
$$p_5 = \tfrac{1}{2}(\xi^{1/2} - \eta\xi^{-1/2})\{(2 + \sqrt{2})(\xi^{1/2} - 1) + 1 - \xi - \eta\}$$
$$p_6 = (1 + \sqrt{2})(1 - \xi^{1/2})(\xi^{1/2} - \eta\xi^{-1/2})$$

Alternatively, the radial displacement variation $u(\xi, \eta)$ may be written

$$u(\xi, \eta) = a_1 + \frac{a_2\xi + a_3\eta}{\sqrt{(\xi + \eta)}} + \frac{a_4\xi\eta}{\xi + \eta} + a_5\xi + a_6\eta$$

As well as including a dependence on $\sqrt{r}$, this form also includes a dependence on r, so that constant strain is permitted.

A three-dimensional form of this element can readily be developed and has already been used in a large number of problems (T.K.Hellen & W. S. Blackburn (1975). The calculation of stress intensity factors in two and three dimensions using finite elements, *Computational Fract. Mech.*, ASME, New York).

An alternative form of the mid-side node compatible special element has been described by Barsoum [14] and Henshell & Shaw [16]. They noted that with quadratic isoparametric elements the $\sqrt{r}$ displacement function can be achieved by moving the mid-side nodes to the quarter position nearest the crack tip nodes. This holds for triangular or quadrilateral elements in two and three dimensions, provided in the latter case only nodal movements in the plane normal to the crack profile are made. In the case of the plane triangular element, the shape function derived by this technique is almost identical to the Blackburn formulation:

$$u(\xi, \eta) = a_1 + \frac{(a_2\xi + a_3\eta)}{\sqrt{\xi + \eta}} + \frac{a_4\xi\eta}{\xi + \eta} + a_5\xi + a_6\eta$$

and gives results of similar accuracy, as shown by Hellen [17].

In the case of post-yield fracture mechanics, the presence of gross crack tip yielding removes the validity of the Westergaard equations. Explicit algebraic equations can no longer be derived. However, assuming power hardening laws, displacement variations can be established. For the case of elastic–perfectly plastic behaviour, special elements have been described by Barsoum [18].

An alternative is to use the linear elastic $\sqrt{r}$ special elements to give a good prediction of initial yield, which will be at a point a finite distance from the tip, then to retain or discard these elements as the plasticity spreads.

REFERENCES

1. Zienkiewicz, O. C. (1971). *The finite element method in engineering science*, 2nd edition, McGraw-Hill, London.
2. Desai, Ch. S., & Abel, J. F. (1972). *Introduction to the finite element method*, Van Nostrand Reinhold, New York.
3. Gallagher, R. H. (1975). *Finite element analysis: fundamentals*, Prentice-Hall, Englewood Cliffs.
4. Hubner, K. H. (1975). *The finite element method for engineers*, John Wiley, New York.
5. Irons, B. M. R. (1970). A front solution program for finite element analysis, *Int. J. Num. Meth. Engng*, **2**, pp. 5–32.
6. Hellen, T. K. (1969). A front solution for finite element techniques, CEGB Report R/D/B/N1459.
7. Turner, M. J., Clough, R. W., Martin, H. C. & Topp, L. J. (1956). Stiffness and deflection analysis of complex structures, *J. Aeron. Sci.*, **23**(9), pp. 805–23.
8. Clough, R. W. (1960). The finite element in plane stress analysis, *Proc. 2nd ASCE Conf. on Electronic Computation*, Pittsburgh (Pa.).
9. Irons, B. M.. R. (1966). Numerical integration applied to finite element methods, Conf. on the Use of Computers in Structural Engineering, Univ. Newcastle.
10. Barlow, J. (1976). Optimal stress locations in finite element models, *Int. J. Num. Meth. Engng*, **10**, pp. 243–51.
11. Gallagher, R. H. (1971). Survey and evaluation of the finite element method in fracture mechanics analysis, 1st Int. Conf. Struct. Mech. Reactor Technol., Berlin.
12. Byskov, E. (1970). The calculation of stress intensity factors using the finite element method with cracked elements, *Int. J. Fract. Mech.*, **6**, pp. 159–67.
13. Benzley, S. E. (1976). Representation of singularities with isoparametric finite elements, *Int. J. Num. Meth. Engng*, **8**, 537–45.
14. Barsoum, R. S. (1974). Application of quadratic isoparametric finite elements in linear fracture mechanics, *Int. J. Fract.*, **10**, pp. 603–5.
15. Blackburn, W. S. (1972). Calculation of stress intensity factors at crack tips using special finite elements, Conf. on Maths of Finite Els. and Appls., Brunel Univ., Apr.
16. Henshell, R. D., & Shaw, K. G. (1975). Crack tip finite elements are unnecessary, *Int. J. Num. Meth. Engng.*, **9**, pp. 495–507.
17. Hellen, T. K. (1977). On special isoparametric elements for elastic fracture mechanics, *Int. J. Num. Meth. Engng*, **11**, pp. 200–3.
18. Barsoum, R. S. (1977). Triangular quarter-point elements as elastic and perfectly plastic crack tip elements, *Int. J. Num. Meth. Engng*, **11**, pp. 85–98.

Appendix 7

Finite Element Formulation of Creep Analysis

T. K. Hellen

CEGB, Berkeley, UK

BASIC THEORY

In addition to time-independent plasticity effects dealt with in the main text of this book, a class of problems exist where creep and plasticity behaviour should be considered simultaneously. This can readily be effected by finite element analysis.

The analysis of creep behaviour is similar to that of plasticity. The main difference is that the creep strain rate depends on the effective stress, the total effective creep strain, the temperature, θ, and time t, by experimentally observed relationships of the type:

$$\{\dot{\varepsilon}_c\} = F(\sigma^*, \varepsilon^*, \theta, t) \qquad (A.7.1)$$

The equivalent stress is defined by eqn. (4.15) and the equivalent creep strain by eqn. (4.18), replacing the plastic strain increments by total creep strains.

The form of eqn. (A.7.1) is derived from uniaxial creep data and is extended to the increments of creep strain rate by use of a flow rule, in a manner analogous to that of plastic strains in eqn. (4.4). Details are given by various authors, for example Rashid [1], and are not reproduced here. Flow rules corresponding to eqns. (4.19) are

$$\frac{\partial \varepsilon_{cx}}{\partial t} = \frac{\Delta \dot{\varepsilon}^*}{2\sigma^*}(2s_x - s_y - s_z)$$

$$\frac{\partial \gamma_{cxy}}{\partial t} = \frac{3\Delta \dot{\varepsilon}^*}{\sigma^*}\tau_{xy} \qquad (A.7.2)$$

For time-hardening, an increment of creep strain rate is given by

$$\frac{\partial \varepsilon_{cx}}{\partial t} = \frac{F_1(\sigma^*)}{2\sigma^*} F_2(t) F_3(\theta)(2s_x - s_y - s_z) \quad \text{(A.7.3)}$$

where F_1, F_2 and F_3 represent functions of stress, time and temperature respectively, derived from experiment. For strain-hardening

$$\frac{\partial \varepsilon_{cx}}{\partial t} = \frac{F_1(\sigma^*)}{2\sigma^*} F_2(T) F_3(\theta)(2s_x - s_y - s_z) \quad \text{(A.7.4)}$$

where T is equivalent time.

SOLUTION TECHNIQUES

Despite the similarity of the plasticity and creep analyses, the finite element solution for creep is somewhat simpler than for plasticity. Several programs were developed relatively early, all using the plane or axisymmetric constant stress triangles, such as those described by Greenbaum & Rubinstein [2], Sutherland [3] and Leech & Ecclestone [4].

The solution of creep problems using finite elements proceeds in small time intervals, dt. The initial strain method is invariably used. The total strain increment is expressed as the sum of the creep strain and the elastic strain, other strain contributions (thermal and plastic) being ignored here since they are represented by their respective load vectors in the equilibrium equations. Thus,

$$d\{\varepsilon\} = d\{\varepsilon_e\} + d\{\varepsilon_c\}$$

Since

$$d\{\sigma\} = [D] d\{\varepsilon_e\}$$

$$d\{\sigma\} = [D](d\{\varepsilon\} - d\varepsilon_c\})$$

$$= [D][B]d\{u\} - [D]d\{\varepsilon_c\}$$

From the material laws,

$$d\{\varepsilon_c\} = f(\sigma^*, \varepsilon_c^*, \theta, t)\, dt$$

and so the equilibrium eqn. (4.24) (regarded here to hold for a time increment as opposed to a load increment) becomes

$$\left(\int_V [B]^T[D][B] \, dV \right) d\{u\} = d\{R\} + \int_V [B]^T[D] \, d\{\varepsilon_c\} \, dV$$

or

$$[K_0] \, d\{u\} = d\{R\} + \int_V [B]^T[D] \, d\{\varepsilon_c\} \, dV \qquad (A7.5)$$

The creep strains are thus accounted for as initial strains, which only affect the right-hand side of eqn. (A.7.5). Hence the initial elastic stiffness matrix can be used throughout the whole solution procedure, unless material properties change radically, because of temperature dependent material properties, or large strains exist.

The combination of plasticity and creep is effected by a sequence of separate analyses of plasticity and creep, each retaining the strains due to the other as initial strains in the constitutive equations. The sequence is dictated by the physical situation. Usually, this entails a plasticity analysis corresponding to the applied loads or temperatures at the moment of load application (time = 0). Then, over a time period, a creep analysis is required to enable stress relaxation from this initial state. The relaxation can occur in yielded as well as elastic regions. At any point in time, a change in the applied loads or temperatures would necessitate another plasticity analysis, after which further time-wise creep behaviour would be required.

In such mixed analyses, each plasticity calculation should be complete in that all the applied load is considered as a set of increments before changing to a creep analysis. In the latter case, however, there is no need for completed analyses over any particular time interval before changing to a plasticity analysis.

REFERENCES

1. Rashid, Y. R. (1973). Analysis of multiaxial flow under variable load and temperature, Int. Conf. on Creep and Fatigue in Evaluated Temp. Applications, ASME/Inst. Mech. E. Paper C183/73.
2. Greenbaum, G. A. & Rubenstein, M. F. (1968). Creep analysis of axisymmetric bodies using finite elements, *Nucl. Engng Design*, 7, pp. 379–97.
3. Sutherland, W. H. (1970). L'AXICRP—finite element computer code for creep analysis of plane stress, plane strain and axisymmetric bodies, *Nucl. Engng Design*, 11, pp. 269–85.
4. Leech, A. J. & Ecclestone, M. J. (1969). Triangular element stress and strain analysis (TESS): Part 2—elastic and creep behaviour. CEGB. Report RD/C/N363.

Index

Index